Jörg-Peter Ewert

Neuroethology

An Introduction to the
Neurophysiological Fundamentals of Behavior

Translation: Transemantics, Inc.

With 171 Figures, Most in Color

Springer-Verlag
Berlin Heidelberg New York 1980

Professor Dr. rer. nat. Jörg-Peter Ewert
Arbeitsgruppe Neuroethologie, FB 19, GHK
Universität des Landes Hessen
Heinrich-Plett-Str. 40, D-3500 Kassel, FRG

Translation:
Transemantics, Inc., Suite 407
1901 Pennsylvania Avenue N.W.
Washington, D.C. 20006, USA

Title of the German edition
Jörg-Peter Ewert, Neuro-Ethologie
Springer-Verlag Berlin Heidelberg New York 1976

ISBN-13: 978-3-642-67502-7   e-ISBN-13: 978-3-642-67500-3
DOI: 10.1007/ 978-3-642-67500-3

Library of Congress Cataloging in Publication Data. Ewert, Jörg-Peter, 1938-.
Neuroethology. Translation of Neuro-Ethologie. Bibliography: p. Includes index.
1. Neuropsychology. 2. Animals, Habits and behavior of. I. Title. QP360.E9313
156'.2 7923144.

Typesetting: Beltz, Offsetdruck, Hemsbach/Bergstr.

2131/3130-543210

Preface

Historically the search for the neural bases of behavior goes back a long way. Neuroethology, which is concerned with the experimental analysis of the releasing and control mechanisms of behavior, is a young discipline. Results from this multidisciplinary branch of research, which uses physical, chemical, and mathematical methods, have not yet been extensively treated in textbooks of neurophysiology and ethology. This book is intended as a first attempt to pose major questions of neuroethology and to demonstrate, by means of selected research examples, some of the ways by which these questions are being approached. Inevitably this cannot be a complete and in depth detailed treatment of all of the neurobiology examples, and I realize that such a selection is of a subjective nature. The overall goal of the book is to present an introduction. After outlining some of the very basic neurophysiological and ethological concepts (Chaps. 2 and 3), neuroethological questions and methods are demonstrated extensively by means of a particular example (Chap. 4). There are two reasons to choose the visually guided prey-catching and avoidance behavior of the Common Toad: (1) it is a system which I have investigated for about fifteen years and therefore know best, (2) the toad story is one of the most comprehensive neuroethological approaches so far. Thus, it is possible here to outline the major concepts of neuroethology and to pose the basic questions. Chapters 5–8 show that the same questions and concepts are also relevant to other animals and to other systems. Not all of the examples are sufficiently developed because of space limitations. For a more detailed account of some of these examples, references are provided in the figure captions, and reading lists will be found at the end of the book (cf. Special References and Suggested Reading). Numerous, in part, strongly schematized two-color diagrams illustrate the text. Essential summarizing statements are printed on a colored background. The reader can be quickly oriented by reading these paragraphs and studying the illustrations. Information about current neurobiological methods is condensed in an appendix (Chap. 9). The English edition *Neuroethology* is a translation of the German edition *Neuro-Ethologie* published in 1976. The English version is more comprehensive, brought up to date and provided in some places with new illustrations. I wish to express my appreciation to Transemantics, Inc. for translating this book. I want to express special thanks to Howard Carl Gerhardt (University of Missouri, Columbia) and Ananda Weerasuriya (National Institutes of Health, Bethesda, Maryland) for careful revision and detailed comments on the manuscript from which I have benefited greatly. I am most grateful to the publisher, Dr. Konrad F. Springer, for the generosity shown in the production of the book.

Kassel, Spring 1980   JÖRG-PETER EWERT

Contents

1 What is Neuroethology?

1.1 Principal Scope

The behavior of animal and man is based on the processing of data which takes place within groups of interconnected nerve cells. Such populations of neurons take the form of relatively diffuse networks in primitive animals, e.g., certain jellyfishes, whereas in more highly organized animals they are concentrated in particular structures (brain, ganglia, ventral or spinal cord) [1]. The results of neuronal data processing are electrical events conducted to particular muscle groups which then contract according to a central program and produce spatially and temporally coordinated patterns of movement, that is, behavior (Fig. 1). The "commands" for the activation of such motor programs can *originate* from the central nervous system (CNS) itself, and they can be *influenced* by the endocrine system. The execution of a "motor command" – the behavioral response – may be *controlled* by sensory systems. "Commands" can also be *elicited* by particular signals of the environment, following appropriate processing of the sensory input (Fig. 1). In the latter case, too, it is important to note that the efficacy of an environmental stimulus – and with it its signal character – depends in considerable measure upon internal conditions; we refer to *drive* or *motivation*.

Let us select an example to illustrate this point: the female American cricket frog is attracted to the male by particular mating calls. However, only the calls of conspecific males are of interest to the female frogs. Furthermore, a female cricket frog from Georgia ignores the mating calls from conspecific males imported from Alabama, who belong to another geographical race and call in a foreign "dialect". Moreover, the female frog replies to the specific mating calls only at a particular time of the year, when she is motivated to mate. As the males utter no mating calls outside the spawning period, the mating behavior of both sexes is coordinated.

What capabilities should the auditory system of the female frog have? It has to select and recognize the biologically important features of the frequency spectrum and temporal patterns of different calls. Further, the brain must localize the source of the call and thereby the position of the male. These computational processes simultaneously start the motor program: the movement toward the male frog. In addition, such a stimulus-response pattern is controlled by the central nervous system of the female and is modulated by the endocrine system so that it occurs only during a particular period of the year.

Neuroethology[1] deals with the experimental exploration of the following main questions (Fig. 1):

1. Which filtering processes of the sense organs and brain – known as sensory systems – are responsible for the differentiation of behaviorally important from unimportant stimuli? In other words, how are signals detected?

2. How can signals in the environment be localized by the CNS?

3. What are the means of acquiring, storing, and recalling information in the CNS?

4. What is the neurophysiological basis for the motivation of a behavioral pattern?

5. In what way is behavior coordinated and controlled by the CNS?

6. How can ontogenetic development of behavior be related to neuronal mechanisms?

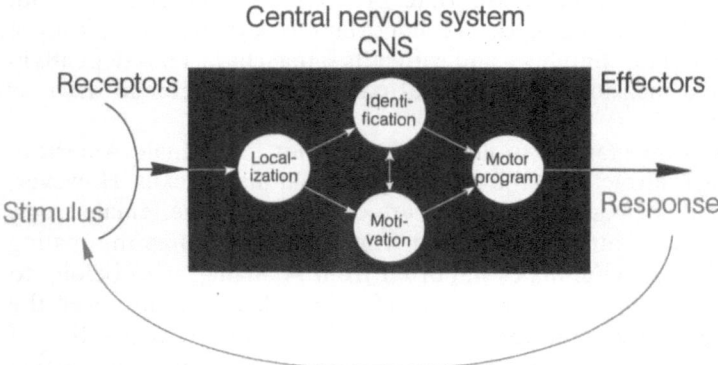

Fig. 1. Important information-processing steps in the central nervous system (CNS), leading to a behavioral reaction in response to a stimulus

----

1 neuro- (Greek-Latin) = related to nerve cells, neurons; éthos (Greek-Latin) = habit, custom.

1.2 History

Subjects of neuroethological research are animals that differ in structural complexity. Answering the above questions is of equally great concern for man. The comparative study of functional principles in lower animals can contribute toward the understanding of functional processes at the highest plane of integration. Seen in this light, neuroethology is as old as the search for the structures and functions of our brain, which is known to go as far back as Prechristian times.

The first brain localization diagrams for specific intellectual capacities (Fig. 2) – based on concepts of Albertus Magnus (13th century) – date from the 16th century. However, they have no scientific value. Leonardo da Vinci (1452–1519) was among the first to experiment with the central nervous system of the frog. He discovered that a frog without a head is viable, but dies after destruction of the spinal cord. Unfortunately his observations remained unknown until quite recently so that they failed to contribute to the historical development of brain physiology. In the middle of the 16th century, with Vesalius' *De humani corporis fabrica libri septem* (1543), the discipline of human anatomy finally emerged, and the prerequisites for physiology were also created. René Descartes (1596–1650) interpreted the human, as well as the animal, organism as a kind of machine. For him the brain consisted of a conglomerate of reflex mechanisms which reacted, as it were, automatically to external stimuli. In the tubular nerves there were supposed to be medullary fibers reaching into the brain; upon stimulation of a sense organ the medullary

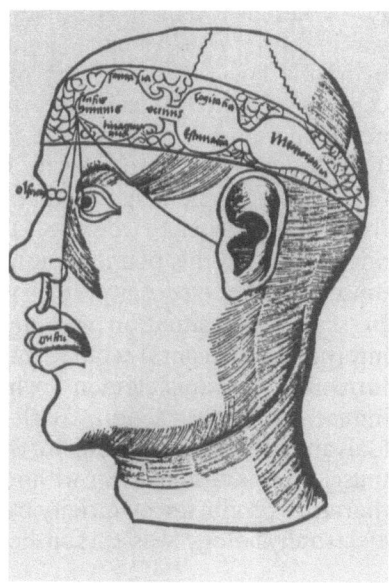

Fig. 2. Speculations on localization of particular mental capacities in the brain, e.g., Sensus communis; Imaginatio; Cogitatio. [By Reisch G (1504) Margarita philosophica. Grüninger, Strassburg]

fibers, set in motion, opened valves in the brain and thereby sent nervous fluids to the muscles. Whereas, according to Descartes' concept the soul had its seat in the brain and controlled the reservoir and valves as a kind of "fountain or engine master", the founder of microscopic anatomy, Marcello Malpighi (1628–1694) granted the brain only a function as a middle-man. The English scientist Thomas Willis (1621–1675) could, however, demonstrate on the basis of numerous, in part localized, brain ablations in various animals that the brain was essential for life. He also came to the conclusion that mental diseases are diseases of the brain. Willis charted brain maps on the basis of his experimental findings in which he assigned specific functions to different brain regions. Thus, for example, instinct was located in midbrain and memory in the cerebral convolutions. The significance of the cerebral motor cortex was first pointed out by Boyle (1691) and Swedenborg (1740). Stephan Hales (1730), with experiments on the decapitated frog, laid the foundations for *reflex physiology,* which was later developed by Hall (1833), Sechenov (1863), and Sherrington (1906). This discipline is important for the understanding of spinal cord functions as well as the integrative mechanisms of the brain. Using precise lesion techniques, the French physiologist, Flourens (1824) systematically investigated the effects of brain ablations on the behavior of pigeons.

Against the background of the violent discussions between the *vitalists* who based life on "life spirits" and the *mechanists* who saw "reflex machines" in all animal beings, Albrecht von Haller (1708–1777) finally succeeded in establishing *experimental physiology* as a scientific discipline. He disproved the doctrine of the "oscillatory movement" of the nerves and showed that muscle contraction is based on "excitability". Whereas Friedrich Hoffmann (1660–1742) still believed that there was in the nervous system a "fluid" which fulfilled functions according to hydraulic laws, the discovery of *bioelectricity* at the end of the 18th century provided a decisive turn in the approach to the physiology of nerve and muscle.

The discovery of bioelectricity came by chance. The Italian Galvani (1786) hung some frog legs on his balcony railings one day and noticed that the legs twitched when they touched the metal fence. Galvani at first misinterpreted this phenomenon, attributing it to the production by the muscle itself of electricity which would be conducted through the railing. In reality the situation was the other way around: Volta correctly interpreted the causal relationship. The fence, made of copper and iron, formed a "galvanic element" which excited the muscle electrically and induced the contraction. With the second important discovery by Galvani (1793), according to whom twitches occurred in the nerve-muscle preparations even without any metal – nerve and muscle were therefore producing electricity by themselves – the actual foundation for electrophysiology was established. It became evident that certain cell membranes can generate electrical potential differences, a phenomenon which later attracted the attention of many famous scientists, among them DuBois-Reymond (1848) and Helmholtz (1850). In 1963 Hodgkin, Huxley, and Eccles received the Nobel Prize for the development of the ionic theory of nervous function and the discovery of synaptic potentials.

Prerequisite for the anatomical investigation of neural tissue was the development of the light microscope by the Dutch morphologist Leeuwenhoek (1632–1723). Camillo Golgi (1844–1926) provided the key to the microscopic investigation of nerve tissue. The breakthrough in the field of neuroanatomy belongs to the Spanish anatomist Ramón y Cajal (1852–1934) for developing the "neuron doctrine": (1) neural tissue is formed by individual neurons, (2) neurons are connected by particular contacts, (3) different morphological features of various neuron types are linked with different functional properties.

Galvani had predicted that brain functions would also be explained by electrical forces, and he was to be proved right.

During the German–French war of 1870–71, Eduard Hitzig *galvanically* stimulated the exposed brain of a wounded, unconscious soldier, with the result that the soldier moved his eyes. Previously Fritsch and Hitzig (1870) had used electrical brain stimulation on vertebrate brain with *constant* direct current – mostly on dogs – and observed that stimulation of different regions of the cerebral cortex produced contractions in different muscle groups of the body. At King's College, London, David Ferrier (1873) performed similar studies on monkeys with *faradic* stimulation, and at the same time Bartholow (1874) in the U.S.A. was carrying out corresponding experiments on the motor cortex of man. Both scientists found that particular motor patterns of movement could be released by stimulation of defined regions of the cerebral cortex and charted such sites on a brain map (Fig. 3). Ferrier also observed another phenomenon: if he locally destroyed a certain area of the cerebral motor cortex which had previously induced a lifting of the arm upon electrical stimulation of this brain, the monkey at first was no longer able to lift his arm after the operation. Surprisingly, the function was regained after a time. Ferrier experimentally excluded the possibility that the return of the function originated in the intact brain hemisphere and he rightly concluded: "It would appear that after destruction of the centre on one side some other part of the *same* hemisphere may take up the functions of the destroyed part."

At the end of the 19th century, Ferrier was able to roughly localize in the cerebral cortex the sensory centers for vision, auditory, gustatory, and olfactory sensations, as well as touch perception. Only several decades later was the method of brain stimulation with *pulsating current* further refined. By stimulation of different areas of the brain stem – particularly the diencephalon – behavioral components, as well as complete

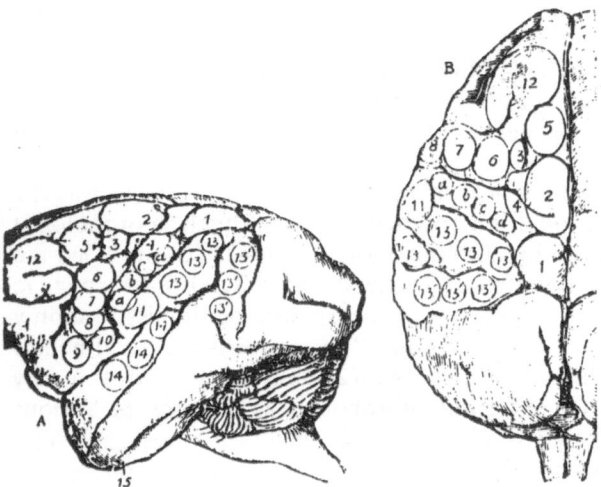

Fig. 3A and B. Areas for electrical stimulation on the surface of the left cerebral hemisphere in the monkey (Ferrier, 1876). **A** Lateral view, **B** view of left hemisphere from above. *1* Advance of opposite leg as in walking; *2* complex movements of thigh, leg, and foot with adapted movements of trunk; *3* movements of tail; *4* retraction and adduction of opposite fore-limb; *5* extension forward of opposite arm and hand, as if to reach or touch something in front; *a, b, c, d* individual and combined movements of fingers and wrists, ending in clenching of fist; *6* supination and flexion of forearm, by which the hand is raised towards the mouth; *7* action of zygomatics, by which the angle of the mouth is retracted and elevated; *8* elevation of ala of nose and upper lip, with depression of lower lip, so as to expose the canine teeth on the opposite side; *9* opening of mouth with protrusion of tongue; *10* opening of mouth with retraction of tongue; *11* retraction of angle of mouth; *12* eyes opening widely, pupils dilating, head and eyes turning towards opposite side; *13, 13'* eyeballs moving to opposite side, pupils generally contracting; *14* sudden retraction of opposite ear; *15* subiculum cornu ammonis, -torsion of lip and nostril on same side. [Encyclopaedia Britannica (1885), 9th ed., vol. 19. Ch. Scribner's Sons, New York]

sequences of action patterns, could be released in the freely moving animal. The breakthrough in that field belongs to the Swiss physiologist W. Hess for his work on cats around 1920. He was awarded the Nobel Prize for Medicine in 1945.

In the literature before the turn of the century one finds in the 9th edition of Encyclopaedia Britannica (1885), Vol. 19, p. 42 the block diagram given in Figure 4. It shows the flow of information in vertebrate brain similar in principle to that represented today:

1. *Sense organs:* eye EY., ear EA., smelling organ SM., taste organ TA., skin receptors SK.
2. *Information processing:* identification and localization S.SE.
3. *Effector systems:* motor centers, muscle M., vascular muscles V., glands G., electrical organ in some fishes EL.O.

Fig. 4. Information processing centers in vertebrate brain. Comments in text [Encyclopaedia Britannica (1885)]

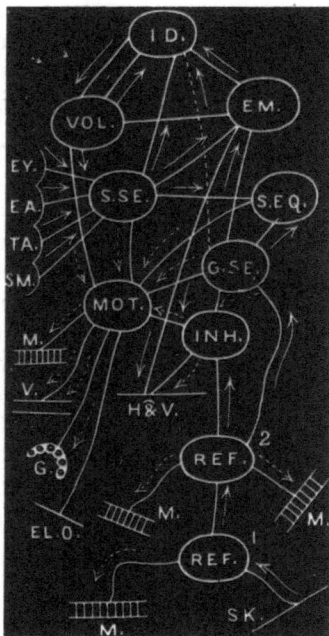

The scheme includes further functional principles such as "ideational" centers ID., volitional centers VOL., emotional centers EM., inhibitory centers INH., and centers of general sensation G.SE. Also considered are centers for equilibrium S.EQ. and reflex excitability REF. Excitatory (EM) and inhibitory (INH) connections to the autonomic nervous system are symbolized by H & V (heart and vessels).

At this late moment in the chapter on the history of neuroethology we should perhaps emphasize two points:

1. The general methodology for the exploration of the neural basis of behavior was, in principle, already provided toward the end of the last century. Today this might surprise the physiologists working in the field and give them cause to reflect. Progress in the fields of physiology and morphology are, however, linked to technical developments such as electronic measuring techniques, chemical tracer methods, and electron microscopy. The possibilities using scalpel and galvanic pincer were more or less exhausted by the turn of the century. Seen in this light, neurophysiology became fully functional only in the last few decades. As we know today, biology and technology are bound together in a kind of dualistic manner. On the one hand, technical developments often have biological origins, while on the other it is evident that the exploration of biological functional processes – as in the brain – depend heavily upon technical support. The most recent past has shown that progress within a

field of research, whose goal is to explore the neural processes underlying the release and control of behavior patterns, goes hand in hand with technical developments, especially in the field of electronics.

2. This rough historical sketch demonstrates the first phase of development of a particular research orientation within medicine, that is neurophysiology, and especially brain physiology. Neurology and psychiatry, because they primarily ask clinical questions, are interested in exploring and understanding the functional properties of our nervous system. As a rule, two paths are taken: (1) after brain damage the function of the specific brain regions can be determined by specific neurological symptoms. In addition the clinician can obtain supplementary knowledge from the history of the patient's deficiencies. A drawback is that site and extent of injury cannot be described without post mortem examination. (2) Fundamental comparative research is done on mammals phylogenetically close to man. Advantage: systematic experiments can later be exploited histologically. Drawback: the animal "answers" only indirectly. Thus, appropriate behavioral tests have to be carefully designed.

Let us now review the most important historical dates in behavioral research, that is in ethology. The exploration of animal behavior also goes as far back as antiquity. Comparative exploration of behavior was born at the turn of the century with the work of the American Charles Whitman (1899) and his coworker Wallace Craig (1918). On the ground prepared by Charles Darwin's treatise *On the origin of species by natural selection,* Whitman was able to ascertain in numerous observations on animals that the theory of evolution applied to particular behavior modes. He found that drinking habits were quite similar in several hundreds of pigeon breeds. Craig showed later that the repertoire of instinctive action consists of two basic kinds of behavior patterns: a variable, *appetitive behavior* which brings the animal into situations where it is likely to encounter a key stimulus, and a rigid, always similar *consummatory act,* which terminates the series of actions elicited by means of the *releasing mechanism.*

Whitman's and Craig's results received little attention at first and were masked by the feud of the prevalent research approaches of the mechanists (Dogma: all types of behavior are based solely on reactions to external stimulation; fault: denial of any spontaneous behavior), and the vitalists (Dogma: guidance of the purpose of behavior by divine natural force; fault: inhibition of any causal-analytical research). Against this background, two research orientations developed which in part still exist today. One of them emphasizes the study of types of behavior which relate to learning processes. The behaviorists belonging to it trace their tradition to the findings of the Russian physiologist Ivan Pavlov (1849–1936). They have obtained important insight into the fundamentals of learning psychology with the labyrinth and the Skinner box experiments. The other approach fixed its attention – surely just as much one-sidedly – on the innate character of behavior modes. Its representatives pushed the study of instinct. Thus, Heinroth (1871–1945) found that there are certain expressive gestures which are specific for each species and, within the species, invariable in form. Modes of behavior which are often expressed as special abilities apparently form an important element upon which the selective forces act in the evolution of a species. The following period was dominated by the studies of ethologists Jacob von Uexküll, Karl von Frisch, Konrad Lorenz, William Thorpe, and Niko Tinbergen. New concepts were born, such as *innate* and *acquired releasing mechanisms, imprinting, drive intensity,* and ideas of *central nervous hierarchies* for sensory and motor functional processes of *fixed action patterns* were also developed. In 1973, von Frisch, Lorenz, and Tinbergen, the founders of comparative behavior research, were awarded the Nobel Prize.

Before strictly descriptive behavioral research was threatened with degeneration, the zoologist Erich von Holst (1908–1962) drew attention to the necessity of analyzing quantitatively the physiological fundamentals of behavior patterns. He discovered the *reafference principle* which states that every command to the musculature is stored in the central nervous system as a kind of copy, which is compared with the execution of the order, and then eventually computed for correction. Furthermore, his most important work was on automatic central nervous processes which play a role in *coordination of locomotion* in vertebrates and invertebrates. Thus, in the spinal cord of fish he demonstrated the separate identity of the motor neurons which activate the musculature of the fins, and autorhythmic neurons which coordinate their beat. Electrical brain stimulation experiments in the freely moving chicken for the first time gave important insight into the *central framework of drives.* The scientific work of Erich von Holst is characterized by three qualities:

1. a wealth of ideas for experimental inquiry,
2. simplicity in the experimentation required,
3. experimental versatility.

Erich von Holst proved with his work that one can also solve certain neurobiological problems without elaborate instruments. His studies on the physiology of behavior secure his place as a founder of neuroethology. Building in part on the ideas and results of von Holst, comparative physiology of behavior has established a wealth of facts and rules. With the introduction of electrophysiology research into the sensory and motor processes of animals had, for the first time, a foundation which was open to physical analysis. The progress in the study of the *cellular analysis of behavior* in invertebrates results from the pioneering work of C. A. G. Wiersma. He confined his attention to crustaceans, starting with the investigation of motor neurons, then turning to the roles of interneurons and currently working on the coding and integration of sensory input. Meanwhile the approaches and strategies of Wiersma are applied to insects, molluscs, and annelids – and in part even to vertebrates.

What is neuroethology?

1. It is a research orientation which seeks to establish types of behavior on neural mechanisms.

2. From a historical viewpoint neuroethology is rooted in the field of brain physiology. Its newest sub-disciplines include the comparative physiology of behavior.

3. Neuroethology draws its problems from the biology of behavior. The foundation of experimental exploration is the quantitative analysis of behavior.

4. Even with our modern methods we cannot immediately expect exhaustive answers to question about the neural basis of behavior in highly developed animals given the complexity of their brain organization. Nevertheless neuroethology strives to explore complex systems, not directly, but through a comparative approach.

5. The principles of neural functions are studied in different animals (including the invertebrates) at various levels of integration. Some fundamental similarities may be expected in all animals whereas inter-species variations may be necessitated by their special ecological and behavioral adaptations.

6. Through common problems neuroethology aims to work as a catalyst to promote interdisciplinary research among scientists from various fields, such as zoology, medicine, psychology, and communications science.

Historical Development of the Concept of Neuroethology [2]:

1939: E. von Holst with his studies on *Die relative Koordination als Phänomen und als Methode zentralnervöser Funktionsanalyse* (The relative coordination as a phenomenon and as a method of analysis of central nervous function) laid the foundation for **Verhaltensphysiologie** (comparative behavioral physiology).

1951: N. Tinbergen used the term **ethophysiology** in connection with studies carried out by Hess and Bruegger in cats on the triggering of motivated behavior by electrical stimulation of the hypothalamus.

1961: J. Segaar studied the effect of frontal brain lesions on innate motor behavior in the three-spined stickleback and proposed the term **ethoanatomy** for this type of investigation.

1963: J. L. Brown and R. W. Hunsperger introduced the term **neuroethology** in connection with their studies on the activation of agonistic behavior modes by electrical brain stimulation in cats.

1970: The term **neuroethology** was more closely defined by G. Hoyle in the context of a review on cellular mechanisms underlying behavior of invertebrates.

1.3 Problems

Emphasizing the "synthetic" aspect of neuroethology is justifiable. It joins scientific disciplines which have had a totally separate historical development up to now. One of them (neurophysiology) should to a certain extent explain the other (ethology), which can result in scientific tensions. For instance the ethologist who is aware of the variability and plasticity of behavior considers it pointless to infer causes of behavior from the recordings of electrical potentials in single nerve cells of the brain. The neurophysiologist, however, has learned that even a few neurons linked together can produce unpredictably complex responses. Thus, correlations with the animal's behavior require great caution. In fact, one should respect neurophysiologists as well as ethologists who raise such objections. However, it is better to indicate rather than generally to forbid new ways. Furthermore, restrictions should be replaced by new ideas for future work seeking to integrate behavior and physiology more closely.

For the neuroethologist conducting neurophysiological experiments the most important factor is to always consider those parameters which are significant for the type of behavior under investigation.

Otherwise there is the risk of seeking the solution to the problem in the wrong direction. An anecdotal illustration (Fig. 5): A pedestrian lost his glasses one night on a dark street. He asked passers-by to help him look for the glasses near a street lamp; after a few moments of fruitless searching they asked him if he were sure that that's where he dropped them. He had to admit that he was not, but that it was a lot easier to look where there was light.

Fig. 5. ... because there is more light here

2 How Does a Stimulus Elicit Muscle Contraction?

2.1 Neural Building Blocks

Man and animal interact with their environment through sense organs responsive to external stimuli and effector organs that cause movement. Sense organs (receptors) and effector organs (effectors) are related through the nervous system. Sensory cells, nerve cells, and muscle cells have one thing in common: at rest, owing to the nature of their cell membranes, they represent charged batteries, negative on the inside, positive on the outside. Their activities are correlated with fluctuations in the potential difference across cell membranes; these changes lead to neurophysiological responses. Sensory cells specialize in transforming physical or chemical energy into a nervous code to which the nerve cells respond and this response they conduct further with or without modification depending upon the properties of the conduction path. Muscle cells alter their resting potential after receiving the nervous information, and this initiates processes which lead to changes in cell shape (contraction).

2.1.1 Neural Tissue

The neural tissue consists of nerve cells (neurons) conducting electrical potentials and of glial cells (neuroglia) which have insulating properties and fulfill supporting and metabolic functions. A neuron is a nerve cell together with all its processes (Fig. 7). According to the number of processes, one distinguishes among unipolar (Fig. 6), pseudounipolar, bipolar, and multipolar neurons. The area of the neuron containing the nucleus is called the soma, and the plasma around the nucleus, perikaryon. Processes which conduct electrical potentials to the nerve cells are called dendrites and that which conduct impulses away from the soma are called axons (neurite). The process of a unipolar nerve cell is always a neurite. Exception: e.g., the amacrine cells of the retina which have only dendrites (see Fig. 27f). Dendrites and axons have different functional properties (see Table 3). One or more of the dendrites with extensive ramifications pick up the information from the sensory or nerve cells in contact with them; the processed information is then

Eye

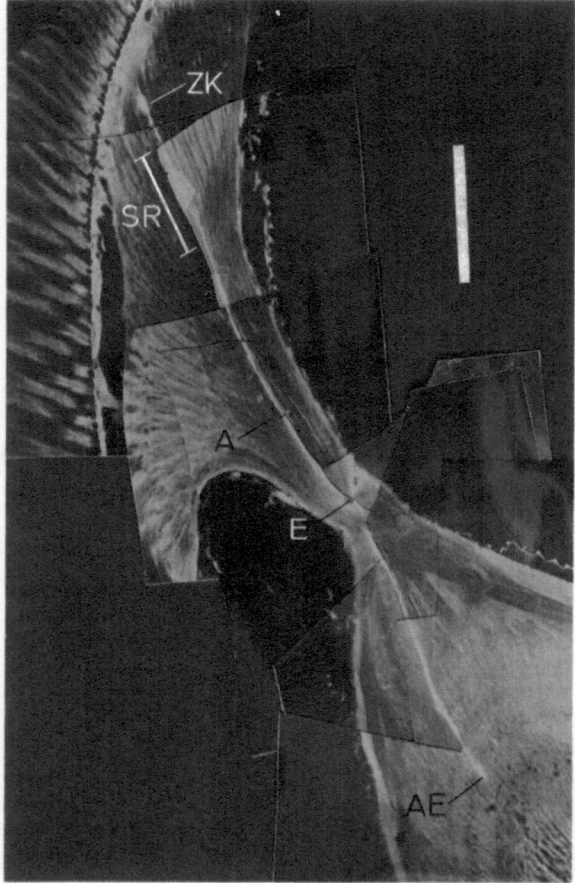

Brain

Fig. 6. Labeling of an axon from a monopolar neuron from the eye of a fly by intracellular injection of a fluorescent dye (Procion yellow, see Chap. 9, Methodological Appendix). Photomontage of 7 μm serial sections, each showing a portion of the stained neuron. *A* axon, *AE* terminal bud of axon, *E* site of micro pipette (tip) serving both as recording electrode and as injection cannula for the marker, *SR* synaptic region with presynaptic photoreceptors; *ZK* cell body; scale mark: 100 μm. [Zettler F (1975) Umschau 75: 118–120, by permission of the author]

conducted by the single axon onto the succeeding neurons or effector organs (muscles, glands).

Morphological measurements indicate that axons of up to 8 μm in diameter often attain lengths of up to decimeters. The axon, after leaving the axon hillock of the cell body (Fig. 7), and after a short bare region, becomes surrounded by glial cover cells which divide the axon into regular segments (Fig. 7). The axons with their sheaths are termed nerve fibers. The cover sheaths may produce myelin and thereby insulate the fiber and increase its speed of conduction. The narrow gap between two

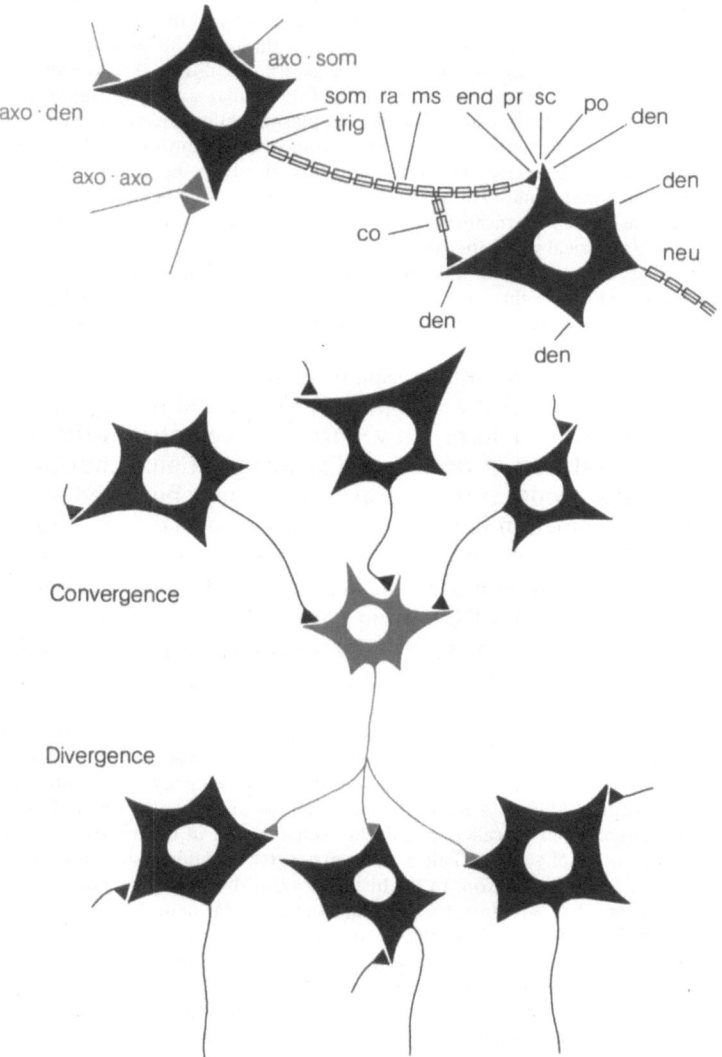

Fig. 7. Functional structures of a nerve cell: *co* axon collateral, *den* dendrite, *end* axonal ending (terminal button), *ms* myelin sheath, *neu* neurite (axon), *po* postsynaptic membrane, *pr* presynaptic membrane, *ra* node of Ranvier, *sc* synaptic cleft, *som* soma, *trig* trigger zone (axon hillock). Examples of synapses: *axo-axo* axo-axonic, *axo-den* axo-dendritic, *axo-som* axo-somatic

neighboring segments is called the node of Ranvier, and the segment between two nodes, the internode.

Apart from the nucleus, Golgi apparatus, and mitochondria the nerve cell possesses a well-developed endoplasmic reticulum which appears under the light microscope as Nissl bodies (Tigroid). Nissl granules are also found in the broadly based dendrites of large nerve cells, whereas they are absent in the axon hillock and the axon. With silver staining the neurofibrils are made visible. They have a threadlike fibrillar organization in the perikaryon and can be followed right into the axon and dendrites. Electronmicroscopically they are seen as fibers (neurofilaments) and fine tubular structures (neurotubules). The protein synthesis in the ribosomes of the endoplasmic reticulum is thought to contribute to the development and maintenance of the neurons which, at the level of the neuroblast have lost their mitotic capacity. The proteins synthesized by the ribosomes also probably mediate functional modifications of shape. The participation of endoplasmic reticulum in the synthesis of neurotransmitter appears probable. Within the axonal ending (terminal button) are spherical structures, the synaptic vesicles, which contain the neurotransmitter. Cell bodies and processes are confined by a plasmalemma (cell membrane). This membrane, which might have a different structure according to location, plays an important role in generation, conduction, and transmission of electrical activity.

The properties of the nerve tissue of vertebrates can best be understood from its embryological development. The neural tissue originates from the ectoderm and in its primitive form is represented by the neural tube and the neural crest. A dorsal, median invagination of the ectoderm, the neural plate, produces the neural groove; this is bordered by the neural crest at the junction between skin and neuroectoderm. Gradually the groove closes itself into the neural tube, detaches itself from the ectoderm, and the neural crests are translocated into the deeper layer. The nerve and glial cells of brain and spinal cord derive from the tube, whereas nerve and glial cells (peripheral glia) outside the central nervous system originate from the neural crests.

Initially the tube consists of a single layer of cylindrical neuroepithelium. By proliferation, displacement, and differentiation, a multiple layered cellular aggregation is formed around a cavity with zonal stratification consisting of the following layers: a ventricular layer of ependymal spongioblasts, a mantle zone of neuro- and glioblasts as well as a cell-free marginal zone. Each so-called radial glial cells located in the ventricular zone send a process to the outer surface. This process may serve as guiding structure for the leading process of a migrating neuron. In the histogenesis of the nerve cell, nuclear growth is the first indication of progressive development. In the apolar neuroblasts fibrillogenic zones arise due to increased density of the plasma at one or both nuclear poles. Out of these are formed the processes of the temporarily bipolar neuroblasts. As the process close to the ventricle degenerates, the unipolar neuroblast is formed. The remaining process grows in the direction of the marginal zone and becomes the axon (neurite). Subsequently, on the soma, additional small plasma growth cones (primordial dendrites) can appear which then characterize the bipolar or possibly multipolar nature of the neuroblasts. For instance in a spinal cord segment the neurite of a multipolar motor neuron of the ventral horn leaves the marginal zone (the further white matter), and thereby the central organ, and approaches its innervation area, participating at the same time in the formation of ventral and dorsal root nerves. The bare axis cylinder is the guiding structure for the glioblasts, originating in the neural crests, which will become the Schwann cells. They "seek" the axons – being capable of amoeboid movement – position themselves in serial order, encircle the axis cylinder and cover it by formation of protein-lipoid lamellae. These structures are visible as segmented myelin sheaths.

In the central organs (brain and spinal cord) the myelin sheath is formed from the oligodendroglial cells. The thickness of the myelin sheath and the length of the sheath segments are proportional to the size of the axon cylinders. The nodes of Ranvier of the central and peripheral fibers lead to *saltatory* impulse propagation, and their intervals determine conduction velocity.

We summarize in a survey diagram the most important differentiation steps in the formation of neural tissue of vertebrates (Table 1).

Table 1. Histogenesis of neural tissue

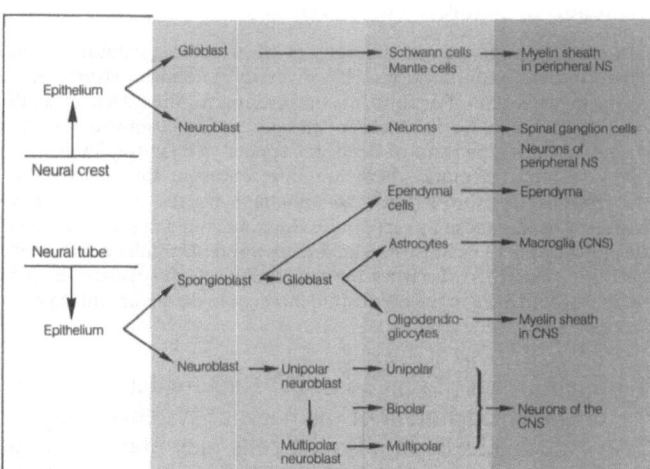

The histogenesis of the vertebrate neuron shows:

1. The neurite (axon) originates as a first process of the neuroblast.

2. The myelin sheath is not a product of the nerve cell.

3. The segmentation of the myelin sheath is due to a serial arrangement of single Schwann or oligodendroglial cells around the axon of the nerve cell.

The neurons communicate with each other via specialized regions of contact known as synapses (Fig. 7). These contact points exhibit morphological and chemical peculiarities. The concept of the synapse was introduced by physiologists and was at first given a strictly functional interpretation. Today the morphological substrate is known. A synapse consists of a presynaptic membrane (generally an axonal ending of a neuron) and a postsynaptic membrane (often dendrite or soma of a subsequent neuron). The two membranes are usually separated by a narrow synaptic cleft of about 200 Å (ranging from 130 to 400 Å among different types; chemical synapses, see Fig. 25 top left). There are also appositions with 20–25 Å spaces named "gap junctions" or low-resistance electrical synapses (see Fig. 25 top middle). The gap consists of an array of 20 Å channels which connect the *intracellular* spaces. The intercytoplasmic channels allow ions to pass between both neurons. There are also appositions without gaps which function as high-resistance seals; they are called "tight junctions" (cf. Fig. 25 top right). Synapses are also classified according to the site of the contact: some of these types are sketched in Figure 7 (top left).

Neurons of vertebrates when compared with neurons of invertebrates exhibit similar morphological components: soma, axon, dendrites. The diversity of dendritic structures is particularly striking in the invertebrates. Possibly this compensates for the relatively small number of neurons in the CNS that are available for the processing of information. In vertebrates, however, the branching patterns of dendrites appear to be more defined for the particular neuron types. Furthermore, there are differences in the neuropil of vertebrates and invertebrates. In the latter, axo-somatic synapses are extremely exceptional [1]. Myelin sheaths like those of vertebrates are not formed in invertebrates; the fibers possess little or no myelin (e.g., loose myelin sheaths). Rapid impulse conduction is ensured by relatively thick, giant fibers. To attain the conduction velocity of a myelinated fiber, the myelin-free fiber would have to have a 50-fold increase in diameter and expend 100 times the amount of energy.

Neuroglia may fulfill a number of functions: (1) Radial glial cells form guiding structures for the arrangement of the nervous system during its development. (2) As mechanical systems glial cells may play a role in regeneration: during proliferation Schwann cells form the so-called Hanken-Büngner bands at a sectioned axonal stump; the bands form guiding structures for cones growing out of the axonal stump (axonal growth cones) and leading to the postsynaptic neuron. (3) The myelin sheaths laid down by Schwann cells provide electrical insulation for the axon. (4) Astrocytes fulfill a trophic function for the neuron and serve as mediators for the transfer of chemical substances between cerebral capillaries and neurons. (5) Apparently neurons need glial cells in their proximity in order to synthesize neurotransmitters. Schwann satellite cells are capable of storing transmitters and they may also be able to form them under certain conditions.

2.1.2 Cell Architecture

With its numerous dendrites a neuron is able to pick up and process *different information* from many presynaptic neurons (convergence principle). The result of such processing however can be conducted further in only one way, that is by the axon (Fig. 7). Through collateral branches originating from the axon, the *same information* can be

Fig. 8. Five different neuron
types from mammalian cerebellar cortex. *1* Granular cell,
2 Purkinje cell, *3* Golgi cell,
4 basket cell, *5* stellate cell.
Neurite *red*, dendrites *black*,
soma *black*, cell nucleus *white*

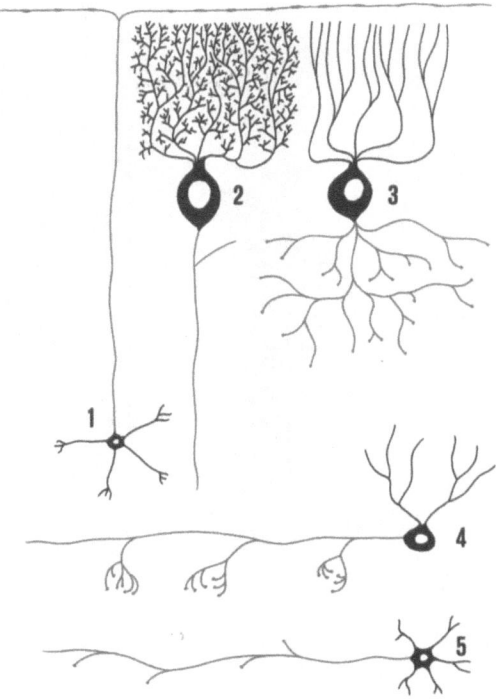

conducted to several postsynaptic neurons (divergence principle).
Example: Figure 8 shows five different types of neurons from the
cerebellar cortex. They are all constructed on the same basic plan:
dendrites, soma, and axon (neurite). By virtue of the different numbers
and structures of the dendrites and axon collaterals they already have
specific patterns of connections that represent the following principles:
types 1 and 4: mainly divergence; type 2: convergence; types 3 and 5:
convergence and divergence. Finally, the structural organization of the
cerebellar cortex is determined by a specific arrangement of neuron
types in the three spatial planes (Fig. 9). We call this *cell architecture*.
(For neuroanatomical techniques for the exploration of fiber connections in the brain, see Chap. 9, Methodological Appendix and Fig. 6).

2.1.3 Neural Function Structures

Several hundred million neurons participate in the composition of a
"simple" vertebrate brain. The human brain consists of about 10^{11}
neurons. A single neuron (e.g., a Purkinje cell of the cerebellum) can
receive inputs over its numerous dendritic ramifications from nearly

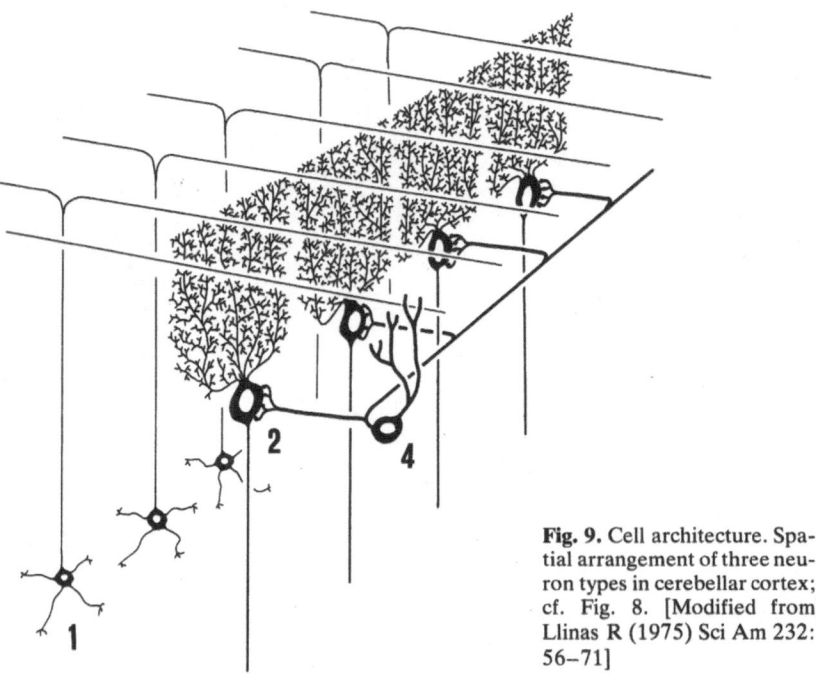

Fig. 9. Cell architecture. Spatial arrangement of three neuron types in cerebellar cortex; cf. Fig. 8. [Modified from Llinas R (1975) Sci Am 232: 56–71]

200,000 synapses with other neurons (convergence). Contrariwise it is possible for a single neuron (e.g., a pyramidal cell of the cerebral cortex) to convey information to thousands of other neurons over many axon collaterals (divergence). The convergence and divergence phenomena – together with the three fundamental arithmetic operations (see p. 41) – give us insight into the means by which the central nervous system processes data. In view of this complexity it appears at first sight hopeless to try to correlate the functions and morphology of the neurons in the brain. However, the central nervous system in vertebrates is not constructed in such a way that processing stages can always be performed by a few neurons or even by a specific type of neuron alone. (But there are exceptions, for example the Mauther neuron, cf. p. 210.) Particular processing steps are often multiplied so that many neurons with certain functional properties work simultaneously in the same manner. Such a redundancy in neuronal populations and functions provide security in several respects, such as (1) a fail-safe guarantee for the function, and (2) compensation for the decay of neurons in many areas, which goes on from birth to death. (The neuron has lost its mitotic capacity; its decay is – for anatomical reasons – an irreversible loss).
Nerve cells which contribute to a certain function are often localized together in defined areas or layers of the brain (Fig. 10). Such regions are

called nuclei, ganglia, areas, regions, strata; for example, nucleus gracilis, dorsal root ganglion, preoptic area, regio prepiriformis, stratum griseum. These areas are connected by fiber tracts formed by the axons

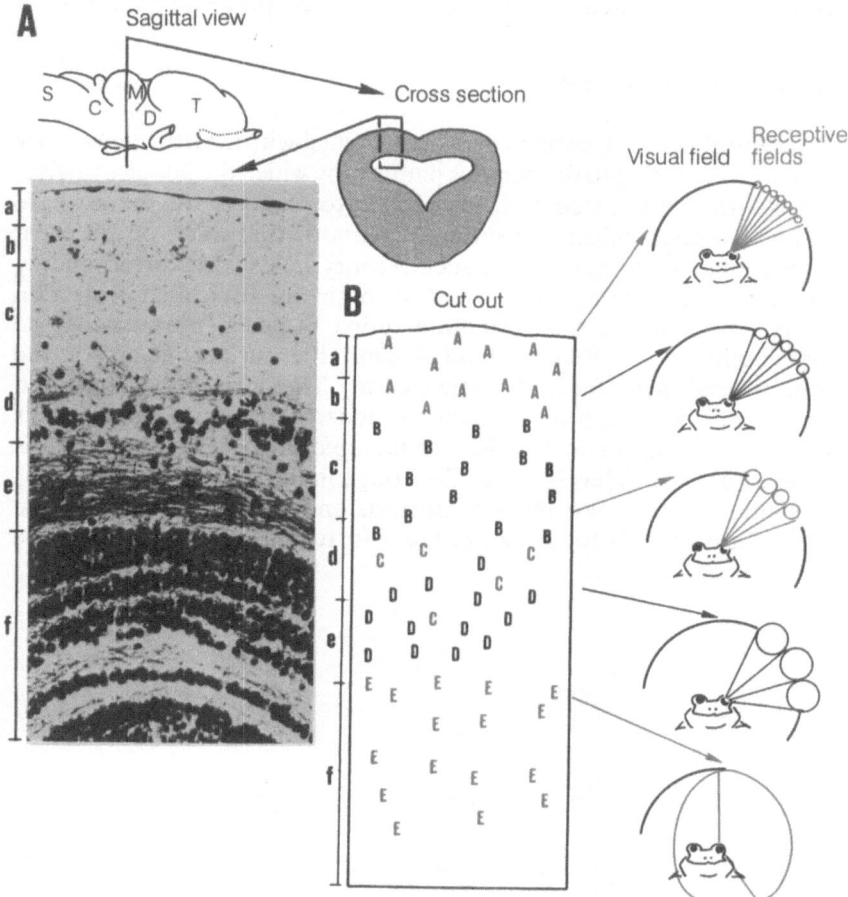

Fig. 10. A Structural patterns in the brain. The *rectangular section* shows the layered histological structure of the optic tectum in the midbrain of a toad. Neurons with identical or similar response characteristics are frequently localized in the same layers. (Combined cell and fiber stain by the method of Klüver and Barrera). *T* Telencephalon, *D* diencephalon, *M* mesencephalon, *C* cerebellum, *S* spinal cord. **B** Distribution of different neuron types in particular tectal layers. *A–C* Sites of retinal fiber terminals according to three classes of ganglion cells projecting to the optic tectum, class R2 *(A),* class R3 *(B)* and class R4 *(C),* distinguished by the size of their excitatory receptive fields. *D* Tectal T5(1) and T5(2) neurons with relatively small excitatory receptive fields; *E* tectal T4 large-field neurons (cf. also Fig. 74). By no means have all response types been included here! The map is based on microelectrode markings; see p. 105. [Modified from Ewert JP, Wietersheim A v (1974) Acta Anat 88: 56–66]

of nerve cell groups; they are named tractus (tract), fasciculus (bundle), funiculus or – as connection between two hemispheres – commissure; for example: pyramidal tract, forebrain bundle, anterior commissure. A discrete area of synaptic interaction between dendrites and axonal arborizations is called a *neuropile;* for example: pretectal neuropile.

2.1.4 Topographic Maps

The central nervous system correlates a signal with a particular sensory modality according to the nervous pathway by which the impulses arrive. Histologically the receptor fields of the eye, the ear, and the skin are mapped topographically in different parts of the brain, on different levels. Fiber pathways linking such sensory fields with corresponding connecting and processing levels of the brain are also called projection tracts. For instance the vertebrate retina projects – preserving the topography of its sensory field – onto the surface layers of the contralateral midbrain roof (optic tectum). The projection is mediated by the optic nerve as projection tract (optic tract) (Figs. 11 and 13). This structural arrangement is called *retinotopic projection.* The retina is mapped at various levels of the CNS (e.g., in the visual cortex).

In similar manner also the skin projects onto different regions of the brain. In mammals for instance, the skin regions are represented via

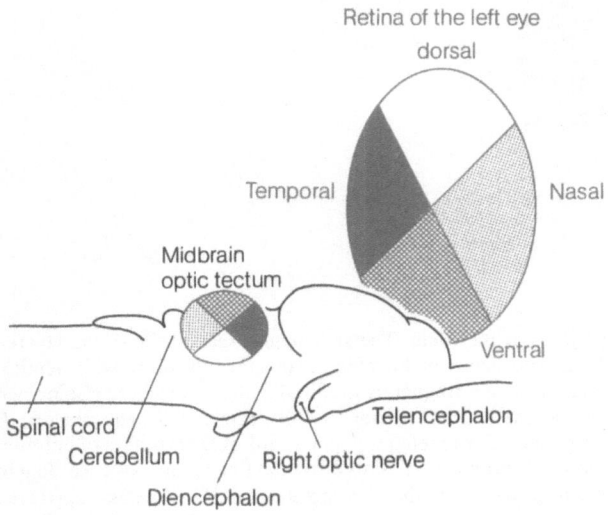

Fig. 11. Projection from retina of left eye to the right optic tectum. Topographical correlations have been roughly schematized for the retinal quadrants. The principle is illustrated by the example of a toad, but is similar for all other vertebrates studied. [Ewert JP, Borchers HW (1971) Z Vergl Physiol 71: 165–189]

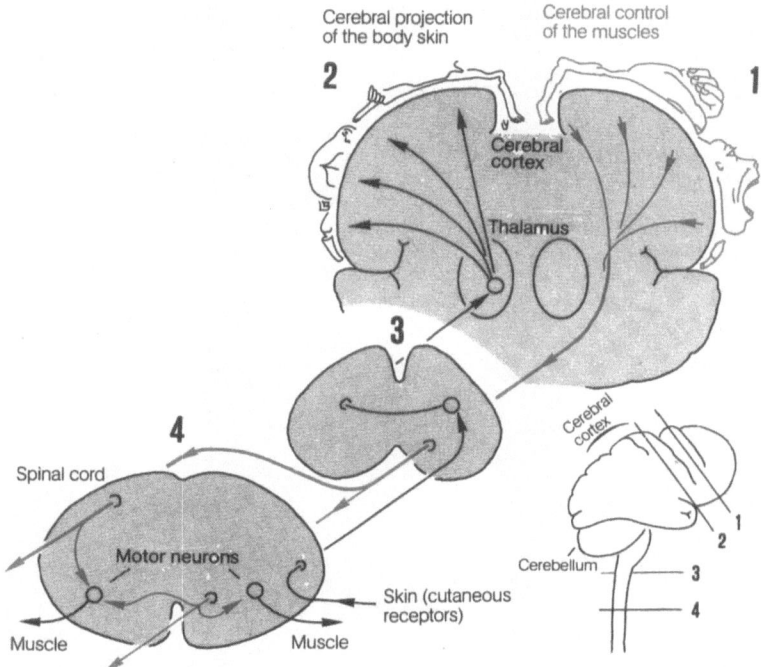

Fig. 12. *Black,* Projection of skin to surface of the sensory cerebral cortex of man. *Red,* topographical relationships for the control of the body musculature from the cerebral motor cortex. The proportions of the two drawings of homunculi correspond roughly to the dimensions of the correlated representation fields in the brain. [Modified from Penfield W, Rasmussen T (1950) The cerebral cortex of man. Macmillan, New York]

"relay stations" in particular surface layers of the cerebral cortex (Fig. 12, black). The spatial extent of cortical representation is correlated with the biological significance of the corresponding skin area. We call this topographic arrangement *somatotopic projection.* In an opposite way, so to speak, the muscle groups of the corresponding region of the body are controlled by a neighboring brain area (Fig. 12, red). By way of these fiber paths, known as pyramidal tracts, the activity of the motor neurons in the spinal cord can be influenced directly or via interneurons. The spatial extent of the muscle representation in the cerebral cortex is correlated to the behavioral importance of the muscle.

With respect to the ear (basilar membrane), there are also *tonotopic representations* in the brain (e.g., auditory cortex, cf. p. 182).

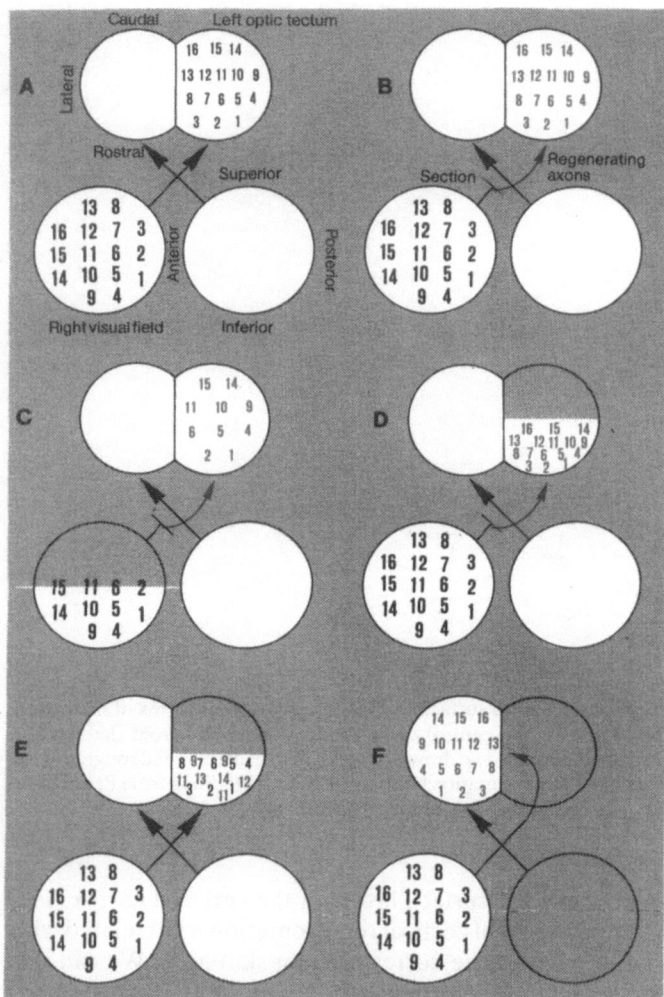

Fig. 13 A–F. Re-arrangements of the retino-tectal projection. **A** Normal projection of the visual field from the right eye onto the left optic tectum. **B** *Reorganization* of the retinotopic map after cutting (or crushing) the right optic nerve. **C** *Expansion* of the projection after partial retinal lesions. **D** *Compression* of the projection after tectal lesions and optic nerve transsection. **E** *Aberrant* projections after tectal lesions if the optic nerve is left intact. **F** *Re-routing* of optic tract axons to the wrong optic tectum after ablation of the contralateral tectum and enucleation of the other eye. Schematized diagrams from results obtained in fish and frogs. [Gaze, RM (1970) The formation of nerve connections. Academic Press, London New York **(A, B, D)**; Horder TJ (1971) J Physiol 216: 53–55 **(C)**; Udin S (1977) J Comp Neurol 173: 561–582 **(E)**. Ingle D (1973) Science 181: 1053–1055 **(F)**]

2.1.5 Neuroplasticity

Topographic connections between sensory areas and central projection fields have a genetic basis. If a projection tract is transsected, processes grow out of axon stumps (axonal growth cones), and under certain conditions they may regenerate the neuronal network by the formation of new synapses. In the CNS, the regeneration capacity is greater, the farther down the animal stands on the "phylogenetic scale". In lower vertebrates regeneration takes place in adults. Among mammals this capacity is present mainly in neonates.

How does a previously sectioned regenerating axon find its right "address" in the projection field?

This phenomenon can be studied with particular clarity in retinotopic maps (e.g., Fig. 13 A). If the projection tract from one eye (optic tract) leading to the contralateral optic tectum has been cut in a frog or fish, the animal is at first blind to stimuli presented in the visual field of this eye. However, after some time, it is again able to respond: processes growing out of the axon stumps have apparently re-innervated *their* corresponding neuronal areas in the central projection field; the topographic map has been restored (Fig. 13 B).

What happens, however, if, after sectioning of the projection tract, also a part of the projection field – and thus some of the "addresses" – is eliminated?

In this case the visual field projection of the whole eye may be re-organized in compressed form in the remaining optic tectum (Fig. 13 D). If, however, the projection tract in the same operation remains intact, there is no compression of the visual field in the residual tectum (Fig. 13 E). Instead the growth cones of optic nerve fibers from the destroyed tectum form an unorganized abnormal projection in the intact residual area which is superimposed on the original unaltered one.

If, in a "reverse type" of experiment, a part of the retina is destroyed, the corresponding retino-tectal fibers degenerate and the projection of the remaining retina may spread over the entire optic tectum (Fig. 13 C).

Let us now go a step further and ask, for example, which way the projection tract of the right eye is directed when their projection field in the left tectum, and in addition, the left eye are eliminated. Now the optic fibers of the intact eye cross the median line of the midbrain and there innervate the "wrong" tectum while retaining the topography of the sensory area (Fig. 13 F).

The results of these regeneration experiments nevertheless pose a number of partly unanswered questions: what mechanisms are responsible for the regeneration and eventual reorganization of the topographic mapping?; how precise is the topography?; is there an exact restoration of synaptic connections?; are all of the neuronal steps of information processing re-organized? Some behavioral correlates of the above experiments will be discussed further in a later chapter (p. 132).

The problems of how a regenerating axon again finds its postsynaptic addressee, how it "knows" in which direction to grow and where to stop cannot as yet be completely answered. From what we know today, a regenerating axon presumably does not find the address in one attempt, but in a sequence of steps, partly preprogrammed genetically. The mechanism for the direction of each step is continually being adapted to the special conditions and situations of the substrate environment. There is some evidence that certain chemical affinities play a role here. If a motor axon of a skeletal muscle is cut, the muscle fibers become chemically hypersensitive; receptors for the neurotransmitter acetylcholine are then spread diffusely over the entire membrane surface. Only when the regenerating nerve re-innervates the muscle will the acetylcholine receptors be concentrated within the areas of synaptic contacts. Another example of chemical effects: from salivary glands of male mice a protein has been isolated which apparently functions as a "nerve growth factor" (NGF) in the development of the sympathetic nervous system [147]. If the protein substance, extracted from the mouse, is injected into a rabbit and the rabbit's serum with the new antibodies is applied to mouse embryos, the sympathetic nervous system is not formed in these animals: the parasympathetic nervous system, including the dorsal root ganglia, however, remains intact. Behaviorally these adult mice respond only weakly to stressful stimuli (p. 249).

To summarize:

1. Functional units in the vertebrate CNS consist of neuronal *cell assemblies*. They take the form of *neural networks*, which may be organized as clusters, layers, columns, or barrels. However, there are examples in vertebrates – and predominantly among invertebrates – where important functions are exerted by single neurons (Mauthner neuron in fish; lateral giant neuron in crayfish, p. 208).

2. Specific developments of dendritic patterns and axonal structures provide a basis for the neuroarchitecture of mutually connected neural networks, and thus determine – to some extent – the nature of physiological interactions among neurons.

3. Important principles of neuronal data processing are based on certain anatomical and functional interaction patterns, such as convergence, divergence, and feedback loop (cf. also p. 41).

4. There are firm connection patterns between sensory fields of sense organs and their projection fields in the brain. These topographic maps have a genetic basis.

5. Maps exist in sensory as well as in motor systems. According to requirements of behavioral functions in an animal species, corresponding "over-representations" may be developed *within* those maps (for example cf. Fig. 104). There are also structures for sensori-motor interfacing.

6. Topographic projections, after severing of the tracts, may have the ability to anatomically re-form their original connections by re-growth of transected axons. Re-mapping depends on pre-existing chemical gradients as well as certain polarity and structural relationships, according to available space in the brain and interaction mechanisms between the axonal terminals.

7. Multiple maps of the same sensory field in different parts of the brain are interconnected topographically. Loss of function caused by partial lesions of connecting tracts may also be "compensated" by *functional recovery*, due to (1) activation of silent pathways, e.g., axon collaterals of adjacent intact tracts, and (2) re-arrangement of functional interactions.

2.2 A Behavioral Experiment

Suppose a toad or a cat sits in front of us. We touch its flank carefully with a bristle. The animal responds to the stimulus with a directed wiping movement of its hind leg. The movement is destined to remove the annoying skin irritation. If the wiping is elicited by the same cutaneous stimulus repeatedly at short intervals the responsivity diminishes until it

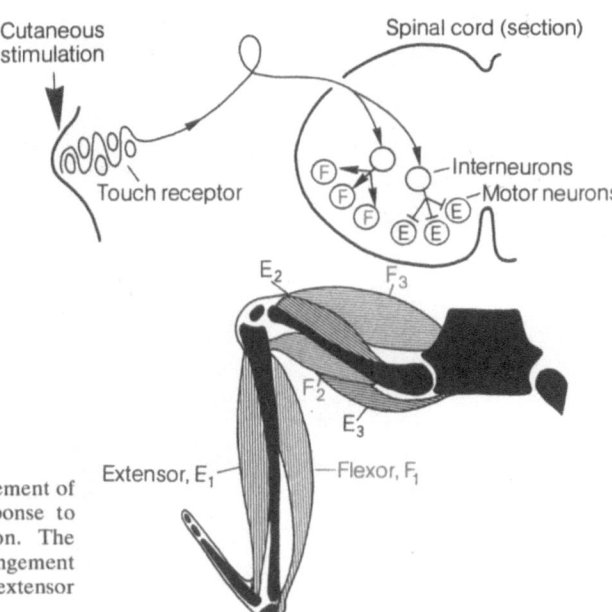

Fig. 14. Wiping movement of a vertebrate in response to cutaneous stimulation. The diagram shows arrangement of flexor *(red)* and extensor *(black)* muscles

is totally extinguished. This relatively simple stimulus response relationship prompts a whole range of questions. We shall examine first the connections among the receptors, neurons, and muscle fibers that are involved in the wiping movement. For this we choose the phase in which the animal activates its hind leg for wiping (Fig. 14). Starting from the cutaneous touch receptors the motor neurons for flexor muscles are activated, and those for extensor muscles are inhibited via specific interneurons in the spinal cord. To understand that the neural circuit triggered by the touch stimulus finally leads to the contraction of the flexor muscle, some neurophysiological fundamentals must first be discussed.

An interconnection between receptors and effectors via neurons of the central nervous system (spinal cord) is represented by the monosynaptic stretch reflex (Fig. 15). Its function is to keep the length of the skeletal muscle constant. If the muscle is extended passively, it can be shortened again to its original length by contraction, via a monosynaptic intrinsic reflex arc. It happens as follows: in the skeletal muscle there is a receptor, the muscle spindle, which responds to stretch. Its excitation is conducted along the sensory nerve fibers (spindle afferent) into the spinal cord to an α-motor neuron which induces – via its axon (α-efferent) – the muscle to contract.

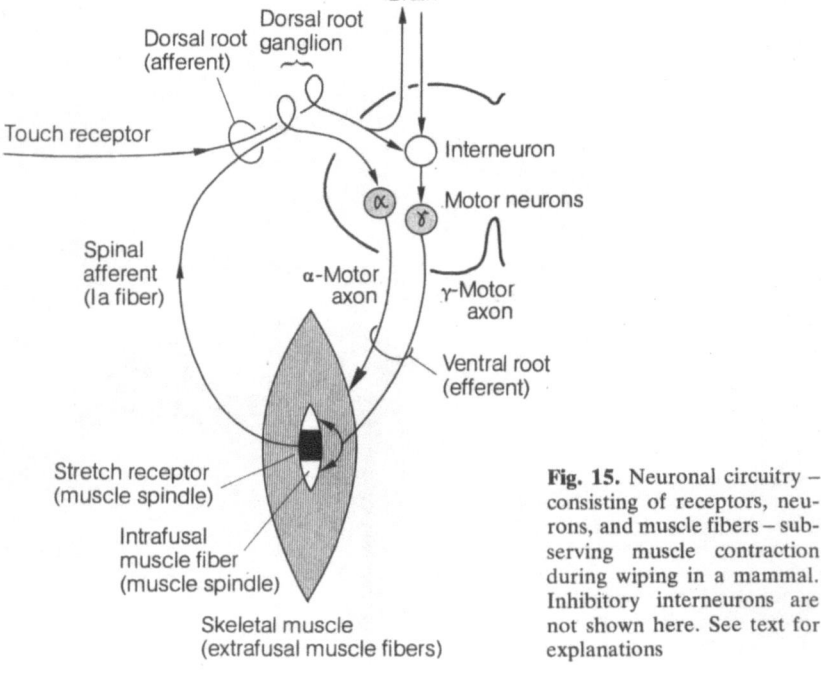

Fig. 15. Neuronal circuitry – consisting of receptors, neurons, and muscle fibers – subserving muscle contraction during wiping in a mammal. Inhibitory interneurons are not shown here. See text for explanations

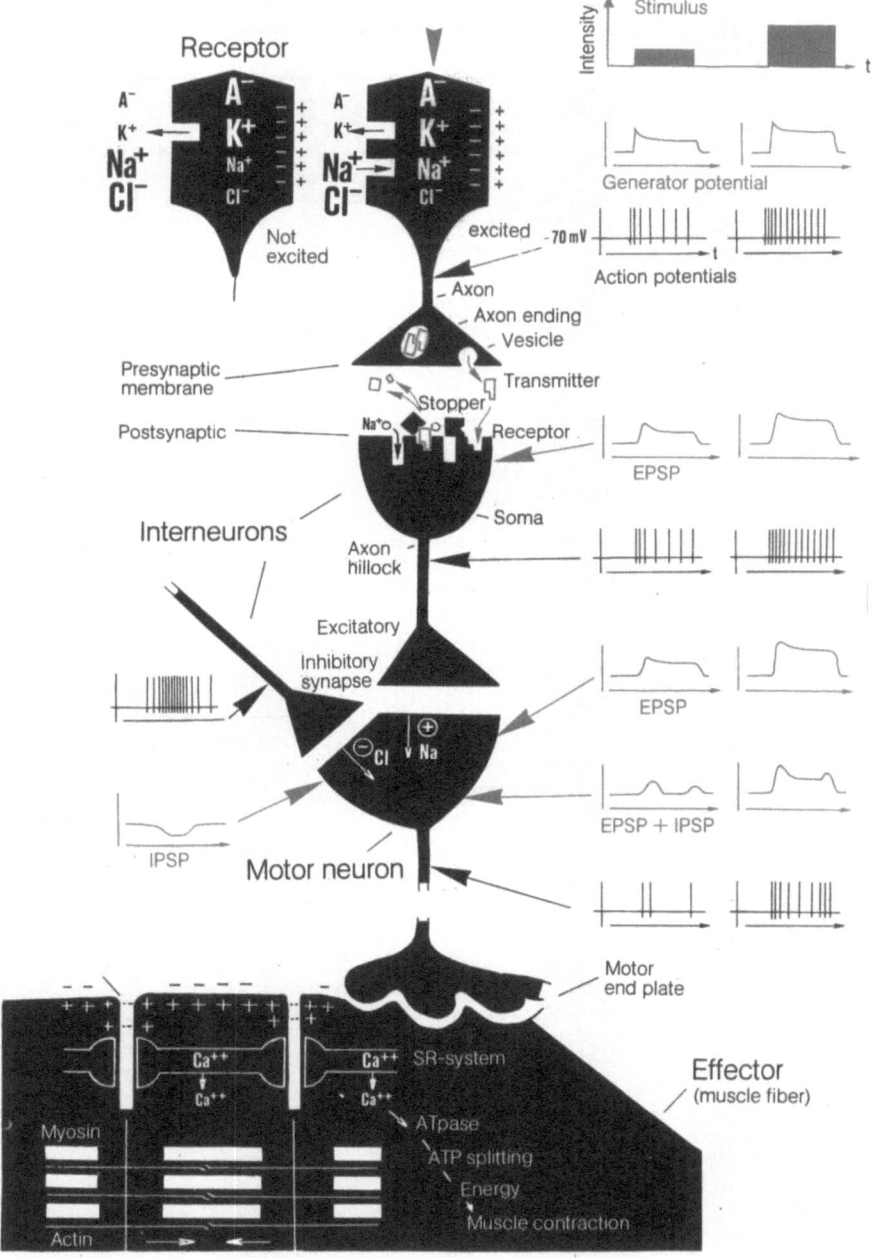

Fig. 16. Basic neurophysiological processes occurring between stimulation (receptor excitation) and behavioral response (muscular contraction). Strongly schematized for striated muscle; see text on p. 33 and Table 2

Table 2. Neurophysiological processes by which information is reaching and leaving the CNS

	Morphological substrate	Physiological process	Function	Task
Sensory cell (receptor)	Receptor membrane areas e.g., dendrites	Stimulus leads to change of ionic conductivity, development of a generator potential (GP)	**Stimulus/ excitation transformation**	Information reaches the CNS
	Adjoining membrane areas	Passive electrotonic spread of the GP		
	Trigger zone of receptor membrane	Suprathreshold electrical excitation (by the GP) triggers action potentials (AP's).	**Coding**	
	Axon (neurite)	Translation of a stimulus-induced depolarization into a sequence of propagated AP's according to the principle of frequency modulation. (There are also axons which in certain situations do not conduct AP's but instead conduct amplitude modulated potentials)	**Conduction of impulses**	
Synapse	Axonal ending, synaptic vesicles, presynaptic membrane, synaptic cleft	AP's release transmitter substance from presynaptic terminals	**Decoding**	Transfer of information from neuron to neuron
	Postsynaptic membrane (e.g., dendrite, soma)	Retranslation of the presynaptic AP-sequence (amplitude modulation). Chemical excitation by neurotransmitter causes postsynaptic potential, PSP. Nonlinear summation of the PSP's		
Neuron(s)	Soma membrane	Passive spread and summation of the EPSP's and IPSP's arriving from different synapses	**Arithmetic operations**	Information processing
	Axon hillock (trigger zone)	Suprathreshold electrical excitation by resulting EPSP causes propagated AP's	**Coding, impulse conduction**	

Table 2 (continued)

	Morphological substrate	Physiological process	Function	Task
Neuromuscular junction (motor end-plate)	Presynaptic membrane of a motor neuron, synaptic cleft	AP releases neuro-transmitter (acetylcho-line)	**Neuro muscular transmission**	Information leaves the CNS
	Post (sub) synaptic membrane	Chemical excitation by transmitter leads to end-plate potenti-al (EPP). Summation of EPP with miniature end-plate potenti-als (MEPP's)		
Muscle fiber (effector)	Fiber and transverse tubuli (T) system mem-brane	Electrical excitation by EPP is suprathre-shold and leads to pro-pagated muscle action potential	**Impulse con-duction**	
	Sarcoplasmic reticulum (SR)	Depolarization of ad-joining T-system mem-brane causes release of Ca^{2+} ions (exchange with Mg^{2+})	**Electromechani-cal coupling**	
	Myofilaments	Ca^{2+} ions activate ATPase (acto-myosin complex). Release of chemical energy by splitting off a phos-phate from ATP		
		Energy for "grip-re-lease mechanism" be-tween actin and myo-sin	**Fiber con-traction**	
	SR-system	T-system membrane is repolarized: SR takes up Ca^{2+} ions in ex-change for Mg^{2+}	**Fiber rela-xation**	
	Myofilaments	Inactivation of ATPase. Separation between ac-tin and myosin by for-mation of an ATP-myosin-Mg^{2+} complex ("plastici-zing" effect of ATP)		

Afferents are sensory nerve paths which conduct impulses (or amplitude modulated potentials) from a sense organ to the CNS. Efferents are motor nerve paths which conduct impulses from the CNS to an effector organ. Generally the terms afferent and efferent may be applied within the CNS in relation to a particular neuronal area.

How is the Length of the Muscle Determined? Obviously the range depends on the degree of stretch of the muscle spindle. The degree of stretch can be altered by contractile intrafusal structures at both ends of the spindle: these motor structures are innervated by a γ-motor neuron (Fig. 15). Thus, the range of the muscle length depends on the γ-efferent. In the monosynaptic reflex it is constant. If it is altered, however, the length of the muscle has to change correspondingly. This is the condition for a *directed movement* of the extremity.

What is the Origin of the Input to the γ-Neurons? In our experiment – the release of the wiping reaction – it comes from the touch receptors via interposed interneurons. The "command" for a wiping movement, however, can also be generated in the brain and executed via excitation of the γ- or α-neurons (Fig. 15). The neural circuit for the spatial-temporal contraction pattern of the muscle groups participating in the wiping movement must preexist in the spinal cord as a fixed motor program. Thus, in the toad, the wiping movement can still be released by a tactile stimulus after ablation of the brain.

In mammals there is evidence that contraction of musculature can be triggered either by the γ-loop or – for immediate response – by directly activating the α-neurons. At least voluntary movements may require α-γ-coactivation. In frogs, unlike in mammals, the intrafusal muscle fibers are innervated by the same motor neurons that also innervate the extrafusal muscle fibers. Thus, spindle muscles cannot be activated independently of the skeletal muscle.

How can the extinction of the wiping reaction after repeated stimulation be explained? There are in principle several possibilities: (1) The cutaneous touch receptors are adapted; (2) the leg musculature is fatigued; (3) habituation to the stimulus is governed by central nervous mechanisms. (*Adaptation* is the cessation of neuronal response following continuous stimulation. *Habituation* is the absence of a behavioral response after repeated presentation of the key stimulus.) These possibilities can be tested experimentally. The first is excluded by neurophysiological results since the response of the sensory fibers remains unchanged by touch stimuli that repeat every 5 s, as in the behavior experiment. The second possibility can also be ruled out since after habituation of the wiping reaction following stimulation of cutaneous area A, the response can be immediately released by stimulation of neighboring area B[3]. This leaves us with the third possibility. Since a touch stimulus also informs the brain via axonal collaterals of the sensory nerve fibers, it is conceivable that the absence of wiping movement after repeated stimulation is due to habituation processes occurring, not only in the spinal cord, but also in the brain. After elimination of the brain, the ability of a toad or frog to become habituated to tactile cutaneous stimulation is greatly reduced [4].

In the coming sections we shall pursue in rough outline the manner in which information from the environment reaches the central nervous system, is processed there and finally leads to the contraction of a skeletal muscle. For better understanding of the principle the following treatment has been greatly simplified and, in part, idealized. (For schematic summaries see Fig. 16 and Table 2).

2.3 The "Language" of the Nervous System

To get a preliminary idea of the excitability of sensory and nerve cells, let us make a comparison with a telegraph system. There the current circuit is open in the rest position and with each depression of the switch, the circuit is closed for a short time producing a pulse of current. The nerve fibers operate according to another principle. In the resting state a potential difference exists between electrolytes inside and outside the cell membrane. With the onset of an impulse the resting potential drops rapidly toward zero, and overshoots, only to return immediately to its starting value. This reversal of polarity of the membrane and its return to the original polarity occurs in about 1 ms.

What Does the Resting Potential Depend on? Here the necessary requisites are charge carriers. The potential difference itself is based on charge separation. Cations (K^+ and Na^+) and anions (Cl^- and protein anions, A^-) act as charge carriers. Their concentrations outside and inside the membrane are different; for K^+ and A^- the concentration is higher inside; for Na^+ and Cl^-, it is higher on the outside (Fig. 16). The

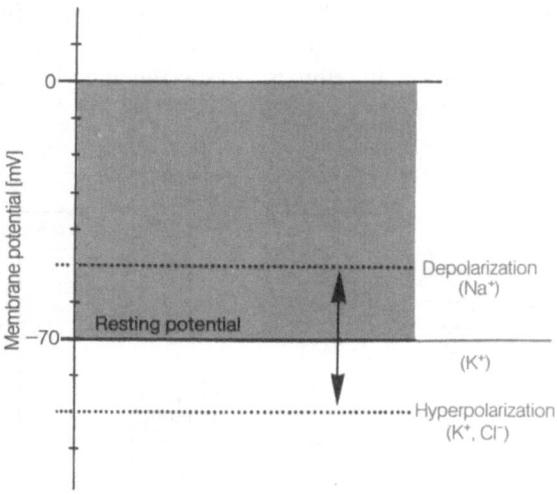

Fig. 17. Membrane states. The potential shifts are mainly due to selective permeabilities for the ions in parentheses

charge separation depends upon the voltage-dependent permeability properties of the membrane. During the resting state the membrane is permeable mostly for K^+ ions. K^+ and A^- ions will try to diffuse to the outside, following their concentration gradient. Whereas K^+ could pass through the membrane, the large protein anions are held back. The ensuing charge separation makes, so to speak, a battery of the nerve cell with the inside being approximately -70 mV with respect to the outside. Conditions are similar with receptor cells and muscle fibers. Some of their functions are related to alterations of this resting potential.

The resting potential is changed when the membrane changes its selectivity for particular kinds of ions (Fig. 17). If the membrane selectively increases its permeability to Na^+, its potential difference is reduced. We then say that the membrane is *depolarized*. However, there are membrane states which make them mostly permeable for K^+ and/or Cl^-. Then the membrane polarity is increased, that is, the membrane is *hyperpolarized*.

How Does Information Reach the Nervous System? A stimulus (e.g., optical, mechanical, chemical) induces a change of ionic conductivity in the membrane of the appropriate receptor (photo, mechano, chemo receptor), via intermediary processes which are not completely understood. The corresponding change of the resting potential is called generator or receptor potential (Fig. 16). It is the first electrically measurable reaction to a stimulus. The generator potential by its amplitude may reflect the intensity and the duration of the stimulus. If the generator potential appears as depolarization, the receptor is excited during that time. The potential spreads passively until it reaches (in primary sensory cells) the electrically excitable region of the axon (axonal hillock). After exceeding a threshold, the stimulus-induced depolarization initially increases the membrane permeability to Na^+

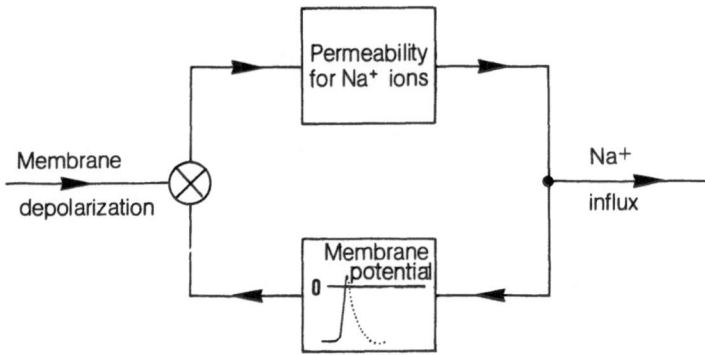

Fig. 18. Interpretation of the action potential (upstroke) as a consequence of positive feedback

ions, causing an increase in the Na^+-influx. Thus, the membrane potential difference is reduced, the depolarization is enhanced, and additional Na^+-"channels" of the membrane are opened. In the course of this self-reinforcing positive feedback process (Fig. 18) the membrane potential drops very fast and overshoots, i.e. the inside of the membrane becomes positive relative to the outside. Subsequently the potential returns to its original value (-70 mV). This fast potential change is termed the *impulse, action potential,* or *spike.* In its *generation,* it is an "all or nothing" reaction.

Specifically, it is assumed that the Na^+-channels of the membrane consist of a selective filter and a gating mechanism. There are approximately 50 Na^+ channels per μm^2. By suprathreshold depolarization these channels are opened, and the Na^+-influx determines the ascending part of the action potential. The membrane depolarization activates, however, not only a Na^+-influx, but also a slightly delayed K^+-efflux system by opening another group of channels. The Na^+-system is again inactivated (sodium channels are closed) before the action potential has reached its peak. The descending limb of the action potential is caused primarily by the increased K^+-efflux.

The Na^+- and K^+-systems can be affected separately by the application of certain drugs. If tetrodotoxin (TTX) – a lethal poison extracted from the ovaries of the puffer fish – is applied to the exterior of the axonal membrane, it blocks the permeability mechanism for Na^+ ions: after membrane depolarization, the influx of Na^+ disappears, whereas the K^+-efflux remains unchanged. The permeability mechanism for K^+ ions can be selectively blocked by tetraethylammonium (TEA).

The return of ions indispensable for the restoration of the resting potential is achieved in the refractory period by means of so-called ion pumps; they are maintained by metabolic energy. Such pumps must also contribute to the upkeep of the resting potential since small quantities of Na^+ ions flow into and K^+ ions out of the cell during the resting state. If Na^+ ions were not actively transported toward the outside against their concentration gradient in exchange for K^+ ions, the resting potential would gradually decrease. Thus, the basis for generation of the impulse in the neuron would be eliminated. This illustrates the direct connection between oxygen requirement, metabolic energy, and neuronal activity.

The coupled Na^+-K^+ pump is based on the concept that – bound by carrier molecules – a K^+ ion can enter for each Na^+ ion transported out of the cell. According to the *carrier hypothesis,* Na^+ forms with a carrier Y an electrically neutral complex NaY in the cell membrane which, following its concentration gradient, diffuses outward and there decomposes. Na^+ is thereby moved outward, whereas Y, inside the membrane, is converted into a carrier X enzymatically. X takes up K^+ from the outside and as an electroneutral complex diffuses according to its concentration gradient into the cell interior and there it decomposes into K^+ and X. Carrier X might be converted into Y with the expenditure of energy and could again be available for the return transport of intracellular Na^+. The energy sources of the cell consist of adenosinetriphosphate (ATP) formed during oxidative phosphorylation. It can be shown experimentally that after loading the cell intracellularly with radioactive isotope $^{24}Na^+$ active ion transport is inhibited by lack of oxygen or by chemicals (e.g., dinitrophenol, DNP) which interfere with the production of ATP.

What Is Neuronal Activity? We saw that suprathreshold depolarization at a particular region of the membrane (trigger zone of the axon) generates an action potential (AP). Continued depolarization can produce several AP's . In accordance with their generation (all or nothing action) they have an amplitude of the same height. The smallest possible interval between two successive AP's depends on the absolute refractory period which may last 1–2 ms. Only after this time interval can a new AP be generated. The magnitude of the latency to the next AP depends then on the strength of the depolarization; the higher the amplitude, the shorter the latency. Thus, the amplitude of depolarization governs the temporal sequence of AP's and thus their frequency. A key to the language of the neuron is its "firing" frequency.

The action potentials are propagated along the axon membrane by electrically exciting neighboring areas and affecting its voltage-gated sodium channels. The conduction occurs on the basis of current flow between the internal and the external electrolyte of the membrane. If the fiber is covered by a myelin sheath, the isolating function of the internodes brings about saltatory excitation spread between adjoining nodes of Ranvier. The conduction velocity is thereby increased. The saltatory principle has another advantage: from the point of view of energy consumption it is economical since active ion transport mechanisms are restricted to the membrane area at the nodes. (On measurement and recording techniques for neuronal responses, see Chap. 9, Methodological Appendix).

How Are Neurons and Glial Satellite Cells Related Electrophysiologically? The membrane of a neuron is separated from that of a surrounding glial cell by tiny extracellular spaces – 20 nm wide – filled with fluid. In the glial cell membrane, too, there is a potential difference between interior and exterior electrolytes. It is somewhat higher than in the neuron and, like the latter, largely determined by the K^+ ion concentrations on either side of the membrane. The glial cell membrane becomes depolarized when the K^+ ion concentration in the extracellular space between glial cell and neuron is raised. That, for example, is the case when action potentials are conducted along the axon. With increasing frequency of action potentials the K^+ ion concentration in the extracellular space rises and the larger is – within limits – the depolarization of the glial cell membrane. Its amplitude reflects the neuronal impulse frequency. Of course, the glial cell membrane itself is unable to generate propagated action potentials.

The function of the glial potential is still unclear. It is possible that the signal flux in the glial cell adjoining a neuron (mutually connected by low resistance pathways) produces a transfer of chemical substances required by the neuron for its own activity. It is assumed that the glial potentials contribute to the brain currents in the electroencephalogram (EEG) (see Chap. 9, Methodological Appendix).

We summarize briefly:

The stimulation of sensory cells leads to the formation of a generator potential whose amplitude can reflect the stimulus intensity. This transformation principle we call *amplitude modulation*. When the generator potential appears as a depolarization, it can be translated in the trigger zone of the axon into the code of the nervous system – action potentials of a corresponding frequency. The amplitude of the generator potential is expressed in the temporal sequence of AP's; the amplitude of AP's remains constant. *Frequency modulation* is due to certain properties of the action potential:

1. The *all or nothing principle* produces action potentials of constant amplitude.

2. The *refractory period* determines the minimum interval between AP's.

3. As a result of the *intensity-latency relation,* the information on the amplitude of depolarization – measured from the baseline of the resting potential – is carried by the frequency of the AP's.

There are also bioelectrical interactions between neurons and adjoining glial cells (satellite cells). Their exact function, however, is still unknown. The "information transfer" between neuron and glial cell is nevertheless fundamentally different from the *synaptic transmission* between neurons.

2.4 Communication in the Nervous System

The precondition for the interaction of neurons – that is the processing of information – is a retranslation of their impulse language. This decoding occurs at those places where receptors are in contact with neurons or neurons with each other, i.e., the synapses. If we start from a 1:1 synaptic transmission, an impulse sequence arriving presynaptically can be retranslated at the postsynaptic membrane into that depolarization which originally gave rise to it (Fig. 16).

How Does This Retranslation Occur? There are small vesicles in the terminal button of an axon. On the arrival of an action potential they move to the presynaptic membrane and empty their content – a neurotransmitter – into the synaptic cleft (Fig. 19). The arrival of an AP at the terminal opens voltage-gated calcium channels. The entry of Ca^{2+} ions into the nerve terminal is required for the release of neurotransmitter. It is supposed that one protein, *stenin,* which coats the vesicle, interacts with another protein called *neurin* located on the microtubules and on the presynaptic membrane. This process leading to the rupture of the vesicles [5] is apparently mediated by Ca^{2+} ions. The transmitter molecules diffuse down their concentration gradient to the postsynaptic membrane and change its ionic conductivity by opening chemically gated channels. For instance, in excitatory synapses, channels mainly for Na^+ ions can be opened by the transmitter (e.g., acetylcholine). One could imagine that the particular ion channel of the postsynaptic membrane is at first closed by a kind of "stopper" molecule (Figs. 16, 19). Near by is a "receptor" onto which the transmitter fits according to the lock and key principle. When the transmitter and stopper combine, the stopper no longer closes the ion channel and so its entrance is freed (Fig. 19). After 1–2 ms the transmitter molecules are split enzymatically, and the components, according to their concentration gradients, diffuse back to the presynaptic membrane which takes them up (Fig. 19). Within the axonal ending the transmitter substance is then resynthetized from the components.

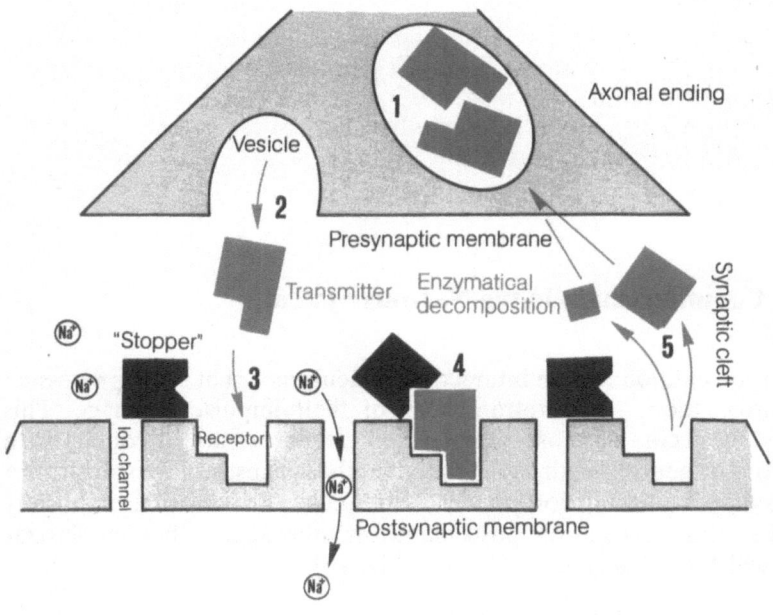

Fig. 19. Simple model explaining chemical transmission at an excitatory synapse with rapidly acting receptors. See text for explanation. [Modified from Eccles JC (1973) The understanding of the brain. McGraw-Hill, New York]

Many neurotransmitters have been identified, e.g., acetylcholine (ACh), epinephrine (adrenaline), norepinephrine (noradrenaline), dopamine, serotonin (5-hydroxytrypta-mine), glutamic acid, γ-aminobutyric acid (GABA), glycine, angiotensin II, enkephalin, vasopressin (the latter three neuropeptides being putative transmitters [147]). It could be shown that norepinephrine, bound to receptor molecules on the outside of the postsynaptic membrane, activates the enzyme adenylate cyclase wich – attached to the receptors – is on the inside of the membrane [6]. Adenylate cyclase then forms cyclic adenosine monophos-phate (AMP) [7] from adenosine triphosphate (ATP). Cyclic AMP activates the enzyme protein kinase and leads to phosphorylation of membrane proteins [8] by linking a phosphoryl group to a side chain of the amino acids serine or threonine. As a result, membrane molecules may alter their dynamic structure, and thereby change the permeability properties of the postsynaptic membrane [9]. Following its action the "first messenger" (norepinephrine) is enzymatically destroyed or recycled back into the vesicles. After relaying its message the "second messenger" (cyclic AMP) is inactivated by phosphodiesterase. Many transmitters utilize rapidly acting receptors in the periphery (as shown in Fig. 19) and longer-acting receptors, involving a second messenger, mainly in the CNS.

At an excitatory synapse a presynaptic spike, by eliciting the release of transmitter, chemically depolarizes the postsynaptic membrane. The time course of the postsynaptic potential (PSP) depends upon the concentration of transmitter substance in the synaptic cleft. It has a relatively rapid increase (commensurate to transmitter release) and slow decline (dependent on enzymatic splitting of the transmitter). A postsynaptic potential differs from an action potential by its relatively slow time course (ca. 15 ms) and by a small amplitude (ca. 10 mV), which can, however, grow by summation with other PSP's. Each PSP owes its

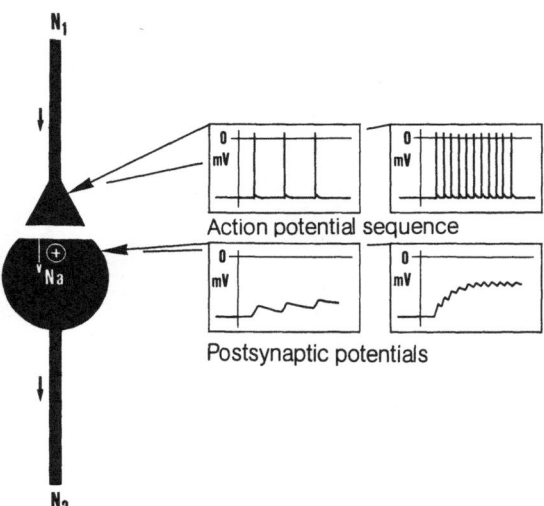

Fig. 20. Principle of decoding at an excitatory synapse. Retranslation of a sequence of action potentials into a nonlinear summated postsynaptic potential. Examples for relatively low and high frequencies of impulses arriving presynaptically

Table 3. Morphological and physiological properties of axonal and dendritic structures

	Dendrite	Axon (neurite)
Functional property	Reception of information (e.g., excitation) from presynaptic neurons	Propagation of excitation to postsynaptic neurons
Physiological characteristics	Predominantly chemically excitable at the synapses. Threshold of electrical excitation is relatively high	Electrically excitable. There are also axons that do not produce action potentials (AP)
	Ionic conductivity of postsynaptic membrane is dependent on neurotransmitter (chemically gated ion channels)	Ionic conductivity for Na^+ (and K^+) is dependent on membrane potential (voltage-gated ion channels)
	Membrane PSP reflects transmitter concentration in synaptic cleft	Potential change (increasing depolarization) leads to increased conductance of Na^+ (and K^+) ions
	PSP's have a slow time course (10–15 ms), a low amplitude (ca. 10 mV) and accumulate (nonlinearly)	AP's have rapid time course (1–2 ms), relatively high amplitude (ca. 100 mV), and according to their mode of generation are "all or nothing" responses
Transformation	Decoding of a sequence of presynaptic AP's into accumulated PSP's; amplitude modulation	Coding of a potential change (PSP) appearing as depolarization into a sequence of APs; frequency modulation
Conduction	Slow electrotonic spread with amplitude decrement	Rapid propagation with constant amplitude
Embryology	Secondary cytoplasm process of the neuroblast	Primary neuroblast process
Histological characteristics	Broad area of origin at the soma	Narrow area of origin at the soma
	Contains Nissl bodies	Contains little or no Nissl bodies

generation to the amount of transmitter released by a presynaptically arriving AP. The amplitude of accumulated PSPs is increased by shorter time intervals between the arrival of consecutive transmitter quanta. In this way the frequency of presynaptically arriving AP's can be expressed postsynaptically by the amplitude of the resulting graded PSP (Fig. 20). The PSP appears as depolarization in excitatory synapses. It spreads passively over the soma and can be transformed into an appropriate sequence of AP's in the trigger zone of the electrically excitable axon membrane. We have already introduced the coding principle.

The smallest physiological unit of neurotransmitter released by a nerve impulse is called a quantum. A quantum may correspond to about 10,000 transmitter molecules (acetylcholine) in the neuromuscular junction (motor end-plate of the frog) and this is able to open up about 1000 Na^+-ion channels for the duration of about 1 ms. About 5×10^4 ions will flow through per channel. At the frog's motor end-plate about 200 quanta of ACh are released by a presynaptic AP. The release of ACh can be blocked by tetanus toxin or botulinum toxin; the effect is similar to removal of Ca^{2+} ions. We can be sure that the synaptic vesicles contain neurotransmitter. However, the proposition that a vesicle contains a transmitter quantum is still hypothetical.

Nervous function is based on the action of various *neural elements*, such as channels, pumps, receptors, and enzymes – all consisting of particular membrane proteins. Neurons provide the *building blocks* of the nervous system. Dendrites and axons form *neural networks* responsible for transmission and processing of information. Using a vertebrate neuron as an example Table 3 summarizes some of their morphological and physiological properties.

2.5 The Three Fundamental Arithmetic Methods

2.5.1 Computations

The transmission of information from one neuron to another takes place at the synapses. Functionally there are two main groups of synapses: *excitatory* and *inhibitory*. In the excitatory synapses, presynaptic action potentials lead to a depolarization of the postsynaptic membrane by the release of specific transmitter; this potential is called excitatory PSP (EPSP). When two neurons N_1 and N_2 each form an excitatory synapse (Fig. 21 A) with a third N_3, the corresponding EPSP can add up in a nonlinear way at the soma membrane of N_3. The result of this *addition* is then frequency-coded in impulses and conducted along the axon to neurons which are postsynaptic to N_3.

At inhibitory synapses the action potentials, by release of another neurotransmitter, lead to a hyperpolarization of the postsynaptic membrane; this potential is called inhibitory PSP (IPSP). It runs as a mirror image of the EPSP. The relationship between the frequency of

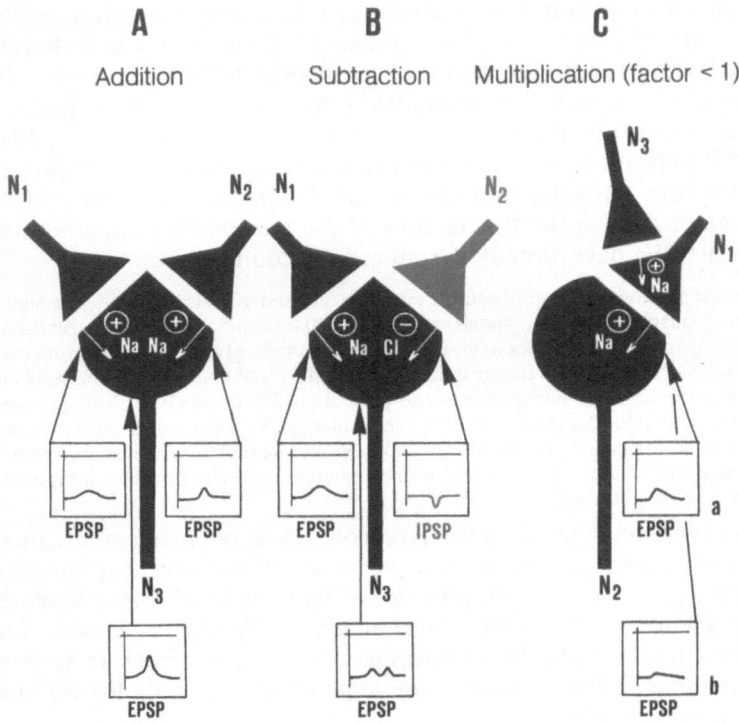

Fig. 21 A–C. Principle of the three basic computations, illustrated by arbitrarily chosen postsynaptic potentials for various types of synapses: **A** two excitatory synapses, **B** an excitatory and a postsynaptically inhibitory synapse *(red),* **C** axo-axonic synapse mediating presynaptic inhibition. The result of the calculations is shown below as the resultant EPSP. *a* Shows the EPSP for impulses arriving in N_1, and *b* shows the modification of the EPSP by additional excitation of N_3. See text for explanation

presynaptic action potentials and IPSP amplitude is in principle the same as that we already met in the excitatory synapses (i.e., a higher frequency of presynaptic AP's leads to a larger IPSP). If a neuron N_1 forms an excitatory synapse and another neuron N_2 forms an inhibitory synapse with a third N_3 (Figs. 16 and 21 B), the IPSP produced by N_2 is *subtracted* from the EPSP produced by N_1. The resulting depolarization is frequency-coded in impulses and conducted along the axon of N_3.

In addition to postsynaptic inhibition, there is another mode, presynaptic inhibition. It requires an axo-axonic synapse (Fig. 21 C). Let us assume that a neuron N_1 forms an excitatory synapse with a sequential neuron N_2 (Fig. 21 C). It is known that the release of transmitter from the axonal ending of the fiber N_1 partly depends on the amplitude of arriving action potentials (AP's); this again is correlated to the existing degree of depolarization of the axon terminal membrane. The greater the depolarization of the membrane potential, the smaller the AP-amplitude, and the weaker the transmitter release. The axonal terminal membrane of N_1 may be depolarized by an excitatory axo-axonic synapse from a neuron N_3 (Fig. 21 C). We now understand the principle: excitation in neuron N_3 always lowers the transmission of N_1 to N_2 by a specific percentage. This inhibition represents a *multiplication* by a factor less than 1, at the synapse between N_1 and N_2.

2.5.2 Neuronal Circuits

For the neuronal computations just described there is a series of examples. Thus, we are familiar with neuronal circuits – called negative feedback loops (Fig. 22 A) – in which a neuron in the transmission of its excitation, silences itself by means of axon collaterals and inhibitory interneurons, such as in the monosynaptic reflex (Fig. 22 C). The reciprocal inhibition of antagonists (Fig. 22 B) demonstrates another principle of neuronal connection; here the activation of the agonist (e.g., extensor) inhibits the antagonist (flexor), and vice versa. It is also possible that neurons could excite themselves via excitatory interneurons (Fig. 22D) and in this way store "circular excitations" for a short while. Short-term memory might be based on such connections.

2.5.3 "Lateral Inhibition"

Lateral inhibition plays an important role in information processing in the nervous system. In this case, a neuron reduces the excitation of its neighbor via interneurons, in proportion to its own level of activity (Fig. 22 A). Lateral inhibition – determined possibly by presynaptic retrograde inhibition – can be found for instance in the spinal cord of vertebrates. It might be developed via inhibitory interneurons among those pathways which conduct excitation to the spinal cord from certain cutaneous receptors excited by tactile stimuli. Excitations from extensive contact stimuli are transmitted at first. However, if the stimulus is maintained for a time, the excitatory input is throttled by onset of feedback inhibition. Excitation produced by a stronger stimulus can pass through the inhibitory system again for a short while. It is certain that the information processing proper takes place at higher levels of the central nervous system. However, simple short-term habituation phenomena might be explained in this way. For example, the habituation to touch stimuli from our clothing. Because of lateral inhibition, changes in stimulation caused by rough clothing, perhaps by poor detergents, can have only a temporary influence on our sensations, contrary to opinions expressed by some companies.

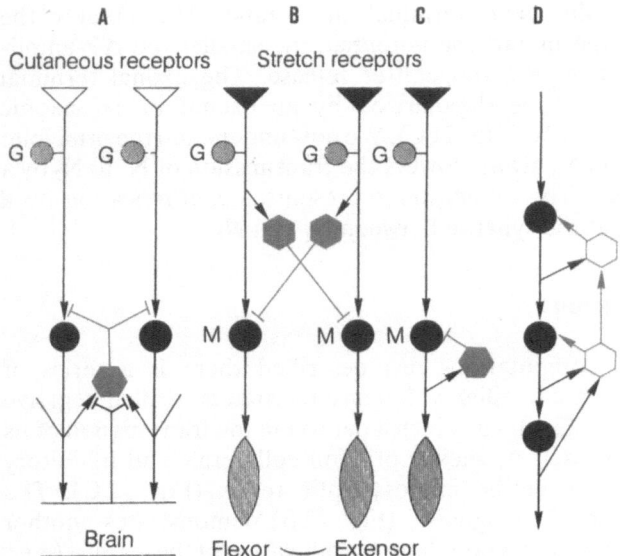

Fig. 22 A–D. Examples of relatively simple neuronal circuits. **A** Principle of lateral inhibition in the spinal cord, illustrated by the example of feedback presynaptic inhibition; *G* spinal ganglion, *M* motor neuron. **B** Reciprocal postsynaptic inhibition of flexor and extensor muscles. **C** Negative feedback inhibition of α-motor neuron in the monosynaptic reflex. **D** Positive feedback. The cells shown in *red* represent inhibitory interneurons in **A–C** and excitatory ones in **D**. (*Arrows* denote excitatory, *lines with cross bars* inhibitory synapses)

The brain is nevertheless able to direct its attention "consciously" to the tactile excitatory processes via certain fiber pathways (pyramidal tracts): lateral inhibition is thereby temporarily reduced – possibly through presynaptic inhibition of the interneurons (Fig. 22 A).

Lateral inhibition is also present in the visual system. Observing the square pattern in Figure 23 we have the impression that "dark stripes" (Mach bands) run in the middle of the white spaces. This sensory illusion is caused by lateral inhibition within our visual system, and it represents for us, in a sense, "visible" information processing. With this we encounter a further effect of lateral inhibition, namely contrast sharpening, which plays an important role in the perception of various degrees of shading. The principle is demonstrated by the rough schematic example in Figure 23. Here a neuron j reduces the excitation of its neighbors j + 1 and j − 1 respectively by 25 % of its own excitation. Result: through lateral inhibition the whole excitation level is slightly lowered, and the excitation differences are exaggerated in the area of contrast borders. Now the "dark stripes" are also understandable. Lateral inhibition has also the function of eliminating possible jamming

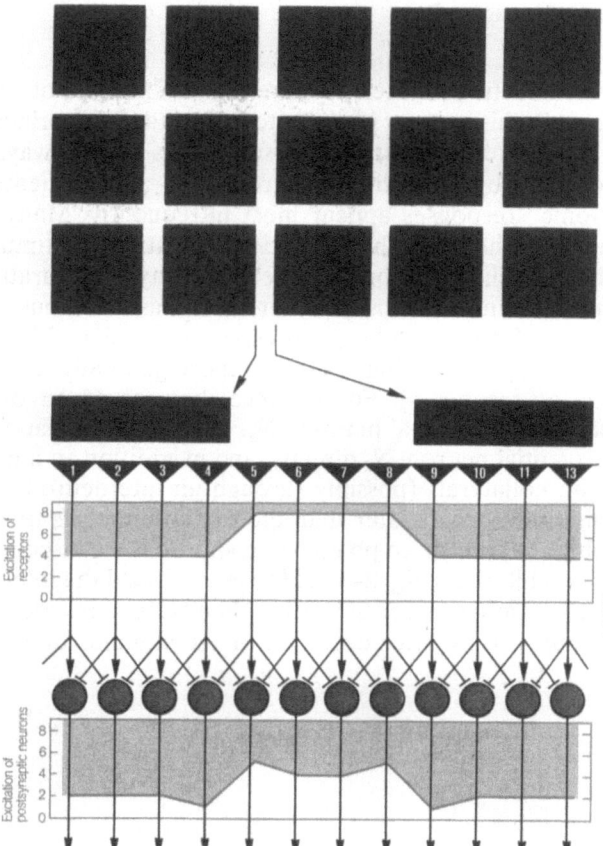

Fig. 23. Contrast accentuation by lateral inhibition in sensory pathways in the visual system. The principle of the mechanism is illustrated by focusing on two contrasting edges extracted from the pattern of squares. The *red* diagram on top represents the spatial distribution of excitation in corresponding sensory pathways *1–13*. The diagram below shows the distribution of excitation after lateral inhibition. It was assumed in this example that each tract subtracts 25% of its excitation from the fibers immediately lateral to it. (*Arrows* denote excitatory, *lines with cross bars* inhibitory synapses). See text for explanation

that might occur when input signals overlap. The principle of information processing by lateral inhibition is found in many sensory systems at various levels of integration and interaction – in vertebrates as well as in invertebrates.

2.6 Complexity

For didactic purposes we have made a number of simplifying assump-
tions in this chapter in order to understand the principle.
1. The intensity pattern of a stimulus is never always truly reproduced
(copied) by the activity of receptors and sequential neurons. Such
"tonic" responses appear most infrequently. Many sensory cells and
neurons show a slight adaptation to continuous stimulation (Fig. 24A);
they give information on onset, intensity, and duration of the stimulus
(phasic-tonic response). Other fully phasic neurons respond only at the
onset of the stimulus (*on*-response, cf. Fig. 24C), at its cessation
(*off*-response, cf. Fig. 24B) or, more generally, to stimulus variations
(*on-off*-responses). For instance, how could an *on*- or *off*-response
occur? Suppose a neuron, N_1, forms an excitatory synapse with a
sequential neuron, N_2 directly, and in addition an inhibitory synapse via
axon collaterals (possibly through an interneuron) (Fig. 24C). If the
inhibition occurs later than the excitation, N_2 is first excited and then
inhibited; the sharp phasic *on*-response is a consequence of postexcita-
tory inhibition (Fig. 24C). However, should the inhibition occur before
the excitation (preexcitatorily), then N_2 remains silent at the onset of the
stimulus; only after cessation of the stimulus does N_2 respond, so to

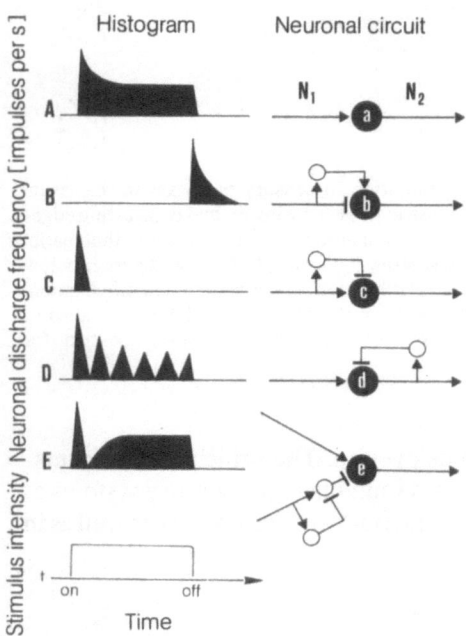

Fig. 24A–E. Various neuronal
response patterns, represented as
schematic histograms, and possible underlying circuits. **A** Phasic-
tonic response *(a)*. **B** *off*- response as result of a delayed excitatory
connection *(b)*. **C** *on*-response as
result of feedforward inhibition
(c). **D** Serial *on*-responses due to
feedback postsynaptic self-inhibi-
tion *(d)*. **E** Paused *on*-responses
as result of two convergent inputs
(e): one is tonic-excitatory, the
other phasic-inhibitory via an inhibitory interneuron. *Arrows* indicate excitatory, *lines with cross
bars* inhibitory synapses. [Modi-
fied from Suga N (1973) In:
Møllner AR (ed) Basic mechanisms in hearing. Academic Press,
London New York]

speak as consequence of the removal of the inhibition (Fig. 24 B). Phasic responses can also be due to certain membrane properties of a neuron.

2. The neurons of the central nervous system do not receive their excitation exclusively by stimulation of the sense organs. Many receptors show weak permanent activity all of the time. This is called spontaneous or background activity. In spontaneously active neurons endogenous changes of their membrane potentials result in impulses (e.g. cardiac pacemaker cells).

3. Particular types of receptors (secondary sensory cells) have no axons. Not all receptor and neuron types which have an axon necessarily produce action potentials. Electrotonic spread of current constitutes their mode of functioning. The membrane potentials of the postsynaptic neurons may then, for example, be controlled by the graded release of transmitter from the nonspiking cell [10].

4. It has been assumed that there is a 1:1 transmission at the synapses, but this is rarely realized. There are mechanisms to insure that a presynaptic action potential leads to an EPSP exceeding the response threshold for an AP at the trigger zone of the postsynaptic neuron. For example (Fig. 25 below): (a) In chemical synapses effective transmission may be secured by an "increase" of the transmitter concentration, e.g., as in large synapses (Fig. 25,1) or by the enlargement of the synaptic surface produced by extensive branching of the axon terminals giving rise to multiterminal synapses (Fig. 25,2). Furthermore, the dendrite may have numerous spines on its surface. These small axo-dendritic synapses exhibit anatomical as well as functional plasticity. Their number and extent appear to depend on use and nonuse (see p. 145) and they are possibly related to learning processes. (b) With a small synaptic cleft the path of diffusion for neurotransmitter is shortened. There are also synapses where information transmission may be produced by electrical coupling of the membranes. They are called low-resistance electrical synapses (cf. Fig. 25, top, middle and p. 18). (c) Localization of the synapse, close to the trigger zone (axonal hillock, Fig. 25,3) insures minimum decrement of the amplitude of the EPSP which must be sufficient to exceed the threshold for spiking.

5. The description of excitatory and inhibitory chemical synapses is in some respects an over simplification. It is firmly established that neurotransmitters produce certain ion-specific permeability changes in the postsynaptic membrane during synaptic transmission. The state of the postsynaptic cell in relation to the immediate substrate environment determines the effect of the process. Therefore, strictly speaking, there are neither definitely excitatory nor definitely inhibitory neurotransmitters. For example, the transmitter γ-aminobutyric acid (GABA) may in some synapses hyperpolarize the postsynaptic membrane. In the presynaptic inhibition described on p. 42, the same transmitter may depolarize the presynaptic membrane (of the synapse formed by N_1 and

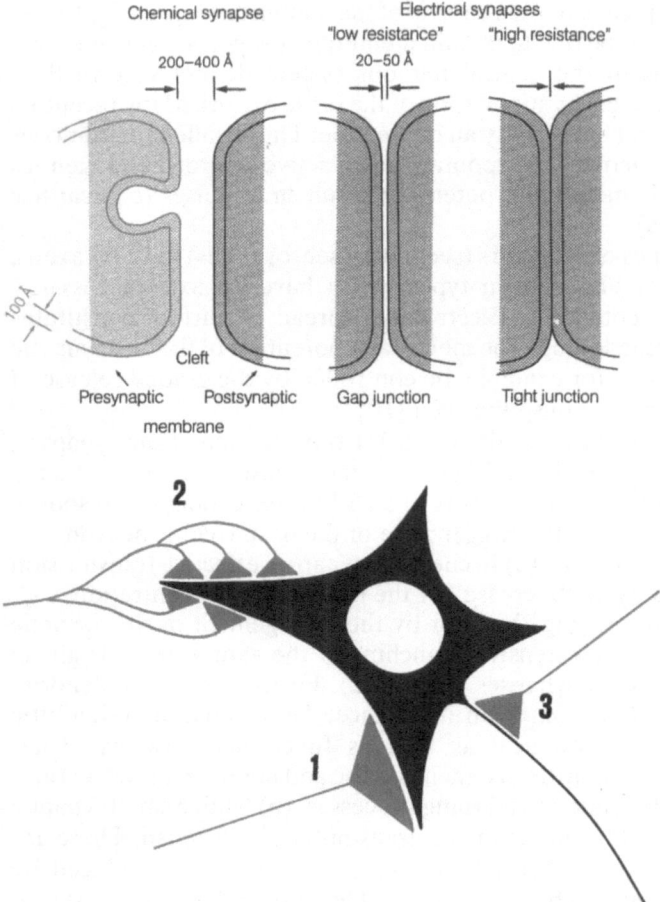

Fig. 25. Synapse types *(above)* and "safety factors" *(below)* for the synaptic transmission. *1* Giant synapse, *2* multiterminal synapse, *3* synapse adjacent to axon hillock (highly schematized)

N_2 as shown in Fig. 21C). Furthermore, it is quite possible that the function of a synapse as either excitatory or inhibitory is *not* fixed once and for all [11].

The permeability changes that cause depolarization of the presynaptic membrane are not well understood. Na^+-influx is often suggested (Fig. 21 C). But other possibilities are also discussed, since the conditions for depolarization appear to be associated with increased levels of extracellular potassium. In some cases presynaptic inhibition may be mediated by 5-hydroxytryptamine (5-HT) or enkephalin as transmitter [12].

6. After repeated activation of an excitatory synapse, its transmission properties do not necessarily remain constant. In particular types of

synapse the efficacy of transmission may be increased by repeated ("tetanic") activation. The reason for this is an increased resting potential of the presynaptic membrane which increases the amplitude of incoming action potentials and thus increases the amount of transmitter released into the synaptic cleft. Such posttetanic synaptic potentiations form perhaps a basis for facilitation and learning processes (p. 239). There are also synapses whose excitability after repeated activation decreases owing to diminished transmitter release. These processes might perhaps form the basis for relatively simple habituation of behavior patterns in response to repeated sensory stimuli.

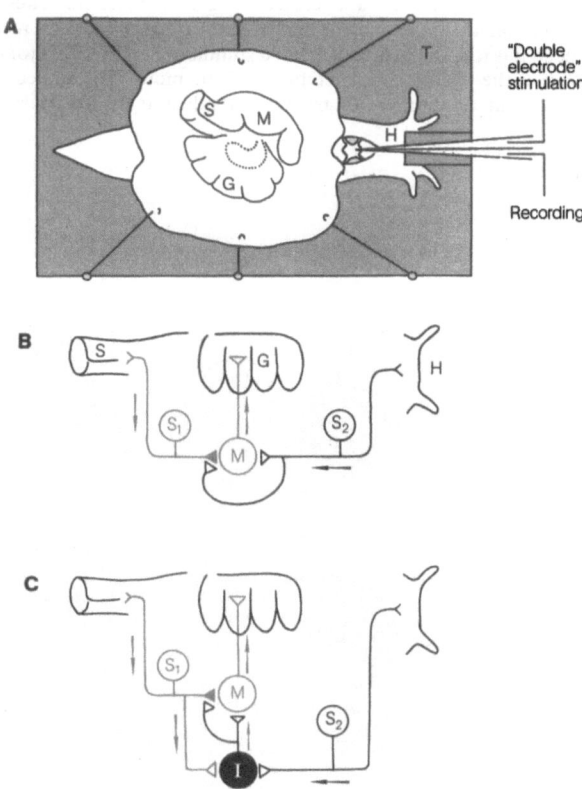

Fig. 26 A–C. Postulated neuronal circuits explaining habituation and dishabituation of the gill withdrawal reflex in the marine snail *Aplysia*. **A** Procedure for combined electrical stimulation and recording experiments; *G* gill, *H* head, *M* mantle shelf, *S* siphon, *T* tank. **B** and **C** Two possible circuits; S_1, S_2 sensory neurons, *M* motor neuron, *I* interneuron. In both circuits the transmission from S_1 to *M* is mediated by a plastic synapse. [Modified from Kandel ER, Brunelli M, Byrne J, Castellucci V (1976) Cold Spring Harbor Lab Symp Quant Biol 60: 465–482]

The following example illustrates plasticity of synapses:

The marine snail *Aplysia* breathes with a gill which is partly covered by a mantle shelf of the body. When the shelf or its prolongation (siphon) is touched, the gill is immediately withdrawn into the protective mantle (gill-withdrawal reflex). This stimulus response sequence is based on a monosynaptic reflex arc (Fig. 26B). A receptor neuron (S_1) conducts excitation induced by stimulation of the siphon to a motor neuron (M) which induces a contraction of the gill muscle (G). If the siphon is stimulated repeatedly, the strength of the gill contraction decreases until finally, it is totally extinguished. Thus, the animal has become habituated to the stimulus. After 30 min, at most, the response is recovered. What kinds of neuronal events underlie the habituation? This question has been studied by a combination of intracellular stimulation and recording techniques with the aid of microelectrodes (Fig. 26 A,B; Chap. 9, Methodological Appendix). Neither repeated electrical stimulation of the receptor nerve fiber (S_1) leads to a comparable diminution of its excitability nor does continuous electrical stimulation of the motor neuron (M) result in a decrease of gill contraction. Thus, behaviorally observed habituation might depend on changes at the synapse between receptor and motor neuron. Indeed it is found that after repeated stimulation of the receptor neuron, the amplitude of EPSP at the postsynaptic membrane of the motor neuron declines, due to a progressive decrease in the amount of transmitter released by the presynaptic terminal. The gill

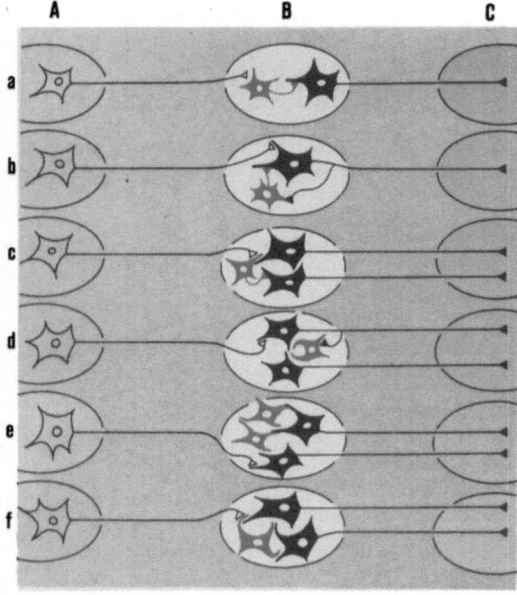

Fig. 27. Examples of local neuronal circuits in a schematic diagram. Cells shown in *grey* or *black* are relay neurons projecting with their long axons from a nucleus *A* via *B* to *C*. Information processing is done mainly by the local circuit neurons shown in *red*; they have short or no axons. *a* Interneuron in spinal cord (cf. Fig. 22 B), *b* the Renshaw cell in spinal cord (cf. Fig. 22 C), *c* the periglomerular cell in olfactory bulb, *d* and *e* the axonless neurons in the thalamus of the diencephalon, *f* the amacrine cell in the vertebrate retina. Note axo-dendritic and dendro-dendritic synapses. [Modified from Rakic P (1975) Neurosci Res Prog 13: 291–446]

withdrawal response can also be released by stimulation of the head region from another receptor nerve fiber (S_2) via the same motor neuron (M). Interestingly, a habituation induced by repeated siphon stimulation may be annulled after head stimulation. Such a dishabituation (i.e., sensitization) is possibly caused by the fact that neuron (S_2) via axon collaterals facilitates the synaptic transmission from neuron (S_1) to neuron (M), perhaps by presynaptic excitation. Interneurons (I) might also be involved in these processes (Fig. 26 C).

Only few (three or four) neuron types participate in the described plastic behavior of *Aplysia*. The involvement of additional interneurons results in more complex systems, such as the habituation of the wiping reaction of the toad (p. 32) [13].

As we know, the substrate for information processing in the central nervous system is highly complex and consists largely of so-called local circuits which are represented by the most diverse types of synapses (Fig. 27) whose functional properties, however, are not known in every detail. (Methods for functional exploration of neuronal circuits, see Chap. 9, Methodological Appendix and Fig. 169).

2.7 Nerve Impulses Induce Muscle Contractions

Looking at the fine structure of striated muscle fiber, we find tightly packed myofibrils (Fig. 28). Each one of them is made of long parallel protein strands, the myofilaments, which consist of actin and myosin. Both are responsible for the "mechanics" of muscle contraction. Today we have evidence that particular molecule structures, coordinated with the myosin complex, pull alongside the actin filaments in a kind of grip–release mechanism (Fig. 28, bottom). The filaments slide like a telescope.

For such a mechanism, energy is necessary. It can be supplied from an energy rich phosphate bond of adenosinetriphosphate (ATP). Energy is available when an enzyme (ATPase) splits off a phosphate from ATP. The acto-myosin complex itself acts as ATPase. However, Ca^{2+} ions are required for its activation (Fig. 29). They are sequestered in an intracellular cavity system of the muscle fiber, the sarcoplasmic reticulum (SR). The Ca^{2+} ions are only released by depolarization of the fiber membrane along the transverse tubuli – an extracellular cavity system (T-system) – bordering on parts of the membrane of the sarcoplasmic reticulum (see Fig. 16). The release of Ca^{2+} initiates a chain reaction which culminates in muscle contraction.

How is Depolarization Achieved? The axon of a ventral horn motor neuron arborizes into a number of axon collaterals (Fig. 30). Each one of them together with a muscle fiber forms a kind of giant synapse, the motor end-plate or neuromuscular junction. The relatively large amount of transmitter released across the large surface area guarantees the transmission of excitation. The end-plates of the striated vertebrate

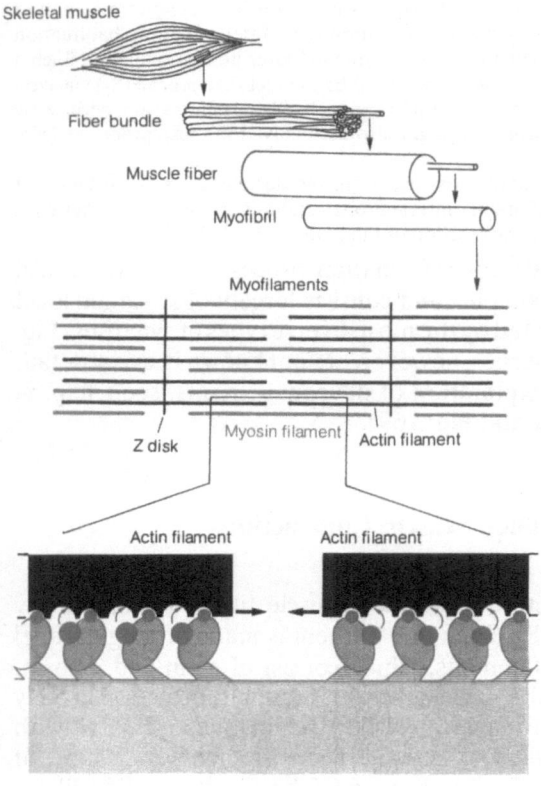

Skeletal muscle

Fiber bundle

Muscle fiber

Myofibril

Myofilaments

Z disk   Myosin filament   Actin filament

Actin filament   Actin filament

Myosin filament

Fig. 28. Organization of vertebrate striated muscle and elementary mechanism for its contraction (see *bottom*). It is assumed that the ovoid heads of a myosin filament form articulated bonds with corresponding molecular structures of the actin filaments and pull them over their tops in a sort of "grip-release" mechanism. A synchronized movement of all myosin heads in the direction of the arrows pushes the two actin filaments shown in the diagram toward each other by 5 nm from each side for a total shortening of 10 nm (see *black arrows*). At a pulling motion frequency of 50 Hz, the whole muscle would be shortened by about one half after 1 s. [Modified from Huxley HE, Simmons R (1971) Nature (London) 233: 533]

muscle are always excitatory; the induced postsynaptic potential – mediated by the transmitter acetylcholine (ACh) – is called the end-plate potential (EPP). The efficacy of synaptic transmission is further enhanced by small quantities of transmitter (ACh) that are released spontaneously – also in an unexcited state. This causes small, irregular depolarizations of the post(sub)synaptic membrane. Upon these miniature end-plate potentials (MEPP's) are superimposed the EPP's in the event of motor axon excitation. The muscle fiber membrane

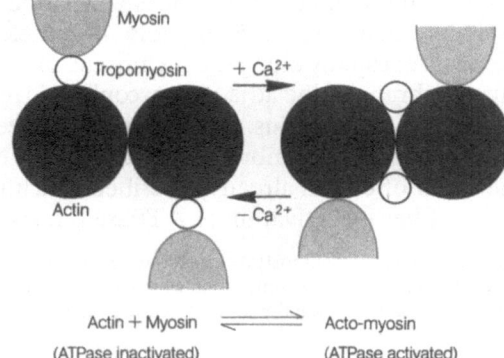

Fig. 29. Schematic diagram explaining the influence of Ca^{2+} ions on the activation of ATPase. *Left,* In the relaxed muscle the myosin and actin filaments are separated from each other by tropomyosin filaments (shown here in a section). *Right,* Under the influence of Ca^{2+} troponin molecules are deformed (not shown here) leading to dislocations of the tropomyosin filaments; by its bonds to actin the myosin becomes activated as ATPase. [Modified from Huxley HE (1973) Cold Spring Harbor Lab Symp Quant Biol 37: 361]

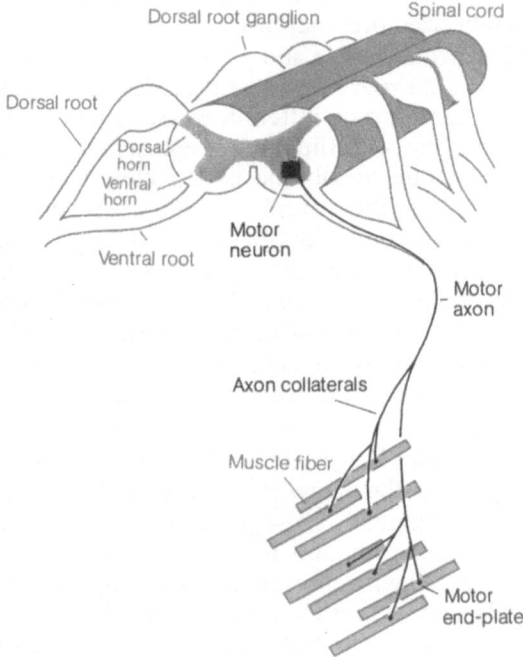

Fig. 30. Components of a motor unit of vertebrate striated muscle

(sarcolemma) adjacent to the subsynaptic membrane is electrically excitable and the EPP triggers a muscle action potential which propagates rapidly along the membrane of the T-system. By this process the membrane of the adjacent sarcoplasmic reticulum (Fig. 16) becomes permeable to Ca^{2+} ions. Following their concentration gradient they can pass through the membrane in exchange for Mg^{2+} ions and then trigger the chain of events already described which leads to contraction of the muscle fiber by activation of ATPase (cf. also Fig. 29).

The experimental demonstration of the release of Ca^{2+} ions from the sarcoplasmic reticulum during muscle contraction was achieved by the injection of aequorin. The latter is a protein isolated from the jellyfish *Aequoria;* it emits light in the presence of free Ca^{2+} ions.

With the relaxation of the fiber, the Ca^{2+} ions are pumped against their concentration gradient again into the sarcoplasmic reticulum in exchange for Mg^{2+} ions. Then the ATPase is inactivated. ATP can no longer be split and exerts on the fiber a so-called plasticizing influence. In this way, ATP, together with Mg^{2+} separates myosin (M) and actin (A)

$$AM + ATP + Mg^{2+} \rightleftharpoons M \cdot Mg^{2+} \cdot ATP + A.$$

How Can Neuromuscular Transmission Be Blocked? The natural transmitter for the motor end-plate is acetylcholine (ACh); it is synthesized presynaptically within the axonal endings of the motor nerve fiber, from choline and Acetyl CoA. The effect of ACh can, however, be inhibited experimentally by particular drugs known as *blocking agents.* This can occur in different ways, depending on the kind of drug. Let us start again with the model representation: the Na^+-ion channel of the subsynaptic membrane is first closed by a "stopper" molecule (Fig. $31 A_1$), and the natural transmitter – resting in an adjacent receptor site – forms a bond with the stopper, a bond which elevates it (Fig. $31 A_2$); shortly thereafter the transmitter is split enzymatically and the ion channel is blocked again by the "closing" stopper. On the basis of their similar molecular affinities, a group of drugs also combines with the receptor and releases the stopper, but the substances themselves can be broken down only very slowly (Fig. 31 C). *Succinylcholine* is such a substance. Because of constant depolarization the membrane becomes inexcitable. Drugs which inhibit the ACh-splitting enzyme (ACh-esterase) have the same effect (Fig. 31 B). To this group belong *eserine* (physostigmine), *prostigmine* and the pesticide *E 605* (nitrostigmine). The latter drug acts irreversibly. There is another group of blocking agents. Their representatives totally inhibit the development of an end-plate potential. *Flaxedil* and the South American Indian arrow poison *curare* are such examples. They fit into the membrane receptors but cannot open the ion channel and therefore exclude the neurotransmitter ACh from the receptor sites (Fig. 31 D).

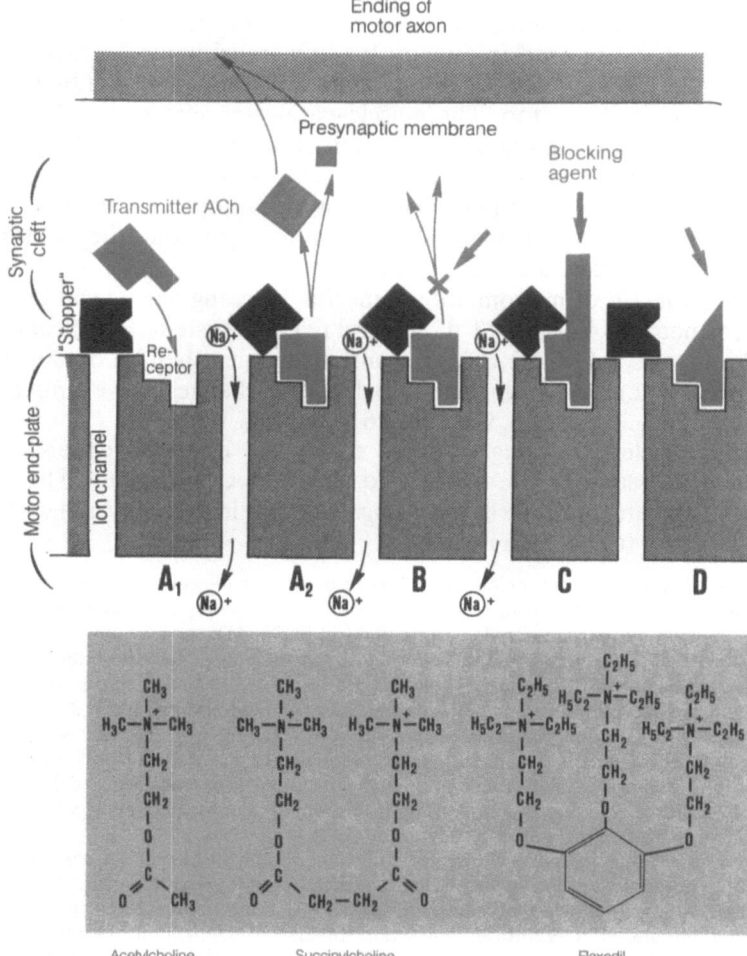

Fig. 31 A–D. Transmitter and blocking agents of the motor end-plate of vertebrate striated muscle. Simple concepts of a model of action mechanisms. **A₁** Normal depolarization of end-plate membrane by acetylcholine (ACh) as transmitter with subsequent hydrolysis **(A₂)** by the enzyme ACh-esterase. **B** The drug prostigmine blocks the enzymatic breakdown of ACh by inhibition of ACh-esterase *(red arrow):* change of membrane properties due to lasting depolarization. **C** Lasting depolarization of the end-plate membrane caused by the drug succinylcholine. **D** Displacement of acetylcholine from receptor sites by the drug flaxedil. Further details in text

The peripheral effect of ACh-blocking agents (which do not pass the blood brain barrier) is a paralysis of the striated musculature. Since the respiratory musculature is involved, death by suffocation is the consequence. Reversible blockers, such as curare and succinylcholine, are used together with artificial respiration to relax the muscles during surgery requiring general anesthesia. In neurophysiological experimentation they are used frequently for immobilizing awake experimental animals (see pp. 100, 271 and Fig. 59). Adequate ventilation is a prerequisite here, too. The muscle-paralyzing effect of curare can be reversed more rapidly by application of prostigmine: this drug inhibits ACh-esterase and thereby increases ACh-concentration in the synaptic cleft of the motor end-plate. The natural transmitter ACh now displaces the blocking agent from the receptor's site and the subsynaptic membrane again becomes excitable.

With the aid of micromethods, specific blocking agents can be placed near neurons in areas of the central nervous system. Clues concerning the identity of putative neurotransmitters may then be obtained from possible neurophysiological or behavioral changes, for example functional deficits (see pp. 98, 223, 264, and Fig. 131).

The muscle-paralyzing effect of snake bite may also depend on the ability of some snake venoms to act as blocking agents. The venom α-bungarotoxin of the snake *Bungarus* binds irreversibly and with great specificity to the ACh receptors.

Striated muscle is not unique to vertebrates. It is widespread in invertebrates and particularly wherever rapid contractions are required: umbrella muscle of the jellyfish (hydrozoa), propulsion muscle of the squid (cephalopoda), and locomotor muscles of the arthropods. In contrast to the vertebrates, muscle fibers in these animals are often innervated by different neuronal types, i.e. slow- and fast-exciting neurons. In arthropods (with the exception of insects) inhibitory neurons might also be involved. Each neuron forms numerous end-plates which spread out over the fiber. What is the purpose of this multiterminal innervation? The muscle fiber membranes do not necessarily generate action potentials. Multiple innervation of the fiber then insures a quick spatial spreading of the membrane depolarization. Starting from a motor neuron, all fibers of a muscle can be joined into a "motor unit" via collaterals.

The flight musculature of insects is distinguished by specialized capabilities. We differentiate two groups. In one of them the flight muscles are situated directly at the base of the wings; their beat frequency is relatively low, namely about 20 Hz in beetles. Each nerve impulse induces muscle contraction. In species belonging to the other group the flight muscles are situated on the chitin shell; they indirectly cause the wings to beat by lifting and lowering the thorax. The beat frequency can reach 500 Hz in the mosquito. One nerve impulse can trigger up to 20 "oscillating" muscle twitches. They are released, not directly by the impulse, but by mechanical stretching of the fiber.

Amoeboid movements (e.g., locomotion of *Amoeba proteus*) appear to be linked with "gliding mechanisms" between F-*actin* (diameter: 50–80 Å) and *myosin* filaments (160 Å). In flagella or cilia (e. g., of protozoa) the peripheral double tubular fibers consist of *tubulin* filaments which are combined with *dynein*. The amino acid sequence of tubulin is similar to that of actin. Dynein has enzymatic properties; it splits ATP in the presence of Ca^{2+} ions. Movement of the flagellum is obviously due to gliding mechanisms between the filaments of adjacent fibers. Application of ATP causes the isolated flagellum to move. Tubulin – which also plays a role in the formation of the mitotic division apparatus – can be inactivated by colchicin.

3 Key-Stimuli and Releasing Mechanisms: Some Fundamental Concepts in Ethology

3.1 Environmental Stimuli and Information Reduction

In the evening when we watch television a flood of information in the form of 5×10^8 binary data (unit = bit/s) bombards our visual system. Among this vast quantity of inflowing information, however, only a small percentage is consciously perceived – about 5×10^1 bit/s. When we recognize the face of an actor, the visual information is reduced to 1 bit/s: this corresponds to a yes/no decision. Information reduction plays a very important role in the classification and recognition of patterns. By this process the characteristic features are separated from irrelevant ones. The drawing of a caricaturist constitutes a good example: a few strokes capture those features, which, in their totality, yield a configuration that we associate, for instance, with a particular image of a person.

Of all the information from the environment which is conveyed to the brain by the sense organs, only a limited portion is biologically important. Some parts merely serve as cues or signals corresponding to, for example, an enemy, a prey, or a social partner. An important function of the corresponding sensory systems is to recognize such stimulus features. Consequently, for an animal, the environment can only be what the system properties of sense organs and nerve nets, which process the signals, allow. Thus, the compound eye of the bee, which is sensitive to ultraviolet and to the e-vector of polarized light, conveys different information to the brain than that conveyed by the camera eye to the vertebrate brain. The environment also "looks" different to the brain of a frog than to the brain of a monkey. Whichever way the world appears to an animal, in order to function, to orient itself, and more generally, in order to survive in it, the sensory systems must perform among other things two important operations with the corresponding input signals: they must be able to localize stimuli in space and to recognize them, that is to answer the questions about the *location* and the *significance* of the signal. Furthermore, the system should be able to store such information. In the following discussion we shall understand signal recognition to mean the classification of stimulus distributions from the environment into innate or acquired categories of functional significance.

3.2 Innate and Acquired Releasing Mechanisms

Animals have a repertoire of fixed behavior programs. Such types of behavior can be innate. We then speak of *fixed action patterns* which are also called "instinctive motor patterns" in the classic sense. Fixed action patterns are hereditary, preprogrammed, so to speak, in the central nervous system, and are marked by rigid components. In comparative ethology the stimuli, that release such a motor pattern, are called *signals, sign stimuli,* or *key-stimuli* just as the key opens a lock. The sensory recognition system which decides whether the key fits the lock is called an *innate releasing mechanism,* IRM, and it functions like a filter.

Innate releasing mechanisms (IRM) can be modified by experience (IRME) and therefore increase and modify their response spectra. Finally, there are releasing mechanisms which are based exclusively on learning. With them the biological relevance of a signal is acquired in the course of an individual life time; we speak then of an acquired releasing mechanism, ARM [14] (cf. also p. 147).

We must be careful to realize that *innate* does not mean that the environment is totally irrelevant in the development of fixed action patterns and releasing mechanisms as well. However, these patterns develop in a completely normal and species-specific way in the total absence of specific information from the environment. For example, many birds can be isolated from all sounds and still produce a completely normal song.

3.3 Releasing Stimuli

3.3.1 Key-Stimuli and Dummies

Key-stimuli may be relatively simple. They "reproduce" the particular object they are supposed to represent as a kind of caricature. Thus, they can be easily recognized as signals and responded to rapidly. Their effectiveness can be quantitatively analyzed in experiments with dummies (models), by omission, addition, and also by the exaggeration of particular components. Some examples:

Part of the distinctive parental care behavior of cichlid perches *Nannacara anomala* is that the young follow their parents. How does the young fish recognize its mother? It was shown in dummy experiments that the breeding garb and the jerky mode of swimming are the most significant stimuli (Fig. 32). The light and dark pattern elements of the breeding garb provide two simple, important features: dark area contrasting against a bright background and a light point structure within the dark area. The shape of the fish and the exact order of the pattern elements are of secondary significance. Comparative studies with different species of *Nannacara* have shown that the key-stimuli are as

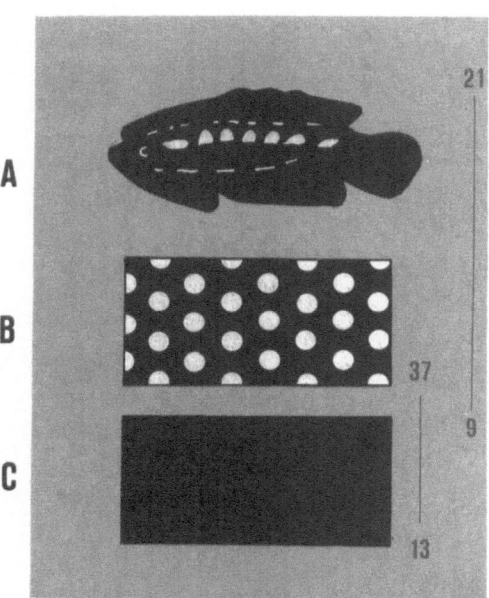

Fig. 32 A–C. Key-stimulus for the "follow the mother" reaction of young perches *Nannacara anomala.* The efficacy of various dummies of the mother compared with each other is indicated by the reaction numbers. [Modified from Kuenzer P (1973) In: Wickler W, Seibt U (eds) Vergleichende Verhaltensforschung. Hoffmann und Campe, Hamburg]

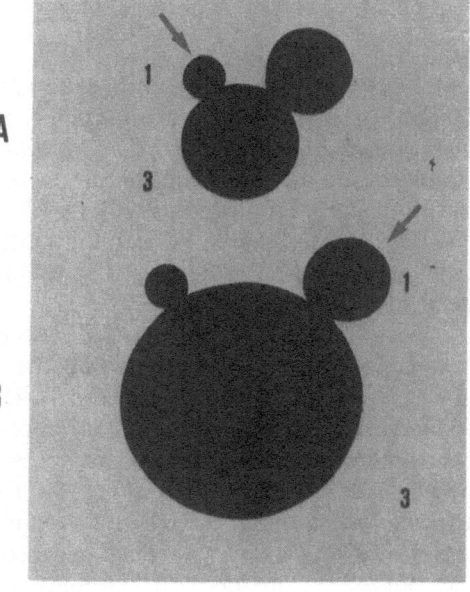

Fig. 33 A and B. Key-stimuli for the gaping of young blackbirds *Turdus merula.* A dummy of the parent bird is maximally responded to when the head-rump ratio is 1:3, see *arrows.* [Modified from Tinbergen N, Kuenen PJ (1939) Z Tierpsychol 3: 37–60]

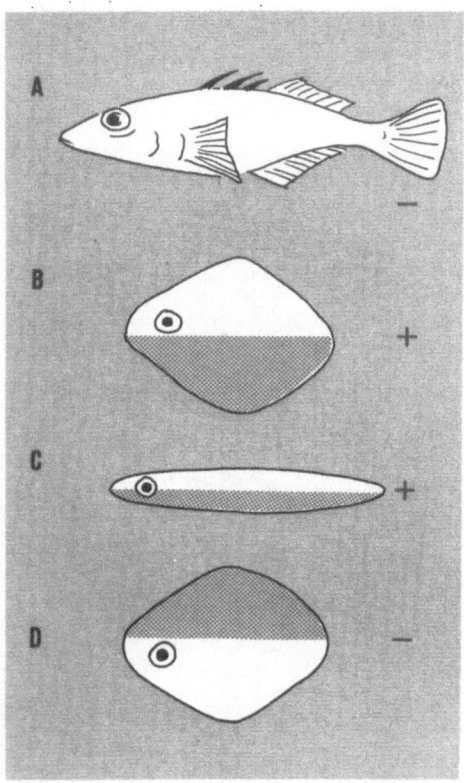

Fig. 34 A–D. Fight releasing (+) key-stimuli for the male stickle-back *Gasterosteus aculeatus.* [Modified from Tinbergen N (1951) The study of instinct. Oxford Univ Press, London]

simple as possible and resemble the appearance and behavior of the response-releasing mother as closely as necessary.

For young blackbirds *Turdus merula* the parent animal may be simulated by a head-rump dummy consisting of two circular disks of different sizes. Experiments with two-headed dummies of various sizes (Fig. 33) showed that as key-stimuli for the gaping of the brood a certain size-ratio of head to rump (1 to 3), is more important than the absolute head size of this dummy.

In male three-spined sticklebacks *Gasterosteus aculeatus,* the red belly represents a key-stimulus exciting the rival to fight (Fig. 34). It is, however, effective only if the red belly is on the ventral side. Here, too, the form of the fish is of secondary importance as a signal. The colored perch *Haplochromis burtoni* possesses characteristic head markings which have a threatening effect on rival males and increase their aggressiveness (Fig. 35 A). The key-stimulus is a black stripe, running lengthwise below the eye, which has a certain orientation in relation to the body-axis of the fish. Dummy experiments showed that the threatening effect of the stripe depends on its angle of orientation (Fig. 35 B).

From these examples it is evident that not only a definite stimulus *pattern,* but also its *configuration* in relation to other features may be of great importance for the releasing value of a key-stimulus. We then speak of "Gestalt" perception. Some further examples:

The common toad *Bufo bufo* responds to small moving objects with prey-catching behavior. Dummy experiments with a small narrow dark stripe contrasting with a white background showed that the key-stimulus "prey" has two main characteristics: (1) the movement, (2) the area dimensions relative to the direction of the movement (Fig. 36). When the stripe moves in a wormlike fashion, parallel to its longer axis, it signals "prey"; but when the longer axis of the same stripe is oriented perpendicular to the direction of movement, the stimulus loses this key feature; it may even signal "threat" (Fig. 36a and c).

Configurational stimuli also play a role as releasers for the food-begging

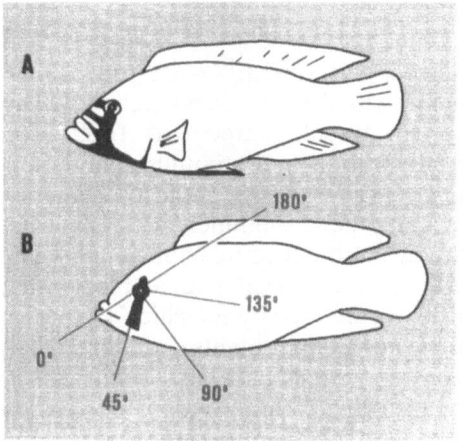

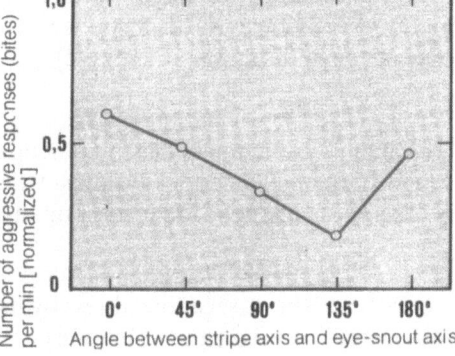

Fig. 35 A and B. Head markings of the perch *Haplochromis burtoni* as key-stimulus with threat effect. **A** Natural markings of the fish. **B** Dummy in which the position of the eye stripe may be changed and related to the level of aggression of a rival male. [Modified from Heiligenberg W, Kramer U, Schulz V (1972) Z Vergl Physiol 76: 168–176]

Number of aggressive responses (bites) per min [normalized]

Angle between stripe axis and eye-snout axis

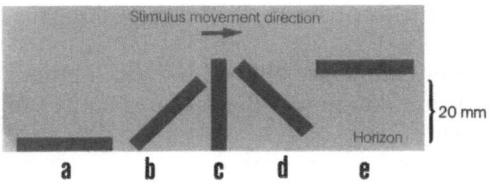

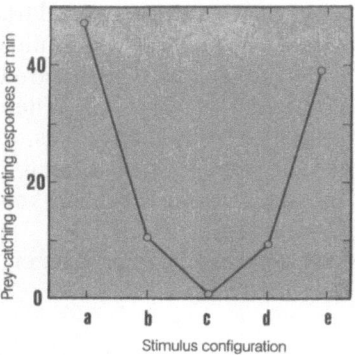

Stimulus configuration

Fig. 36. Key-stimulus for the release of prey-catching in the common toad *Bufo bufo*. The efficacy of a worm dummy depends on the orientation of its longitudinal axis relative to the direction of movement. [Modified from Ewert JP (1968) Z Vergl Physiol 61: 41–70; Ewert (1975)]

behavior of young herring gulls *Larus argentatus*. In dummy experiments with different head models we can test and compare the effects of certain visual components (Fig. 37). Thus, principles of stimulus configuration and stimulus summation can be studied. As shown in Figure 37 it is even possible to exaggerate stimuli characteristics, thereby forming head-dummies with supernormal releasing characteristics. Studies of the feeding behavior of sea gulls indicate that action patterns are not fully developed at birth. Their development appears to be strongly affected by the chick's experience [15].

Behavioral experiments with human babies suggest that during recognition of visual patterns, *stimulus summation* phenomena can be "transformed" into *Gestalt perception* phenomena in the course of ontogeny. For example (Fig. 38) in young babies the intensity, R, of the orienting reaction to a face model equals the algebraic sum of the orienting activities (R) in response to each component of this model (circular disk, d; points, p; cross, c):

$$R_{dpc} \approx R_d + R_p + R_c.$$

After half a year at most, Gestalt perception develops in babies. Now the response to the face model is almost twice as high as the algebraic sum of the responses to each single component:

$$R_{dpc} \gg R_d + R_p + R_c.$$

The pattern is now recognized less by its single components than by the "integrated" configuration (Fig. 38).

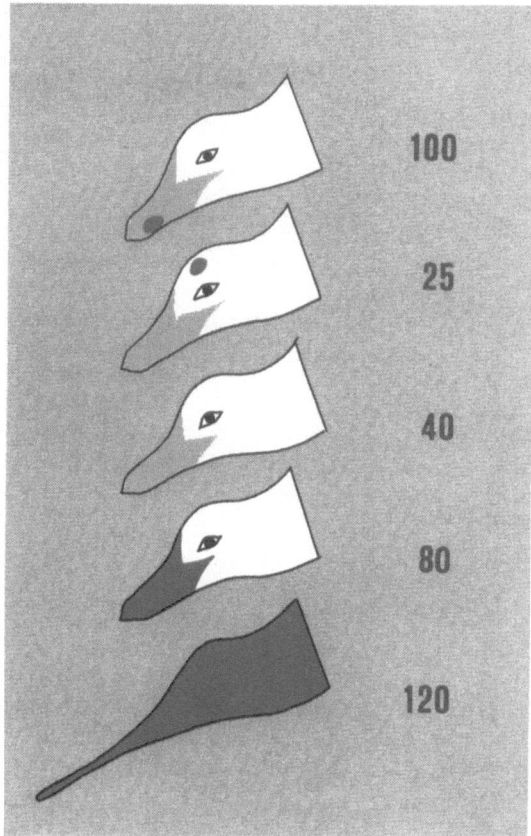

Fig. 37. Key-stimuli for the begging of young silver gulls *Larus argentatus*. The efficacy of the head dummy of a parent bird depends on definite characteristics: shape of bill, color of bill (extent of yellow or red), topography of red spot. The numbers in each case indicate the responses to the dummy as a percent of the response elicited by natural parent bird. [Examples from Tinbergen N, Perdeck AC (1950) Behaviour 3: 1–38]

The ability to classify stimuli into learned categories of meaning is seen especially in higher organized animals – in vertebrates as well as invertebrates. At school we learn to associate symbols of script with categories of meaning. It is important here that unambiguous classification of a configurational stimulus to a category of meaning is largely independent of variations of certain other parameters. For example, our visual system is able to recognize the letter "B" independently of the (1) contrast direction with the background, (2) visual field position, (3) absolute size, (4) script type (Fig. 39, top). Evidently the recognition system exhibits *invariance*. The construction of invariants appears to be

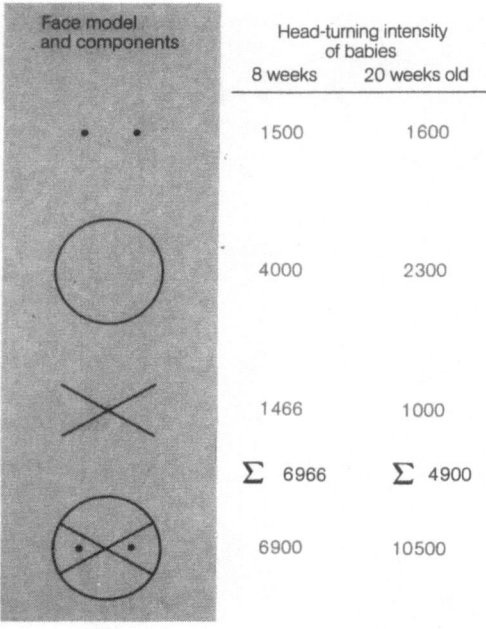

Face model and components	Head-turning intensity of babies	
	8 weeks	20 weeks old
	1500	1600
	4000	2300
	1466	1000
	Σ 6966	Σ 4900
	6900	10500

Fig. 38. Phenomena of stimulus summation and Gestalt perception in the recognition of face and partial-face models by the infant at different ages. [Modified from Bower TGR (1966) Anim Behav 14: 395–398]

a precondition for *Gestalt perception*. In the course of experience, however, more meanings may be added (Fig. 39, bottom).

3.3.2 Releasers and Signal Falsification

Whereas the terms key-stimulus and signal are usually employed synonymously, they may be distinguished from the term releaser. What is the difference? With regard to the releaser the interest on information transmission is shared between the "transmitter" and the "receiver", i.e., there is communication. With the key-stimulus, however, the interest is restricted to the receiver. A worm, for example, is not interested in forming a prey-stimulus for the toad!

Releasers are employed mainly in *intra*specific communication. Social releasers often consist of visual, auditory, or olfactory components. There are also social releasers, which have developed in the course of evolution from particular behavior patterns, such as cleaning and feeding. These ritualized movements are strongly simplified; we may call them *signal actions*. Social releasers may be involved in different kinds of behavioral functions, such as recognizing conspecifics, uniting sexual partners, and demarcating territory.

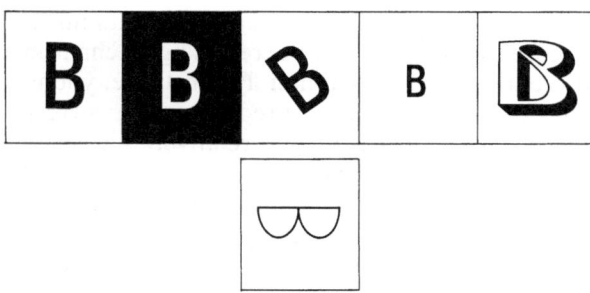

Fig. 39. Classes of invariants in the recognition of the visual symbol "B" *(top):* sign of contrast, position, size, script type. *Bottom,* Ambiguity. See text for details (cf. Fig. 55)

Releasers are also used in *inter*species communication. The butterfly *Automeris memusae* startles and chases away its potential predators by unfolding its wings and thus presenting conspicuous and menacing eye markings. We must also mention in this context "signal falsification" (mimicry). An animal imitates certain characteristics of another in order to procure for itself the biological advantage of the latter. The defenseless syrphidae wear the black-yellow rings of the wasp (wasp-mimicry). To catch its prey the North American turtle *Macroclemis temminckii* uses natural dummies as bait: it lies at the bottom of the water waiting for prey, motionless and with open mouth. Its bait consist of two thin red skin protuberances which contrast against the dark tongue and are moved by it like small worms. As soon as a smaller fish approaches and touches this worm dummy, the turtle snaps its mouth shut.

3.4 Principles of Stimulus Selection

3.4.1 Habituation Phenomena

The efficacy of a key-stimulus depends on the frequency with which it is repeatedly offered to the animal. Habituation phenomena, moreover, are often stimulus-specific: if, for example a behavioral reaction has been habituated in response to a stimulus *A,* it may be released again by a slightly modified stimulus *B.* The reason for stimulus habituation could be stimulus-specific after-effects in the CNS which accumulate during successive stimulation and influence the selectivity of the releasing mechanism. The biological significance apparently is the protection of the releasing mechanism from senseless reactions and to maintain its alertness toward "new" stimuli.

Long-term habituation for particular stimulus configurations may increase the selectivity of the releasing mechanism and induce a further differentiation of key-stimuli. For instance: young turkeys show escape behavior toward airborne predators. The corresponding key-stimulus filtered by an IRM is at first relatively unspecific: Any large shadow (from a bird) moving over them releases escape. The shape at first plays no role. In the course of time, following some kind of learning the IRM becomes selective to configurational stimuli. The turkeys become habituated to goose-like birds (long neck, short tail) flying overhead repeatedly in their environment whereas the less frequently seen birds of prey (short neck, long tail) continue to be avoided. This can be tested most convincingly in a dummy experiment (Fig. 40): When one and the same bird dummy is moved with the short end leading, it symbolizes an airborne predator and releases avoidance behavior. Moved in the opposite direction it symbolizes with its long neck an inoffensive goose and hence elicits no response. In this instance the stimulus *configuration* plays an important role for decision making by the modified IRM (IRME).

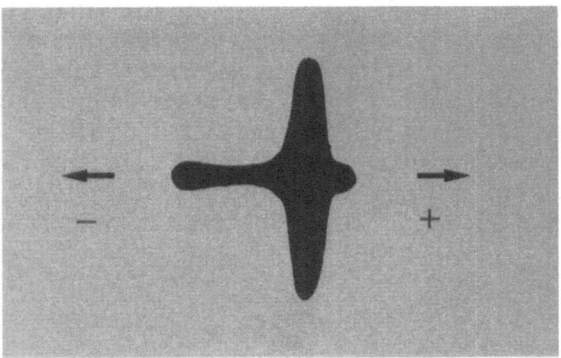

Fig. 40. Effect of bird dummy on the escape response of turkeys. The releasing character (+) depends on the direction of movement of the dummy and thus, its configuration. Further details in text. [Modified from Tinbergen, N (1948) Wilson Bull 60: 6–52; Schleidt WM (1961) Z Tierpsychol 18: 534–560]

3.4.2 Recognition with Searching Images

The abstraction of a particular pattern known to us and which we want to distinguish from many others is called a searching image. It may reproduce our idea of the original by characteristic form or by color and contrast properties. The production of a false searching image may lead to a perceptual psychological game of hide and seek.

The recognition of prey objects by means of searching images has been analyzed in titmice *Parus major*. At first sight the principle appears to be the opposite of the habituation just described. The final objectives, however, are similar in that it is the selection of appropriate stimulus features. When titmice repeatedly find a conspicuous and tasty prey in their environment, they neglect all other prey-objects. Evidently the animal has formed a searching image for the preferred object which facilitates its detection. This image, however, presupposes that it may be rapidly changed if need be: if the preferred prey is found more rarely, a new searching image of the new goal objects is required [16].

3.5 Motivation and Change of Meaning

Releasers are not necessarily permanently associated with their carriers. Their development and releasing values may be limited to certain life periods or to particular seasons. Also the meaning of a key-stimulus does not represent a constant factor for activating a behavior. It may depend on changes of the physiological state which regulate the motivation of the animal; this can be subject to diurnal, as well as seasonal variations. When a sparrow is motivated to build a nest, straw has significance for it. During mating it "overlooks" the straw since at this stage its partner plays a decisive role. In some animals stimuli can be associated with specific meaning during a *sensitive period* in the individual's early life – a phenomenon called "imprinting".

When the motivational level is very low the releasing value of even key-stimuli can be strongly diminished. Conversely, the releasing value of a weak stimulus may increase in the presence of very high motivational level. The corresponding behavior is then activated with inappropriate objects or even in the absence of an external stimulus. We term this behavior vacuum activity. The releasing mechanism may be unlocked, say, by "picking it" rather than with a specific "key".

The following points should be kept in mind:

1. Releasing mechanisms, RM's, are specific sensory recognition systems which activate fixed patterns of behavioral responses.

2. Recognition systems can be *innate* (IRM) and thus have been acquired during phylogenetic adaptation to a certain environmental situation. As a result of an individual's *experience* an IRM may be modified (IRME), thus extending or reducing its filter properties. There are also releasing mechanisms which are *acquired* by an individual during ontogeny (ARM). The filter properties of these ARM's are determined by learning.

3. The corresponding signals (key-stimuli or releasers), which are to be recognized by a RM, usually have simple configurational cues.

4. The efficacy of key-stimuli, and with it also the selectivity of a RM, depends on the motivation of the animal.

What chain of neural events connects a key-stimulus with a specific pattern of behavioral responses? What do we know of neuronal processes which underlie recognition and localization of key-stimuli, and how could an innate releasing mechanism function? We shall first attempt to answer these questions in detail with the aid of a well-studied example.

4 Neurobiological Basis for the
Recognition and Localization of Environmental Signals:
How Does a Toad Brain Recognize Prey and Enemy?

When a toad sits motionless at twilight in front of its hiding place, very little or even no information is fed to its brain from the static visual environment (via retino-tectal projection paths, see p. 101) [17,35]. This has been demonstrated in experiments where, with the aid of a relatively sophisticated electrophysiological recording technique (Fig. 161), the neuronal responses from single optic nerve fibers were recorded in the optic tectum of a quietly sitting animal facing a stationary pattern. Result: responses were absent. Neuronal responses were observed as soon as the pattern started moving, and they continued in certain neuronal classes (R2) for a couple of seconds after the movement was stopped. In this important respect we, for example, differ from the toad. Although the static images on our retinae are also extinguished by local adaptation, we are able to counteract the adaptation by moving the image over an area of the retina with the aid of involuntary saccadic eye movements, thereby transforming it into a moving stimulus.

Frogs and toads do not have involuntary saccadic eye movements [18]. Many behavioral responses, therefore are mainly to those stimulus patterns which pass through sections of their visual field. Thus, a moving beetle releases prey-catching; an approaching hedgehog or a bird of prey, flying overhead, however, activates escape behavior. If both, prey and predator, do not move, the toad generally will not react (cf. p. 88). Sound and smell are of secondary significance. Obviously the toad can afford to restrict its attention to moving objects. A motionless dead worm is not of use to it, nor is a motionless enemy dangerous. The toad may, however, perceive static visual patterns when its head movements cause retinal images to shift positions; this information can be "stored" for a few seconds. For example, very small head and/or eye movements are correlated with strong breathing movements [19].

When a toad is motivated to catch prey and a small moving prey object appears in its visual field, it responds with a sequence of successive behavioral reactions (Fig. 41): (1) orienting toward the prey, (2) stalking up to the prey, (3) binocular fixation, (4) snapping, (5) swallowing, and (6) wiping the mouth with the fore limbs. This sequence of motor patterns may be described as a stimulus-response chain, set in motion by the prey stimulus, and in which each reaction of the toad provides the stimulus situation for the next response (Fig. 41). Single patterns of this

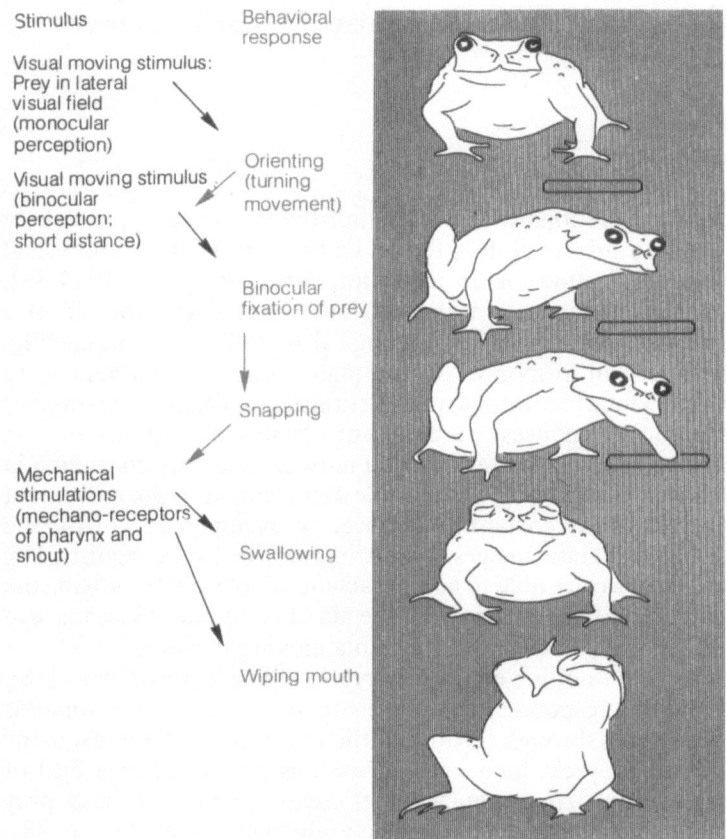

Stimulus

Behavioral response

Visual moving stimulus: Prey in lateral visual field (monocular perception)

Visual moving stimulus (binocular perception; short distance)

Orienting (turning movement)

Binocular fixation of prey

Snapping

Mechanical stimulations (mechano-receptors of pharynx and snout)

Swallowing

Wiping mouth

Fig. 41. Stimulus-reaction chain during prey-catching of the common toad *Bufo bufo*. [Based on results by Ewert JP (1967) Z Vergl Physiol 54: 455–481]

response sequence, however, might be linked together by central nervous mechanisms: If the prey object is quickly removed during the binocular fixation, the subsequent partial actions – snapping, swallowing and wiping – may be carried out automatically in the absence of the prey, and in the correct sequence [20].

If a large moving enemy object confronts the toad, it responds with escape movements that allow it to avoid the stimulus. These may consist of ducking with inflated lymphatic sacks or jumping away (airborne enemy), standing up and inflating itself, dodging, or running away (ground enemy). Common to prey and enemy key-stimuli are the parameters contrast and movement; the distinction between prey and enemy is probably made mainly on the basis of configuration and size.

We may then ask the following questions:

1. Which features of size and shape of a moving object are linked with prey or enemy for the toad? Common toads are active during twilight; the stimulus parameter *color* can be neglected for the time being [21].

2. Are there centers for prey-catching (orienting) and escape (avoidance) in the toad brain?

3. To what extent does the neuronal system of the retina serve as the first information filter?

4. Which areas of the brain process the visual input from the retina and to what extent do they contribute to the localization and interpretation of prey and enemy objects?

5. What is the basis of decision-making – prey/nonprey – in terms of neuronal events?

4.1 The Key-Stimuli "Prey" and "Enemy"

Let us consider first the brain (more properly the whole organism of the toad) as a "black box" and study the input-output relations, i.e., the stimulus-response relationships for visual behavior. The strategy of these experiments is to find out, from the viewpoint of the toad, the critical features and dimensions of the caricature of its prey and enemies.

4.1.1 An Experimental Measure of Stimulus Effectiveness

The effectiveness of the stimulus in releasing a behavior may be reflected in the response activity of the animal. In order to measure quantitatively the prey-catching behavior in response to a simulated prey object, we do not require by any means the complete behavioral sequence. It can also be measured in terms of an "isolated" component. However, it is necessary to know which behavioral component reflects unambiguously the differences in simulated prey objects. Such a component is the *orienting movement.*

Whereas both orienting and fixation adapt to a given stimulus situation, snapping is relatively rigid: When the prey-object is removed during the binocular fixation, the toad snaps, so to speak, in a vacuum. The toad consequently expresses its initial interest in a prey-object by the target-oriented turning reaction. The greater the resemblance between visual stimulus and prey, the faster the toad turns toward it. *Since the toad holds its eyes in a rigid position, it has made a "prey/non-prey/ enemy" decision already prior to the orienting movement. If a stimulus*

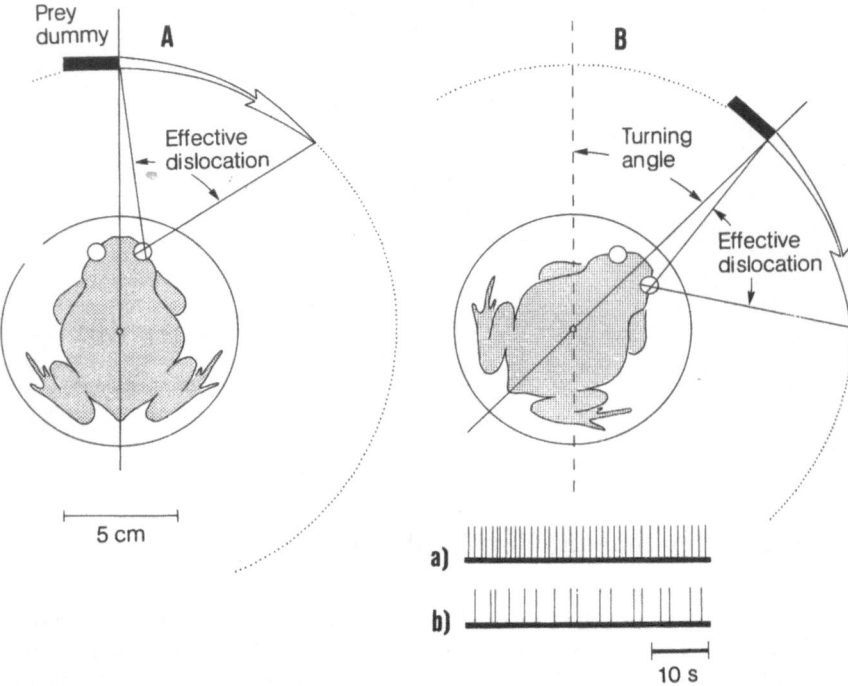

Fig. 42 A and B. Principle of experimental set up for quantitative measurements of prey-catching orienting activity of a toad. Correlation of effective displacement of prey dummy **(A)**, turning angle of toad **(B)** and prey-catching activity; *a* time sequence of turning reactions to a strongly and *b* to a weakly effective prey dummy. The angular velocity was kept constant at 20°/s. [From Ewert JP (1969) Pflügers Arch 308: 225–243]

pattern possesses no prey features the probability of prey-catching orientation is very low. There are no "tracking eye movements" in frogs and toads. Prey recognition is feasible through images formed on any part of the retina.

How Can Orienting Activity Be Measured? Let us use the following experimental arrangement (Figs. 42 and 151). The toad sits in a cylindrical glass vessel around which a piece of cardboard contrasting with the background is mechanically moved in a horizontal plane at a constant angular velocity (20°/s) while keeping a constant distance of 7 cm from the animal. The toad may interpret such a small object as a prey animal and try to catch it in the binocular fixation field by successive turning reactions of head and body and in the process it turns around jerkily. We can take the number (R) of the orientational turning reactions as a measure of its prey-catching activity, and this is expressed as the number of turns which accompany the circling dummy during an interval of 1 min: R [turning reactions/min]. As the releasing value of a prey dummy becomes smaller, the effective displacement, D [degrees of visual angle], of the stimulus in the visual field necessary for releasing an orienting reaction lengthens, the angle, T[degrees], of the turning movement increases, and the orienting frequency becomes smaller (Fig. 42, a and b). If the toad can follow the prey continuously, the product of orienting activity (R) and the mean turning angle (T) is constant:

$$R \times T = c \text{ [degrees/min]}$$
$$R = c/T \text{ [turning reactions/min]},$$

where c is the angular velocity of the dummy in relation to the center of rotation; c is held constant. Thus, R is a reciprocal measure of the average effective stimulus displacement, D. The R of a particular stimulus – in relation to R values of other stimuli – can be taken as a measure of resemblance between that stimulus and prey.

4.1.2 What is Large for a Toad?

There are two criteria by which an animal can judge the size of a visual object. The simplest way is to determine the size based upon the size of the retinal image, in degrees of visual angle. However, since the size of an object remains constant, and the size of the visual angle measured by the animal changes with the distance, reliance upon this cue alone could result in misinterpretations – provided configurational cues are not available or are neglected. An automobile, for instance, moving close to the animal may realistically elicit escape, but at a distance of 15 m the same object may produce a retinal image corresponding to a "bug". It is therefore important to judge the absolute size of a visual object by also taking into account an estimate of its distance. This ability is called "size constancy".

In behavioral experiments it is easy to investigate how the toad judges prey dummies of different sizes (Fig. 43 A). When in a series of consecutive experiments, the orienting activity is measured in response to black squares (or circular disks) of different sizes moved around the toad against a white background at a distance of 7 cm, the optimal prey dummy size is between 4° and 8° visual angle (Fig. 43 A). An object measuring 20° usually does not elicit a response. Larger objects cause the toad to turn away. What happens when the distance between toad and dummy is changed? Figure 43 B shows the result: the optimal value does not remain constant over the range of visual angles. Rather it shifts in the direction of higher values for short distances and vice-versa. Accordingly, it would preserve its position over the absolute (metric) scale. Therefore, during prey-catching toads show size constancy [22]. In

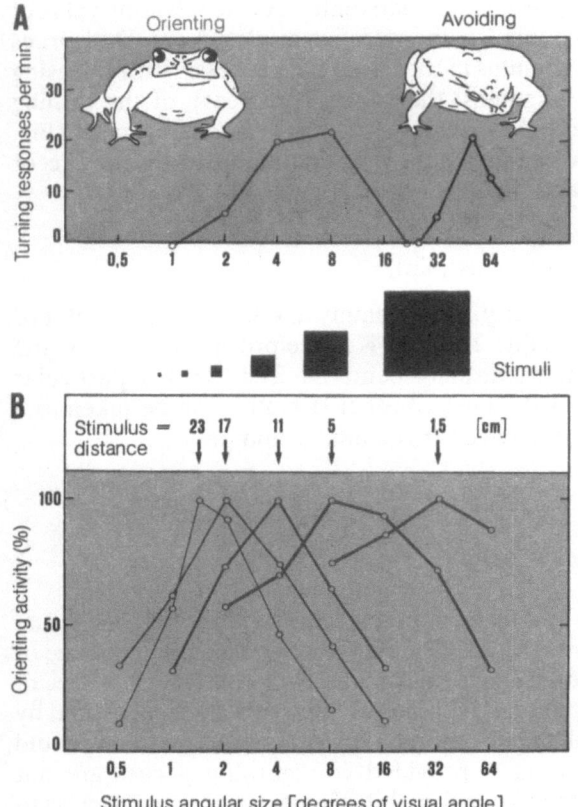

Fig. 43 A and B. Effect of square-shaped prey dummies of various sizes, moved at 20°/s, on the turning activity of the common toad *Bufo bufo*. **A** Transition from prey-catching (orienting) to escape behavior (avoiding) with increasing size of a square object at a constant distance of ~7 cm between stimulus and experimental animal. **B** The size constancy phenomenon. Change of prey-catching activity (normalized) in response to square-shaped prey dummies of various sizes at various distances of the stimulus. Curves of the means from a total of 20 **(A)** or 15 **(B)** experiments with different animals. Further details in text. [Modified from Ewert JP (1976) In: Fite KV (ed) The amphibian visual system. Academic Press, New York **(A)**; Ewert JP, Gebauer L (1973) Z Vergl Physiol 85: 303–315 **(B)**]

judging a stimulus area the decisive factor is the absolute size of the object and not its angular size. Although the attractiveness of a dummy decreases with increasing distance from the animal, in a series of squares of variable size one with sides of 10 mm is maximally effective at distances varying from 1.5 to 23 cm.

The question of how toads obtain information about the distance of the object, which is required for estimating size, is so far unanswered. Since one-eyed animals show similar behavior, the mechanisms of lens accommodation [23] and trigonometric evaluation by short-term shifting of the retinal images during a head turning – a process known as parallax – should be considered apart from the binocular distance estimate.

Do Toads Learn to Recognize the Correct Size? If So, When? These questions were investigated in greater detail in young midwife toads

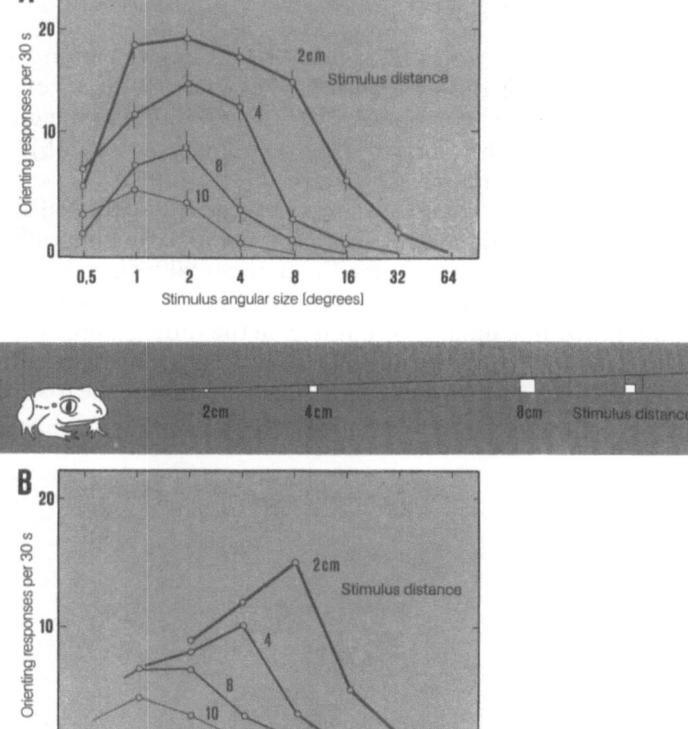

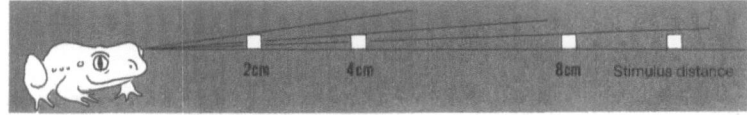

Fig. 44 A and B. Measurement of visual angle **(A)** and estimation of absolute size **(B)** of a square-shaped prey dummy in freshly metamorphosed **(A)** and older midwife toads *Alytes obstetricans* **(B).** The angular velocity of the dummies was 20°/s. Curves of means from 10 experiments in each case, using different animals. Further details in text. [Modified from Ewert JP, Burghagen H (1979) Brain Behav Evol 16: 99–112]

(*Alytes obstetricans,* Laur.). It appears, surprisingly, that young postmetamorphic juvenile animals at first prefer, almost without regard to the distance, a certain angular size of a square prey dummy. Nevertheless a trend for estimating the absolute size can be seen already (Fig. 44 A). Six months later, the size constancy phenomenon during prey-catching is clearly evident. Presumably the ability matures during ontogenesis.

4.1.3 The Effect of Gestalt Features

Since toads perceive prey and enemy objects in terms of *moving* visual stimuli, the analysis of key-stimuli suggests at first the investigation of those configurational components which are related to the direction of the movement of the dummy (x-y coordinates) (Fig. 45). Our first experimental program thus consists of three groups.

Starting with a small square dummy, the side is lengthened stepwise in the horizontal direction in the first group of experiments with constant vertical dimension; in the second experimental group the side is now lengthened perpendicularly to the direction of the movement (with constant horizontal dimension); finally in the third group, both sides are equally enlarged, consequently different large squares are offered. All objects are black and are moved mechanically around the toad at 20°/s in front of a homogeneous white background.

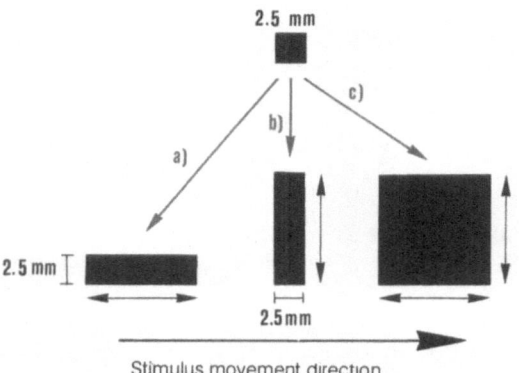

Fig. 45. Three different possibilities of changing the configurational characteristics of a moving two-dimensional visual stimulus. Starting with a 2.5 × 2.5 mm² square, **a** the vertical edge is kept constant at 2.5 mm and the horizontal edge is elongated stepwise in the movement direction in successive experiments, **b** the horizontal edge is kept constant at 2.5 mm and the vertical edge is elongated perpendicular to the movement direction, **c** squares of different sizes are displayed. Further details in text. [From Ewert JP (1969) Pflügers Arch 308: 225–243]

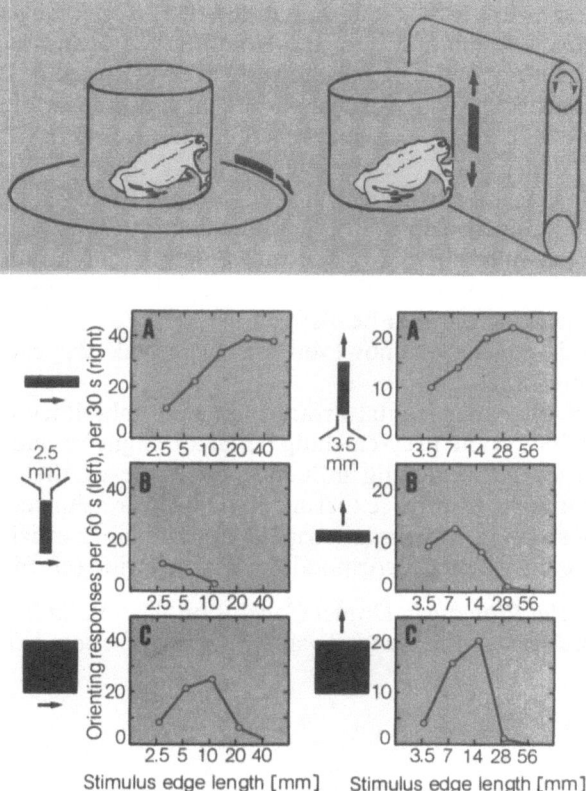

Fig. 46 A–C. Stimulus efficacy of prey dummies on prey-catching activity of the common toad as a function of configurational features varied according to Figure 45 for horizontal *(left)* and vertical movements *(right).* Stripes of varying length, oriented in the direction of movement **(A),** and perpendicular to the direction of movement **(B),** and squares **(C)** of varying sizes. The angular velocity of the dummies was 20°/s, the distance: ~7 cm. Curves of the means of 20 *(left)* or 15 *(right)* experiments per value, using different animals. [Ewert JP (1968) Z Vergl Physiol 61: 41–70; Ewert JP (1976) In: Fite KV (ed) The amphibian visual system. Academic Press, London New York; Ewert JP, Arend B, Becker V, Borchers HW (1979) Brain Behav Evol 16: 38–51]

Since it has been demonstrated that toads show size constancy in catching prey, the data previously given in degrees of visual angle are now convertible to absolute units (mm).

Figure 46 (left) shows the effectiveness of the three series of dummies for releasing prey-catching behavior, as measured by the orienting activity. Lengthening of the stripes in the horizontal movement direction enhances – within limits – the prey effect of the dummy (Fig. 46 A). Lengthening of the stripes across the movement direction decreases the prey simulation and may even signal "threat" (Fig. 46 B). With squares it results in a kind of addition of the two effects: the prey-signal enhancing effect originating from the horizontal edge and the signal-diminishing effect originating from the vertical edge (Fig. 46C). With some animals black squares or vertical stripes moved in front of a white background induce no prey-catching reaction whatever (Fig. 48B).   −

These results can be interpreted as follows. Worm and beetle-like objects are particularly attractive to the toad; such objects turned by 90° and moved with their longitudinal axis perpendicular to the horizontal movement direction constitute stimulus configurations which apparently do not fit the toad's idea of a prey: there are no worms – "antiworms" – walking on their heads!

This conclusion, however, has two weak aspects which need further consideration.

In this experimental arrangement the inhibitory effect of the vertical stripe on the prey-catching behavior might be due to the fact that they exert a threatening influence by virtue of their extension over the horizon. Control experiments, however, showed that, even in this position, a horizontal stripe is a much more efficient stimulus as prey dummy than a corresponding vertical stripe (cf. Fig. 36c and e).

Is the Previously Drawn Conclusion Also Valid If the Pattern Is Moved in Front of the Toad, To and Fro, in a Vertical Direction? As shown in

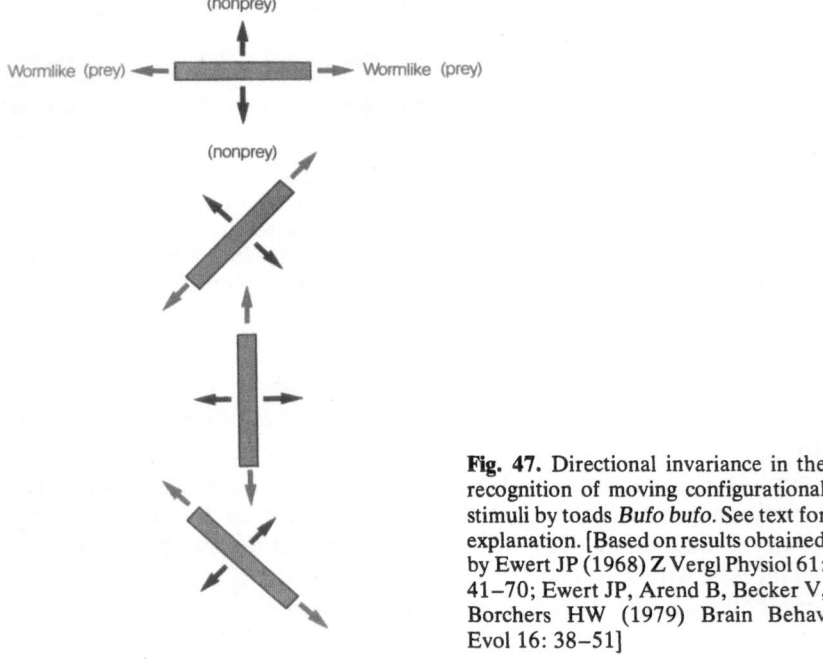

Fig. 47. Directional invariance in the recognition of moving configurational stimuli by toads *Bufo bufo*. See text for explanation. [Based on results obtained by Ewert JP (1968) Z Vergl Physiol 61: 41–70; Ewert JP, Arend B, Becker V, Borchers HW (1979) Brain Behav Evol 16: 38–51]

Figure 46 (right), the study on the effect of Gestalt factors revealed that the prey-catching activity is basically uninfluenced by the orientation of the movement direction of the dummy. Further experiments showed, for instance, that the one and the same 2.5 × 30 mm² stripe releases prey-catching behavior during "wormlike" movement (stripe axis parallel to movement direction) and elicits no response or causes retreat during "antiwormlike" movement (stripe axis perpendicular to the movement direction). The ability of the toad to distinguish between worm and antiworm configuration is independent of the movement direction of the pattern (x-y coordinates, cf. Fig. 47), even if the pattern is moved toward or away from the toad (z-y coordinates). The visual system of the toad thus possesses a directional invariance in this domain of Gestalt perception.

4.1.4 Ontogenetic Aspects and Species Differences

The prey-catching behavior of various species of amphibia, including its ontogenetic aspects, has been studied with respect to differences of shape of moving stimulus patterns. Shortly after its metamorphosis, the midwife toad *(Alytes obstetricans)* behaves in a manner similar to that of the adult common toad *Bufo bufo* toward patterns of the three series of

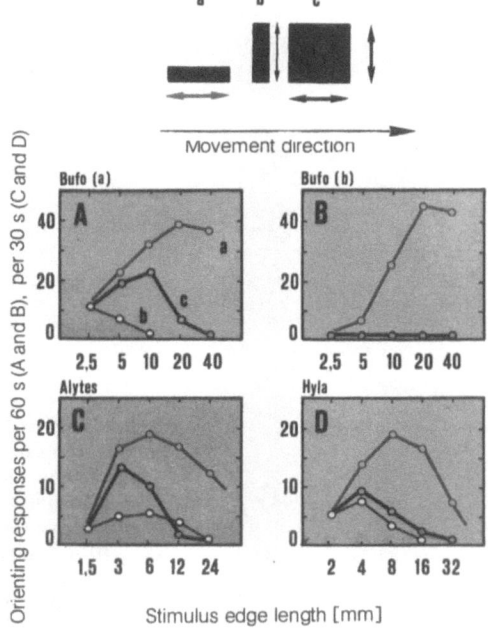

Fig. 48 A–D. Stimulus efficacy of a prey dummy – as a function of various configurational features – on the prey-catching activity of *Bufo, Alytes,* and *Hyla. Red curve,* stripes of different lengths elongated in the direction of movement ("worms"). *Thin black curve,* stripes of different lengths elongated perpendicular to direction of movement ("antiworms"). *Thick black curve,* squares of varying sizes. (This way of representation will be adhered to in subsequent analogous figures). The stripe width was 2.5 mm in **A** and **B**, 1.5 mm in **C** and 2 mm in **D**. Angular velocity: 20°/s; stimulus distance: ~7 cm (**A** and **B**), ~4 cm (**C**) and ~3 cm (**D**). Curves of the means of 20 (**A, B**) or 10 (**C, D**) experiments per value, with different animals. [Modified from Ewert JP, Burghagen H (1979) Brain Behav Evol 16: 157–175]

dummies (Fig. 48 A and C). The main difference between the common toad and the midwife toad is that the efficacy of stripes expanded in the direction of the movement and squares are restricted to a relatively small range of sizes (Fig. 48 C). Similar results were obtained in tree frogs *Hyla arborea* and *H. cinerea* (Fig. 48 D) and in the toad *Bombina variegata*. The ability of these animals to prefer wormlike moving objects is, as far as can be ascertained, an innate characteristic. The general worm/anti-worm phenomenon is in common toads established immediately with transition to terrestrial life. However, the acuity of configurational prey selection increases during the first week, even though in these experiments the postmetamorphic toads had never caught any prey (the tadpoles are not carnivorous). Of course, this does not exclude the possibility that (early) experience may influence prey selection (cf. also pp. 83 and 89). Figure 48 C shows the responses of young postmetamorphic midwife toads which indicates that the worm preference is already present. About six months after metamorphosis, the preference is even sharpened [61]. Whereas the average prey-catching activity in response to wormlike stripes remains about the same as immediately after

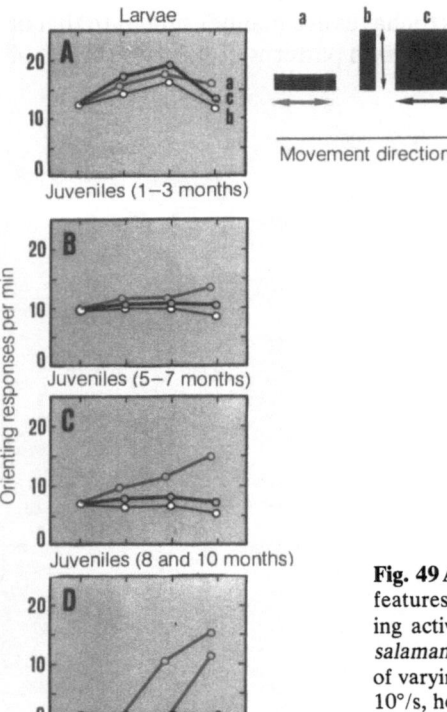

Fig. 49 A–D. Effect of different configurational features of a moving dummy on the prey-catching activity of the fire salamander *Salamandra salamandra*. **A** Larvae and **B–D** juvenile animals of varying ages. Angular velocity of the dummy: 10°/s, horizontal movements. Stimulus distance: ~3 cm. Curves of the means of 20 experiments per value. [Modified from Himstedt W, Freidank H, Singer E (1976) Z Tierpsychol 41: 235–243]

metamorphosis (Fig. 48 C), the activity in response to squares and correspondingly long antiworm-like stripes is now decreased by about 50%.

The fire salamander *(Salamandra salamandra)*, an urodele, is particularly suitable for ontogenetic studies in this context, especially since the larvae also feed by prey-catching. In the larval stage the salamander is hardly able to distinguish square-shaped objects from equally long stripe-shaped stimuli, that are expanded in or perpendicular to the direction of the movement (Fig. 49 A). A pattern size of about 8 mm length in either square or rectangular configuration is always preferred. However, after metamorphosis the juvenile animals begin to differentiate among moving configurational stimuli (Fig. 49 B-D). After 10 months at the latest, the releasing mechanism has obviously become selective for wormlike patterns. The prey recognition system appears to be subject to a "maturation process" in the fire salamander. The biological relevance of such a recognition system is easy to see. For the aquatic larvae those prey animals which do not fit the category "creeping worm" are also important: e.g., *Daphnea, Tubifex,* mosquito larvae, and others. In addition, because of passive transport in water, constantly changing water currents can influence the direction of propulsion of a prey object, and thus alter its configuration. After metamorphosis the salamander – like the toad – mainly finds prey animals whose longitudinal body axis coincides with the direction of movement (caterpillars, earthworms, slugs, bugs etc.). This feature of the new environment has perhaps been preprogrammed into the recognition system of the salamander, quite comparable to that in the toad [61].

4.1.5 Structural Stimulus Patterns

Let us ask, again in the case of toads, if the stimulus effectiveness of structured patterns may also be explained with the aid of the simple Gestalt scheme. A 2.5×2.5 mm^2 black square moved at 7 cm distance at $20°/s$ against a white background represents a prey stimulus that is just suprathreshold for the release of prey-catching:

1   direction of movement $\rightarrow$

If 2 mm away from it another square of the same size is added in the same plane, both moving simultaneously, the signal value is increased:

1   2

If this chain is prolonged by two further links in the direction of the movement, the attractiveness continues to increase:

1   2   3   4

The stimulus efficacy would be even further increased if a fifth member next to the fourth were moved with the others, but it would lose its signal effect as a prey object almost completely, if the fifth member is placed at a distance of 10 mm above one of the other members:

5

1 2 3 4

The signal-extinguishing effect depends on the distance of the object perpendicular to the direction of movement (Fig. 50). The Gestalt recognition scheme is therefore also valid for discontinuous patterns.

Investigations concerning the natural enemy image of the toad suggest that similar "dot-stripe patterns" have a threatening effect and perhaps reproduce the enemy image of a snake, as the toad responds to the dummy (Fig. 51 B) with the same specific defense posture as it does to the natural enemy (Fig. 51 A): it blows itself up, assumes a stiff-legged avoidance posture and offers its flank. A leech walking jerkily on land apparently also fits this enemy image if it lifts its frontal sucker (Fig. 51 C); however, as soon as the sucker is on the ground in the direction of

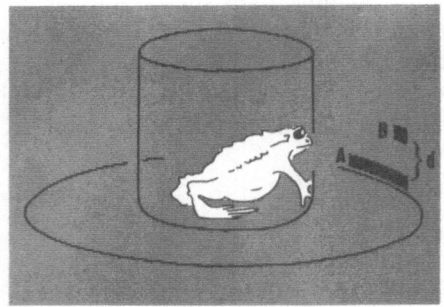

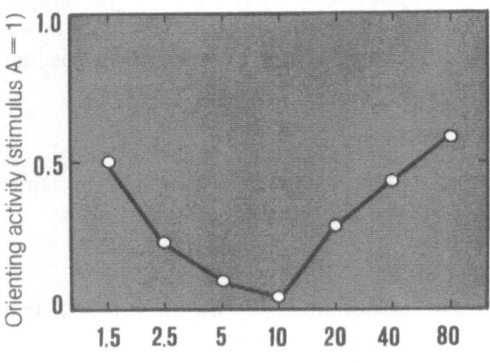

Inter stimulus distance, d [mm]

Fig. 50. Inhibitory effect of a 2×2 mm^2 square *(B)* on the efficacy of an optimal 2×20 mm^2 prey dummy *(A)*. The efficacy of the double stimulus *(AB)* depends on the interstimulus distance, d. Angular velocity: 20°/s; distance between stimulus and toad: ~7 cm. Curve of the means from 20 experiments with different common toads. [Modified from Ewert JP, Speckhardt I, Amelang W (1970) Z Vergl Physiol 68: 84–110]

the movement, the leech is taken for a prey (Fig. 51 D). The South American caterpillar *Pholus labruscae,* if attacked, raises its rostral end and assumes a snakelike posture ("snake mimicry").

4.1.6 Pattern Selection by Habituation

The assumption of the relatively simple Gestalt recognition scheme does not deny the ability of the toad to distinguish finer details of a visual pattern and to exploit these details in its behavior as a consequence of individual experiences. The variety of shapes and patterns of natural prey objects should allow toads to link individual experience to particular visual cues and to store this information in order to re-call it when faced with the appropriate stimulus. This is indeed the case. In the following we use the "habituation method" [24] to quantitatively investigate the toad's ability to select prey dummies of different shapes by experience. With the repeated release of the orienting reaction by the same prey dummy form *A* during a long-term series of presentations (Fig. 52 A), the toad habituates to the stimulus, finally leaving it without responding. If immediately afterwards a prey object of another shape *B* appears (Fig. 52 B) and the toad again reacts, it must have been able to differentiate the latter stimulus from the preceding one. Finally, it also habituates to stimulus *B*. If in the reverse type of experiment the stimulus *A* releases no prey-catching following habituation of the

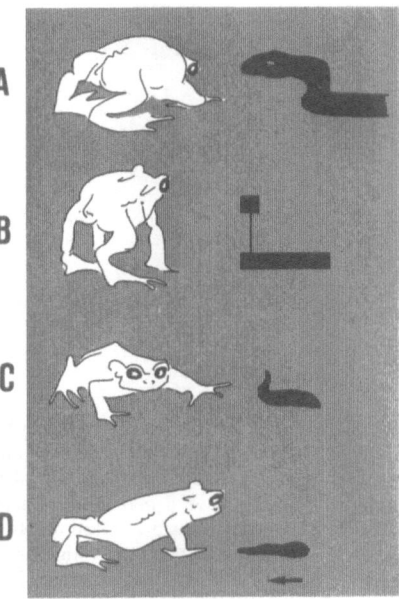

Fig. 51 A–D. Enemy images of the common toad. **A** Ringsnake, **B** head-rump dummy, **C** leech with raised frontal sucker, **D** sucker in the plane of body movement ("worm schema"). The stimulus objects move in the *direction of the arrow* past the toad. The behavioral poses of the toad were drawn from photographs. [Modified from Ewert JP, Traud R (1979) Behaviour 68: 170–180]

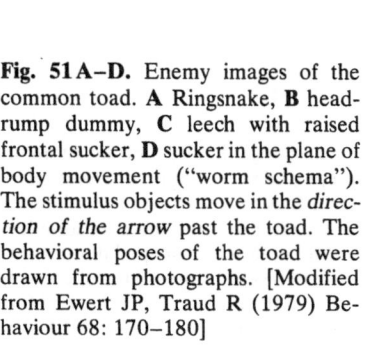

response first to *B*, we can conclude that toads prefer dummies of the pattern *B* as prey to those of the pattern *A*. However, if in both habituation series the second dummy "renewed" the toad's prey-catch-

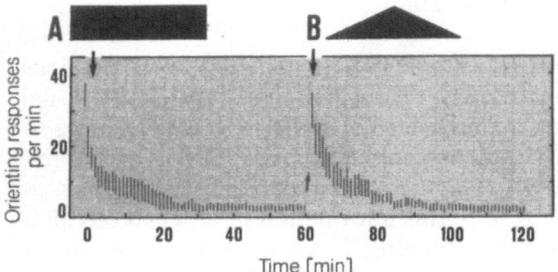

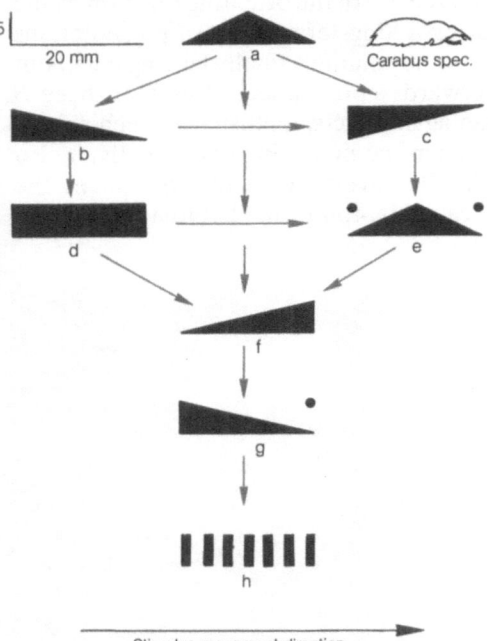

Fig. 52 A and B. Prey-selection resulting from stimulus specific habituation in common toads. **A** Decline of the efficacy of a prey dummy on repeated presentation in a long-term stimulation series (experimental set-up of Fig. 46 left). **B** A differently shaped stimulus may subsequently have a maximum releasing effect, but it, too, soon loses its stimulus efficacy upon continuous presentation. By paired testing of different shaped dummies a kind of hierarchy of features activating prey-catching may be established. The efficacy of the prey dummies shown here decreases in the direction of the arrows *(red.)*. See text for further details. [Modified from Ewert JP, Kehl W (1978) J Comp Physiol 126: 105–114]

ing activity by equal amounts, we can conclude that both dummies are distinguished and that they are almost equally attractive.

The stimulus habituation paradigm is suitable for studying *in detail* the ability of the toad to discriminate patterns. Figure 52 shows a compilation of different shapes in the hierarchy of the stimulus-effectiveness as prey. Such an order was determined by paired testing in habituation series as described above. If $b>c$, $b>g$ and $c>g$, then logically $b>c>g$. The main features governing prey selection by habituation are, as far as has been investigated: (1) area components, (2) leading edges of the stimulus in the direction of movement, (3) isolated dots, (4) striped patterns. Indeed those features are "components" of natural prey objects. The capacity of the toad to distinguish subtle differences is hard to explain with the aid of the simple Gestalt recognition scheme alone.

Stimulus-specific habituation after a long term stimulus series may last up to 24 h. The toad habituates selectively to the location in the visual field if a prey object continuously appears in the same place. Movement of the same prey to another part of the visual field is again effective in eliciting prey-catching. The biological significance of the locus and stimulus-specific habituation consists in alerting the IRM for prey-catching to "new" stimuli. The prey-catching mechanisms per se (e.g., orienting, snapping) remain unaffected by habituation [25].

4.1.7 The Movement Parameter of a Dummy

If in the experimental arrangement of Figure 46 (right) a prey dummy is moved at an angular velocity of 2°/s, it releases little activity in the toad. The prey-catching rate increases if in successive experiments the angular velocity is increased until it reaches a maximum at 20°/s and declines again at velocities in excess of 50°/s; dummies moved at 100°/s are usually ignored. The ability of the toad to distinguish configurational features of a moving dummy (e.g., worm, antiworm) is not affected by the movement velocity (v) within the range investigated, $2.5° \leqq v \leqq 60°/s$ [26].

Corresponding experiments with dark disk-shaped enemy dummies measuring 40° and moved over the toad result in largely similar correlations for avoidance behavior, but the entire reaction range is somewhat shifted in the direction of faster velocities: therefore, a predator would "have to" move faster to release an escape reaction than a prey animal would "have to" to be chased.

The prey-catching activity of the toad may also be affected by certain movement dynamics of a stimulus, especially if the stimuli have an "indifferent" configuration. Dummies, such as small squares or disks, gain in efficacy if they move stepwise. In such cases step frequencies of 1–2 cps are optimal. Indeed, natural prey objects do move jerkily rather than smoothly. The worm/antiworm discrimination in common toads is invariant for different movement patterns.

The prey releasing value of an object having an "indifferent" configuration may also be affected by the direction of its movement: if a giant ant *Camponotus herculeanus* moves toward the toad in the $z(+)$ axis of the visual field it may release avoidance, however, if the ant moves past a toad ($z-$) it signals prey. Furthermore, a prey dummy traversing the toad's visual field in $y(+)$direction (from inferior to superior) is somewhat more effective than the same stimulus moving in $y(-)$direction. At present we do not know whether this directional preference can be explained by an adaptation to natural stimulus situations, e.g., negative geotaxis of prey insects.

4.1.8 The Stimulus-Background Contrast

If we now keep the angular velocity and size of a prey dummy constant, within the optimum ranges and vary its contrast with the background, the orienting activity increases as the contrast of the dummy with the background increases.

The same applies to the effectiveness of an enemy dummy. There are, however, differences concerning the direction of the stimulus-background contrast. Whereas in prey-catching small white objects moving in front of a black background are strong releasers, the escape behavior is mainly activated by large dark objects moving in front of a bright background.

There is also a relationship between the shape of a dummy and the nature of its contrast with the background, provided the amounts of the contrast are held constant. If a dummy is clearly in the prey category because of its shape (worm configuration), its stimulus effectiveness is hardly affected by the contrast direction. If the shape of the dummy does not fit so well (e.g., small square dummies), the direction of its contrast has a bigger effect. In this case small white objects moving against a black background are more effective than appropriate black ones against a white background; the latter may even be ineffective. However, the stimulus efficacy for prey-catching is dependent not only on the direction of the stimulus background contrast. The background luminance is also taken into account. Relatively high intensities inhibit the response to prey.

The worm/antiworm preference of the toad is not generally influenced by the direction of the stimulus-background contrast. However, the accuracy of differentiation with regard to configuration is better in the case of a black pattern on white background rather than for a white pattern on black background – during summer time (cf. also Fig. 53). In this context it is interesting to note that the optimal prey size (of square or disk shaped stimuli) is smaller when the stimulus is white against black rather than black against white.

4.1.9 Background Structure

Obviously, in the natural environment prey objects very seldom move in front of a homogeneous background. Let us test if the discrimination ability of the toad for moving configurational stimuli is influenced by a

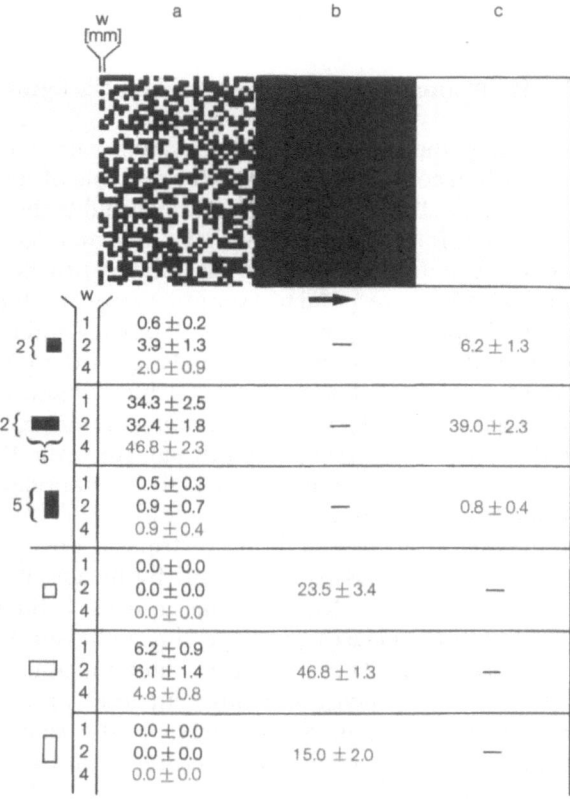

Fig. 53. Influence of background structures on the discrimination of small square, wormlike and antiwormlike stimuli; size: 2×2, 2×5, 5×2 mm^2; angular velocity: $20°$/s; stimulus distance: ~ 7 cm; luminance: black or white; horizontal movement direction. The stimulus was moved (in procedure of Fig. 46 *left*) against different stationary backgrounds: *a* computer-generated noise (Julesz pattern), *b* homogeneous black or *c* homogeneous white background. The structured background consisted of 50% white and 50% black components. Three different structures were tested: minimal width w = 1, 2 or 4 mm. The table shows the average prey-catching orienting activity of 20 different common toads. [From Ewert JP, Albrecht L, Kepper J (1980) in prep.]

bright-dark structured background. For this purpose, we choose a "Julesz Pattern" consisting of a spatial frequency distribution corresponding to visual noise (Fig. 53). Photographs from backgrounds of the natural environment (biotope) can also be used. The effect is surprising. Black or white wormlike stripes and small squares, as well as worm and antiworm configurations are properly distinguished when moved against such structured backgrounds. The discriminative values are in part (black objects) even better than with a homogeneous background of a luminance corresponding to the average luminance of the noise pattern. Small white objects appear to be masked by the background structure (Fig. 53).

4.1.10 Nonmoving Prey Dummies and Background Effects

Following the release of prey-catching with a moving worm dummy toads often continue to respond for a couple of seconds after the motion of the prey stimulus has been stopped suddenly. This after-effect does not occur (following activation of prey-catching with a moving worm) when the stationary stimulus has an antiworm configuration. Common toads also fail to respond to a stationary worm without prior movement.

What happens if a prey object is stationary and the toad itself passes it while walking? The result depends upon the environment. Toads do orient and snap at a stationary stripe stimulus, if it is located on a *homogeneous background* – providing the stripe axis is oriented parallel to the movement direction of the toad. However, the moving animal fails to respond in the same situation if the stationary background of the stimulus is structured.

If the toad, facing a stationary stripe against a homogeneous background, itself is passively rotated (by means of a modification of the procedure in Fig. 46, left) the moving retinal image of a worm configuration releases prey-catching behavior whereas an antiworm configuration elicits no response. Similar results are obtained when the toad is moved stepwise with different accelerations, in order to increase the vestibular input. Surprisingly, in the same situation the worm stimulus, if it is attached to a structured stationary background, does not release any response from the passively rotated toad.

We conclude that stationary prey objects located on a *homogeneous background* elicit prey-catching, if appropriate *moving* retinal images are produced (either during prey movement, or passive displacement of the toad itself). A potential prey object is totally masked by a *structured background,* as long as it does not move. A prey object is also masked if it is moved together with a structured background (at the same velocity). Briefly, the background plays an important role in helping the visual system to classify the origin of moving retinal images [27]. This evaluation is not solely due to "efference copies" in sensori-motor systems (cf. p. 197).

4.1.11 Seasonal Dependencies

There are indications that the visual world does not remain constant for the toad during the course of the year. In the summer months white small square objects moving on black background are more effective releasers for prey-catching than black ones against a white background. Interestingly, this relation is reversed in autumn and winter; in these periods a small black prey dummy is more effective than a white one of the same size, even though the amount of the stimulus background contrast is unaltered [46].
Common toads hibernate. During the winter their prey-catching activity, under constant laboratory conditions, is, on the whole, much less than during the summer months.

4.1.12 Motivation and Learning

The stimulus efficacy of a prey dummy also depends on the motivation of the animal at that time. Well-fed toads are slow to react in experimental

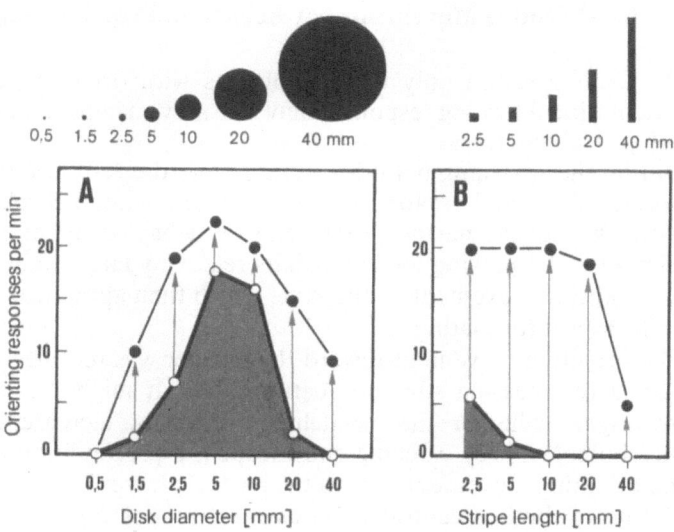

Fig. 54 A and B. Influence of motivation of the common toad on efficacy of various prey dummies. **A** Circular disks, **B** vertical stripes (horizontal movements; angular velocity: 20°/s; stimulus distance: ~7 cm). In the presence of prey odor, the stimulus efficacy may in part be greatly enhanced (see *arrows*). Curves of means from 15 experiments per value, using different toads. [Modified from Ewert JP (1968) Z Vergl Physiol 61: 41–70]

situations whereas hungry animals may be very active. Such differences are generally reflected in the quantitative experiment by the average number of responses. The stimulus response *relationships* are maintained. However, after extremely long hunger periods toads may show some preference for larger prey objects (this effect was tested with square stimuli).

It is also possible to manipulate the prey-catching motivation level by presenting the familiar odor of mealworm excrements. Toads are exposed to the odor during feeding in the laboratory ("self-training"). If this odor is then offered in the experimental situation along with a visual dummy, the stimulus effectiveness can thereby be increased, even for those stimulus configurations which previously had poorly or never resembled a prey (Fig. 54).

Another example: if toads are allowed to eat prey objects out of the experimenter's hand, animals may in the course of time come to associate the hand of the experimenter with food. This is later generalized to include even large objects as prey [28].

We must keep in mind that the accuracy of differentiation of the innate releasing mechanism (IRM) may be modified in relation to individual experiences (IRME).

4.1.13 Orienting Movements not Related to Prey-Catching

Do toads respond only to prey objects with orienting movements? Orientational turning responses may also play an important role in other behavioral functions.

During the spawning period in spring time, the response threshold for prey-catching and escape reactions to visual stimuli is relatively high. Now the sexual partner represents a behaviorally effective visual stimulus. The moving female with its relatively large body may release an orienting movement of the male which then approaches the female and clasps it for mating.

The orienting movement toward the partner was recently studied with regard to releasing stimulus features [29]. It might be important to investigate whether the "switching off" (i.e., increase of response thresholds) of prey-catching and escape is a precondition for releasing the orienting movement toward the partner. It would also be of interest to know whether central nervous mechanisms for prey and enemy detection are utilized (or "occupied") to solve this particular detection problem.

The toad shows a definite kind of turning responses in optokinetic nystagmus (cf. p. 134). These head-pursuit movements are, however, totally different from the prey-catching orienting reaction. This is true for the releasing key-stimuli as well as for the dynamics of the motor patterns [30].

Orienting movements may also be used to find dark hiding places or to avoid stationary obstacles. Head movements also appear spontaneously, e.g., to provide information about the stationary visual environment.

By what does the toad recognize its prey and enemies?

1. Prey and enemy key-stimuli have two basic parameters in common: movement and contrast. These are the prerequisites for classifying the respective objects into the functional categories of prey-catching or enemy avoidance. The movement direction and the direction of the stimulus-background contrast may endow an object with certain prey or enemy characteristics. In addition the nature of the movement, smooth or jerky, and the background structures are also taken into account.

2. The classification of a stimulus pattern (moving in the x-y coordinates) either as a prey, nonprey or enemy occurs on the basis of area and Gestalt features related to the direction of the movement: *wormlike* expansion of a small object *in the* direction of the movement (within limits) signifies "prey", expansion of the object *perpendicular to* the direction of movement (*antiworm* configuration) reduces the releasing value and may signify "threat" or even "enemy" (Fig. 55). Large, extended areas are optimal enemy key-stimuli, especially when they appear in the upper visual field.

3. The prey category is not specific to *wormlike* objects. Other *small* stimuli with "indifferent" configuration (disk or square-like shapes) also generally fit this category, but they are not optimal releasers; relatively long *antiworms*, however, do not fit the category at all.

4. The ability of the common toad to distinguish between a wormlike and an antiwormlike moving, sufficiently long stripe pattern is *invariant* with respect to changes of other stimulus parameters, such as (a) the direction of movement within the x-y and z-y coordinates,

Antiworm:

Worm:

Fig. 55. Classes of invariants in the discrimination of moving worms and antiworms by the common toad: sign of contrast, movement direction, size, stimulus structure; cf. also Fig. 39. [Modified from Ewert JP (1968) Z Vergl Physiol 61: 41–70. Ewert JP, Arend B, Becker V, Borchers HW (1979) Brain Behav Evol 16: 38–51]

(b) the direction of contrast with the background, (c) background structures, (d) the distance to the object from the eyes of the animal (investigated range: $4 \leqq d \leqq 16$ cm), (e) the size of the pattern within certain limits, (f) the movement velocity (investigated range: $2.5 \leqq v \leqq 60°$ /s, d $= 7$ cm), and (g) the movement pattern (cf. Fig. 55).

5. Toads show "size constancy" in prey-catching. Their ability to estimate the absolute size of a prey animal is possibly subject to a "maturation process" and not necessarily based on experience (learning).

6. Certain filter properties of the prey recognition system (innate releasing mechanism) can be modified by learning.

7. On the basis of individual experience (e.g., habituation) toads are able to distinguish detailed structures of a moving visual stimulus, even within an already classified category of prey. This ability cannot be easily explained on the basis of simple Gestalt classification.

8. Ontogenetic studies show that the configurational prey recognition system is not completely developed in the toad on completion of the metamorphosis. In the fire salamander, too, it has to mature during ontogeny: During the larval period it appears to be geared to aquatic prey and it shows no configurational selectivity. Several months after metamorphosis it appears to be "programmed" predominantly to terrestrial prey animals with wormlike configuration.

9. Experiments with different amphibian species hitherto studied show that a prey recognition system seems to have developed in their brains which possibly uses similar basic detection mechanisms. Different emphases and characteristics of their functional components, however, permit variations in individual species according to their special ecological and behavioral adaptations [26].

4.2 Prey-Catching and Avoidance "Areas" in the Toad Brain

Let us now pursue the optic nerve into the toad brain and ask whether there are regions responsible for triggering prey-catching and enemy-avoidance behavior. The fibers of the optic nerve that project to the diencephalon decussate almost completely. Some fibers terminate in defined areas of the diencephalon, the caudal dorsal thalamus and the pretectal region (TP-region), whereas the majority of them project to the surface layers of the midbrain, the optic tectum (Fig. 57). There are still other projection fields of optic nerve fibers. (For neurophysiological and neuroanatomical techniques, see Chap. 9, Methodological Appendix).

Caudal dorsal thalamus and pretectal region are anatomically distinct. But owing to their close proximity it has not been possible so far to unequivocally classify, on physiological criteria, this area of the posterior thalamus into one area or the other. We shall therefore for the time being call this area the thalamic-pretectal (TP) region. However, there is some evidence that cells of the postero-lateral nucleus of the thalamus are relevant in the present context.

4.2.1 Electrical Brain Stimulation

The retina of one eye projects via optic nerve fibers in a topographical manner – "point to point" – to the surface layers of the opposite optic tectum (Fig. 11). Thus, an area of the visual field is associated with a corresponding area in the optic tectum (cf. also Fig. 60).

What would happen, if, in the freely moving toad, a point in the optic tectum is electrically stimulated via an implanted electrode?

As shown experimentally, the toad turns toward an area of the visual field corresponding to the retinotopic map (Fig. 56A). The animal behaves as if this tectal area had been stimulated from an appropriate region of the visual field by a prey object. The toad is obviously able to spatially localize visual stimuli with the aid of the retino-tectal projection map. The function is probably based on fixed motor programs (for the corresponding orienting turn), whose inputs originate in the retinal projection field of the optic tectum and which may be called into play according to the different peripheral situations.

We can illustrate this with an example. For a pattern w pictured on the retina outside the fixation area x, there is a corresponding projection point w' in the optic tectum. If the object is recognized as prey, an "address" is "dialed" corresponding to the site difference x–w. There are predetermined programs in the central projection fields for each possible site difference in stimulus and fixation positions. These programs can be called into play by address (namely x–w) either by visual or by electrical stimulation and then translated into corresponding motor commands. The resulting turning reaction is adjusted to bring the prey w into the binocular fixation area x, so that x–w = O, and thus the retinal image is further processed in the central projection area x'. The snapping response is then triggered at an optimal distance. Orienting toward the prey object (once it is triggered) happens "blindly", without peripheral control *during* the movement. Subsequent corrections require feedback.

Toads program the angle of the orienting turn *and* – within limits – also the route of the approach before they start to move. However, programs can be changed, for example, if barriers arise between the toad and its prey. The toad then modifies the motor program [31]. Using depth estimation, it can reach the prey by detours.

Thus, in brain stimulation experiments, snapping cannot be triggered by all areas of the optic tectum, but rather predominantly by those which are linked with the fixation and snapping zones of the retina (Fig. 56B). Depending on the site of stimulation in the optic tectum, stepwise increases of the threshold current can activate complete or incomplete sequences of prey-catching behavior, while maintaining the correct sequence of components (cf. Fig. 41).

What happens when the retinal projection field in the thalamic-pretectal region is electrically stimulated?

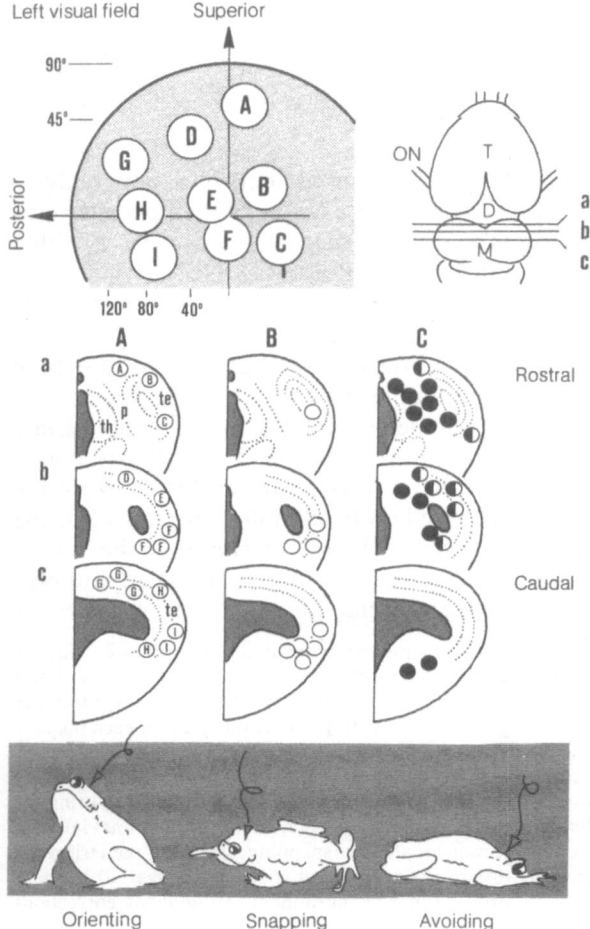

Fig. 56 A–C. Activation of prey-catching and avoidance behaviors by electrical point stimulation of the brain in freely moving common toads. *Above right,* dorsal view of brain; *T* telencephalon, *D* diencephalon, *M* mesencephalon. *a–c* Three (semi)-cross sections through the brain corresponding to the section levels shown; *te* optic tectum, *th* caudal dorsal thalamus; *p* pretectal region. **A** The toad orients toward the areas of the visual field identified by letters when correspondingly designated sites of the optic tectum are electrically stimulated. **B** Points of stimulation from which, in addition, snapping could be elicited. **C** Points of stimulation for the triggering of escape reactions *(black).* From *half-black, half-white* points orienting, as well as avoidance movements could be activated. [Combined from Ewert JP (1967) Z Vergl Physiol 54: 455–481; Ewert JP (1967) Pflügers Arch 295: 90–98; Ewert JP (1968) Z Vergl Physiol 61: 41–70; Ewert JP (1974) Sci Am 230: 34–42]

As has been shown experimentally (Fig. 56 C) the toad responds with avoidance behavior. According to the site of the stimulation, it can turn away from an invisible enemy, or duck or jump away in panic. Escape reactions may also be induced from tegmental midbrain areas.

4.2.2 Brain Lesion Experiments

The experimental finding that there are zones in different areas of the toad brain responsible for the triggering of prey-catching (optic tectum) and avoidance behavior (TP-region) is supported by ablation experiments.

Thalamic-Pretectal Lesions. When, while carefully preserving the optic tectum, the TP-region is destroyed by radio frequency coagulation (Fig. 57 D), or TP is separated from optic tectum by a discrete knife cut (Fig. 57 C), escape behavior fails to occur. Surprisingly the prey-catching behavior is "disinhibited", that is, the animals now respond with prey-catching to anything that moves (Fig. 58), including enemy objects. Their ability to correctly classify *configurational* visual patterns is severely impaired (Fig. 64 A). Moreover, the phenomena of stimulus-specific habituation of prey-catching may be greatly diminished.

The failure of *prey/enemy recognition* after TP-lesions is not necessarily linked with a *deficit of habituation.* We may conclude therefore that the two syndromes are associated with different (adjacent) structures, which have not as yet been separately identified by lesion experiments [32].

When the TP-region is destroyed in one hemisphere only, an animal "divided within itself" may develop: If an enemy dummy is shown to the eye whose optic nerve leads to the intact hemisphere, the toad responds with avoidance. When the same object is shown to the eye whose optic nerve leads to the lesioned hemisphere, it immediately reacts with prey-catching. However, in some preparations the responses toward objects moved in the visual field of the intact hemisphere were suppressed, thus indicating the existence of interhemispherical links. According to the extent of the lesion in the TP-region, the behavior of the toad may become normalized after hours (very small lesions) or days (larger ablations). However, the configurational selectivity of the normal animals will not be regained. After certain small TP-lesions the "disinhibition" of prey-catching is localized to restricted areas of the visual field. When the TP-lesion is completely bilateral, this effect is seen when objects are moved in any part of the visual field, and it may last several days or even weeks. We assume that the phenomena of fast recovery after small TP lesions are based on functional plasticity in the thalamo-tectal wiring system (activation of "silent" pathways) and are not the result of neuroplasticity (e.g., sprouting of axons).

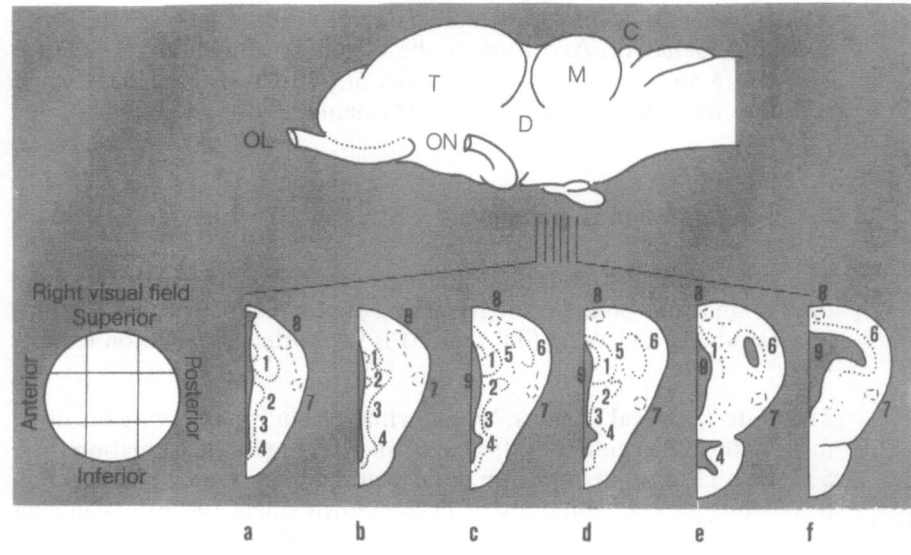

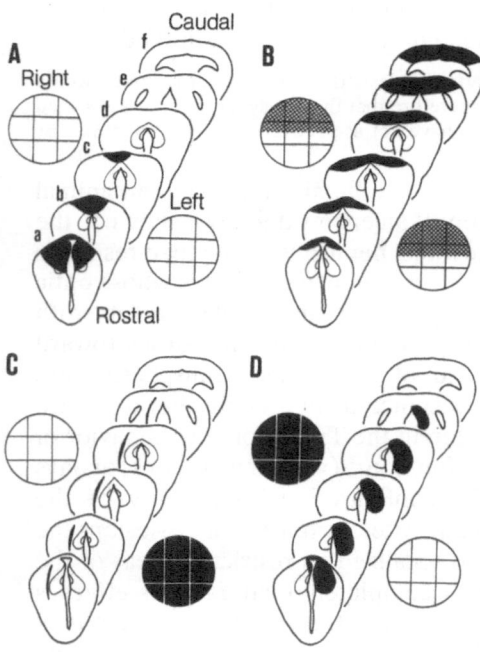

Fig. 57. Brain lesion studies in common toads. *Top,* brain, lateral aspect: *T* telencephalon, *D* diencephalon, *M* mesencephalon, *C* cerebellum, *OL* olfactory nerve, *ON* optic nerve. *a–f* series of (semi)-cross sections corresponding to the levels shown. *1* Dorsal, *2* medial, *3* ventral thalamus, *4* hypothalamus, *5* pretectal region, *6* optic tectum, *7* ventrolateral and *8* dorsomedial branch of optic tract, *9* third ventricle. *A–D* Examples of various brain lesions *(black).* Visual field areas of the left and right eye, respectively, *(large circles)* for which the prey-catching behavior in response to visual patterns is disinhibited as a result of the brain lesion, are *black,* those in which the toad is blind, are *cross-hatched.* Responses to prey or enemy objects moved in white visual areas appear to be normal. [Modified from Ewert JP (1968) Z Vergl Physiol 61: 41–70]

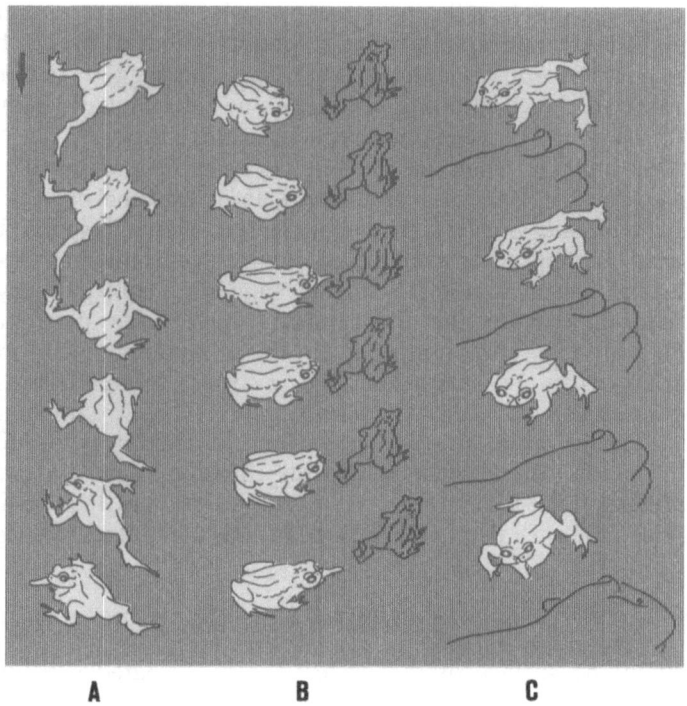

Fig. 58 A–C. Examples of disinhibited prey-catching behavior of a common toad with lesions in the caudal dorsal thalamus and the pretectal region (TP-lesion). The toad turns toward, and snaps at, its own moving extremities **(A)**, a passing conspecific **(B)**, the moving hand of the experimenter **(C)**. Drawn from film sequences, analyzed frame by frame. [Modified from Ewert JP (1967) Z Vergl Physiol 57: 263–298]

Ablation of the Optic Tectum. After destruction of the entire optic tectum *moving* visual stimuli elicit no response. Nevertheless a directed snapping reaction may be induced by tactile stimulation of the dorsal body skin [33]. Accordingly, a possible "snapping center" cannot be localized exclusively in the optic tectum. The adjacent subtectum is a site for multisensory integration (cf. Fig. 74) [34]. Moreover, it has been shown that the TP-region is unable to trigger escape away from moving visual enemy objects without inputs from the optic tectum. "Atectal" animals are, however, able to avoid *stationary* barriers [35].

Removal of the Telencephalon. The telencephalon seems to exert modulating effects on the two behaviors (cf. Fig. 69). If one hemisphere is destroyed, the prey-catching responses to prey objects moving in the visual field of the contralateral eye fail to occur. Prey objects appearing in the visual field of the ipsilateral eye (whose optic nerve therefore leads to the intact hemisphere) elicit responses, although the catching activity is not as vigorous as in the intact animal. This suggests that normally the prey-catching facilitating effects of one telencephalic hemisphere are "distributed" between the two sides of the brain.

What happens when the whole telencephalon is removed? In that case prey-catching behavior cannot be released by visual objects; in contrast, the avoidance behavior may be increased, even in response to relatively small moving visual stimuli. (A similar syndrome was observed after lesions rostral to the TP-region, suggesting that pathways from the telencephalon to the diencephalon have been destroyed). If the TP-region is also removed, in these animals, then prey-catching behavior, as previously described, is hyperactivated; the escape behavior seems to be eliminated irreversibly. Thus, we might suggest that after removal of the telencephalon the inhibitory influence of the TP-region overrides the tectal prey-catching trigger system. Following additional TP-lesions the prey-catching system is then "disinhibited" – quite similar to that in TP-lesioned toads with intact telencephalon.

4.2.3 Pharmacological Effects

When the surface of the optic tectum is treated with dilute curare solution, the prey-catching behavior in response to moving visual objects is "disinhibited" in a way similar to that seen after ablation of the TP-region. If, however, acetylcholine is applied instead of curare, prey-catching fails to occur for the duration of the drug effect, whereas avoidance behavior seems to be normal [36].

How can the pharmacological effects be interpreted?

There are many possibilities to explain the inhibitory effect of acetylcholine on prey-catching: One such explanation is that specific neurons – possibly from the TP-region – form inhibitory synapses with tectal-neurons that might participate in the triggering of prey-catching. If presynaptic inhibition by acetylcholine as transmitter is assumed, the inhibition might be increased by application of acetylcholine and decreased by acetylcholine blockers (such as curare). As a matter of fact the dorsal half of the optic tectum contains very high concentrations of cholinesterase [37]. Anatomical evidence for axo-axonic synapses in the optic tectum was obtained by electron microscopy studies [38]. However, presynaptic inhibition with ACh as transmitter has not yet been described (cf. p. 48).

What conclusions can we now draw about areas in the toad brain that mediate prey-catching and escape behaviors?

1. The optic tectum contains a system for the localization of visual stimuli.

2. Components of the prey-catching behavior sequence (such as orienting, snapping, swallowing, wiping) are represented in the toad brain as spatially and temporally coordinated programs. These may be initiated from the optic tectum. The correct sequence of the single components is presumably maintained by threshold differences in the corresponding neural systems.

3. The tectal prey-catching system apparently receives inhibitory inputs from the TP-region (Fig. 69). Such connections might protect the toad from orienting toward irrelevant objects. After TP-lesions the prey/enemy recognition system is damaged and the prey-catching behavior, "disinhibited" toward anything that moves.

4. Pharmacological experiments suggest that the assumed inhibition of the tectal prey-catching system is mediated by cholinergic transmission.

5. The thalamus/pretectum (TP-region) may initiate motor programs which have a protective function for the toad, as, for example, the escape behavior.

6. In the absence of excitatory inputs from the optic tectum, the TP-region alone, with just the retinal afferents, is apparently unable to trigger escape responses toward appropriate visual stimulus patterns.

7. The central prey-catching and avoidance systems may be modulated by the telencephalon, possibly via inhibitory connections to the TP-region (cf. Fig. 69). For example, a function of the forebrain is to mediate the facilitation of prey-catching brought about by concurrent olfactory stimulation.

8. The recognition of prey and enemy stimuli appears to be a complex process based on interactions between neuronal populations located in tectal, thalamic-pretectal and telencephalic regions.

4.3 Are There "Prey" and "Enemy Neurons" in the Toad Brain?

Having become acquainted with brain areas participating in the triggering of prey-catching and avoidance behavior, we shall now ask if there are specific neurons which take part in the evaluation of the

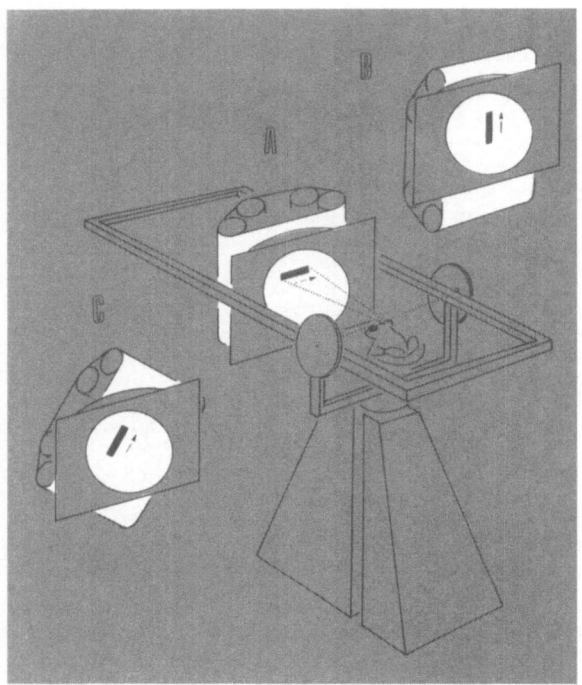

Fig. 59. Perimeter-like experimental set-up for stimulations with moving visual patterns in the neurophysiological experiment. The dummies used in the behavior experiments may be moved in various directions (*A–C*) behind the aperture on an illuminated belt in front of an immobilized toad. Further details in text. [Modified from Ewert JP, Krug H, Schönitz G (1979) J Comp Physiol 129: 211–215]

corresponding key-stimuli. (Concerning neurophysiological measuring and recording techniques, see Chap. 9, Methodological Appendix and Fig. 59).

4.3.1 What Does the Toad's Eye Tell the Toad's Brain? Information Processing in the Retina

This question was first formulated for the frog and dealt with in the fascinating pioneering research of Jerome Y. Lettvin and his colleagues at the Massachusetts Institute of Technology [39]. The neuronal system of the retina represents, embryologically, a protuberance of the diencephalon. We may therefore assume that visual information is being *processed* there.

Neuronal Organization of the Retina. The retina consists of receptor cells and different sequentially connected neuron types (Fig. 61a):

bipolar, amacrine, and ganglion cells [70]. Among these cells there are cross connections via amacrine and horizontal cells. A ganglion cell is connected with numerous receptor cells via amacrines and bipolars. Each ganglion cell therefore responds to a particular section of the visual field. This section is called the *receptive field* (RF); the sum total of the associated neuronal contact elements is called the *receptive unit*. Figure 61a shows an example.

Visual Field Projection. The visual world is "scanned" with such overlapping receptive fields. The preprocessed signals are conveyed as corresponding impulses via the ganglion cell axons in the optic nerve, mainly to the surface of the opposite midbrain (the optic tectum), but also to different areas of the diencephalon. As previously mentioned the visual world is systematically projected "point to point" onto the optic tectum, a genetically fixed retinotopical order being observed (Fig. 60) [40]. This can be easily demonstrated by inserting a recording microelectrode into the optic layer and moving it close to the terminal of

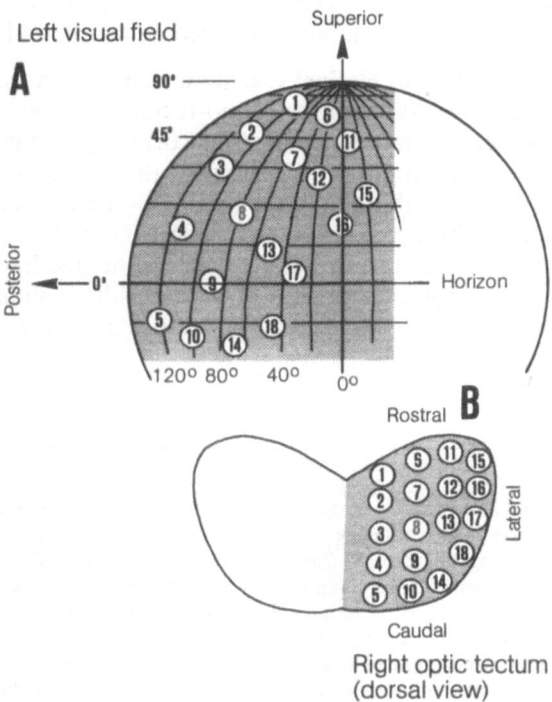

Fig. 60 A and B. Projection of the visual field of the left eye on the surface of the right optic tectum of the common toad. [Modified from Ewert JP, Borchers HW (1971) Z Vergl Physiol 71: 165–189]

a ganglion cell fiber (Fig. 60B,8). Only an object movement in the corresponding part of the visual field, the receptive field of this ganglion cell, leads to neuronal activity at the recording point (Fig. 60A,8). By systematic displacement of the electrode on the optic tectum (Fig. 60B) and the localization of the corresponding receptive field from a recorded neuron (Fig. 60A) a topographical map for the visual field projection may be constructed.

Organization of the Receptive Field. In toads and frogs the receptive field (RF) of a ganglion cell is organized in the following manner (Fig. 61a). That part of the field from which the ganglion cell may be activated by a moving stimulus is called the excitatory receptive field (ERF); the shape of the ERF is slightly ellipsoidal and on the average it is approximately radially symmetrical. The ERF is located centrally within the receptive field and is surrounded by an inhibitory receptive field (IRF) [41]. The IRF is defined this way: an object moving in this area inhibits the activation of the ganglion cell by a visual stimulus, moving simultaneously through the ERF. We may conceive of the distribution of excitation and inhibition in the whole receptive field (Fig. 61a) as the nonlinear sum of two Gaussian functions for a spatial excitatory (lateral excitation) and an inhibitory process (lateral inhibition) (Fig. 61b) [42].

Classification of Retinal Ganglion Cells. As in frogs there are different ganglion cell types in the retino-tectal projection of toads. Three types have been identified so far; they correspond to classes R2, R3, and R4 (R, for retina) described earlier in the frog and may be recorded consecutively in this sequence by dorso-ventral penetration of the optic tectum with a recording electrode (Fig. 10B, A–C). On the basis of different sizes of ERF diameters, on the one hand, and the *on* and *off* response-criteria (to brief change of diffuse illumination in the entire visual field), on the other hand, these neurons can be easily distinguished (Table 4):

Fig. 61A–C. *Right,* Responses of retinal ganglion cells (classes R2, R3, R4) of the ▶ common toad to moving configurational stimuli, *a–c* used in behavioral experiments (visual angular velocity: 7.6°/s; horizontal movement. Curves of means from $n = 10$ different neurons of the same class). *Left,* histological structure of retina: *R* receptor cell, *B,* bipolar cell, *A* amacrine cell, *H* horizontal cell, *G* ganglion cell. *a* Division of the receptive field into a central excitatory *(ERF)* and a peripheral inhibitory area *(IRF). b* Spatial distribution of excitation in the receptive field after addition of two Gaussian distributions, one for an excitatory *(E)* and one for an inhibitory *(I)* process. *c* Original records of a class R3 neuron. In successive experiments, squares with equal visual angular velocities (7.6°/s) and different edge lengths (s = 2, 8, 20°) were moved through the receptive field center *(a)* in a horizontal direction. The width of the stripe stimuli in *A–C* was always 2°. [Modified from Ewert JP, Hock FJ (1972) Exp Brain Res 16: 41–59]

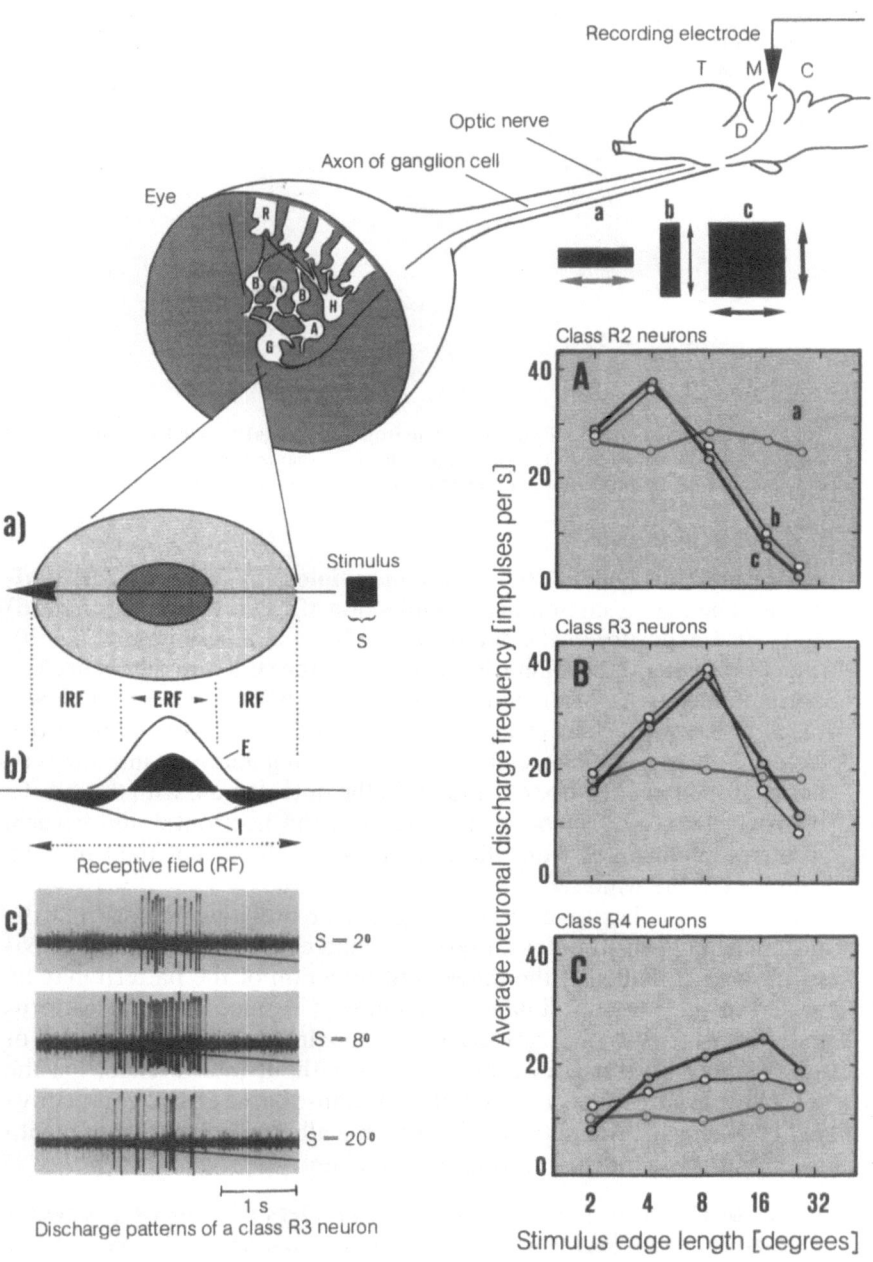

Fig. 61 A–C

Table 4. Characteristic functional response properties of retinal ganglion cells in the common toad

Class	ERF diam.	IRF strength	on	off	
R2	~4°	+ + +			[*on* response]
R3	~8°	+ +			*on-off* response
R4	12–16°	(+)			*off* response

The receptive field of a retinal ganglion cell in frogs (and toads) has small *on-*, *on-off-*, and *off* zones which are irregularly distributed. The RF organization is different from that obtained in *on-* and *off*-center neurons in the mammalian retina [43].

A Method for Quantitative Electrophysiological Recording Experiments.

The toad is immobilized by injection of succinylcholine (cf. p. 56) and, after exposure of the brain surface fixed in a perimeter (Fig. 59). One of the eyes of the animal is positioned at the center of the set up. The perimeter consists of an illuminated screen behind whose window a black pattern attached to a white moving belt traverses the receptive field center of a ganglion cell; the corresponding neuronal impulses can be recorded from the fiber terminals in the optic tectum with the aid of a microelectrode. By pivoting the screen in the horizontal and vertical axis, receptive fields from different parts of the visual field may be centered in the window.

This apparatus has several advantages over previous models: (1) the distance between stimulus pattern and toad eye is variable in the z-axis of the visual field; (2) the movement direction of the pattern may be varied in the x-y-coordinates as required; (3) since different patterns may be attached consecutively to the moving belt, a whole series of stimuli may be tested in random order without interference; (4) the whole program sequence – including the pause between two consecutive field traverses of the pattern – is electronically controlled together with the electrophysiological recording equipment.

Effect of Different Configurational Parameters of a Moving Stimulus Pattern on the Activation of Retinal Ganglion Cells.

A cursory examination of Table 4 might lead one to assume that the problem of prey-enemy signal processing had been solved on the basis of the identification of such neuron types: class R2 neurons with their relatively small ERF's would have to be interpreted as "prey" and class

R4 neurons as "enemy" detectors. However, by ascertaining the intensity of the neuronal activation (impulses/s) dependent on configurations of the moving dummies used in behavior experiments, the

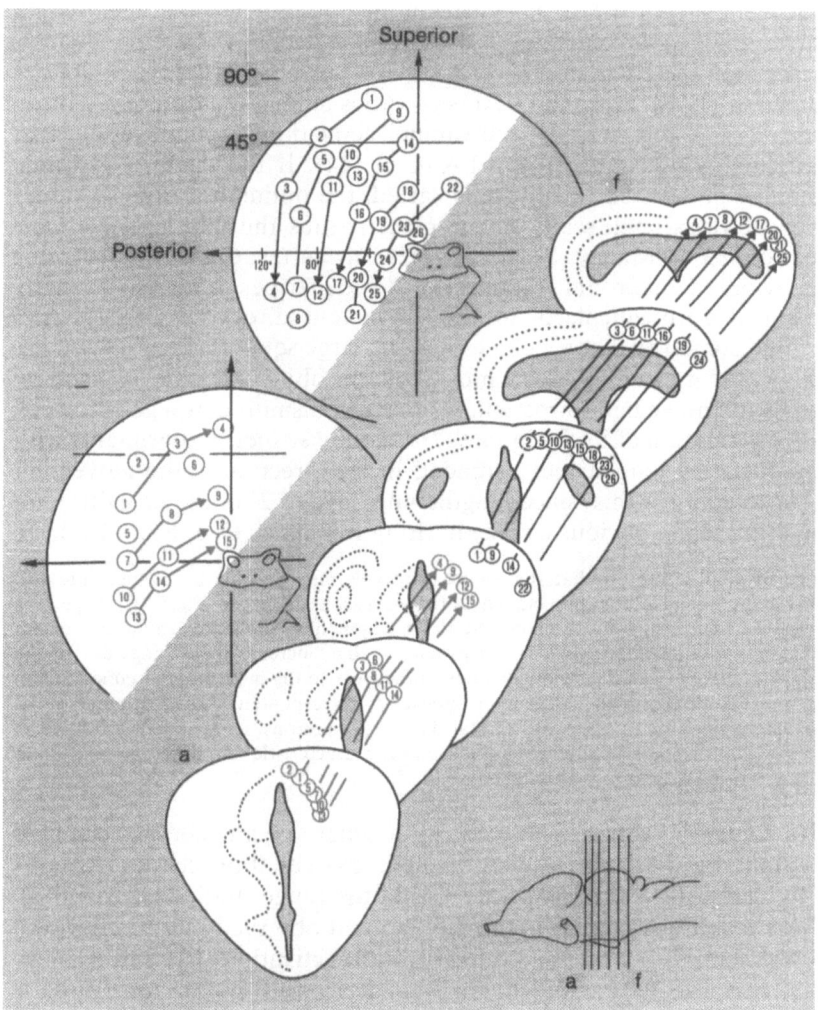

Fig. 62. Projection of the visual field on the contralateral optic tectum *(black)* and the thalamic-pretectal region *(red)* in the common toad: Topographic relations between receptive field positions (centers) of tectal T5 (1, 2) and thalamic-pretectal TH3 neurons and the sites of their recording positions. Series of brain sections through diencephalon *(a–c)* and the mesencephalon *(c–f)*, running from rostral to caudal. Recording positions of the stainless steel microelectrode tips – after passing anodal DC current – were identified in histological brain sections *(a–f)* by staining the iron deposit. [Modified from Ewert JP, Hock FJ, Wietersheim A v (1974) J Comp Physiol 92: 343–356]

following results are obtained (values for angular velocity and contrast being held constant):

Square Objects. The firing rate of all neurons generally increases with increasing edge length of a square object moving through the receptive field center (Fig. 61 c): it reaches a maximum when the edge length corresponds to the diameter of the ERF and declines when the square is larger than the ERF and therefore stimulates portions of the inhibitory surround (IRF). Since the ERF size varies among the different neuron types (cf. Table 4), the maximum activation depends upon that particular visual area (Fig. 61 A–C,c) [44]. If the distance between stimulus pattern and toad eye is varied, the optimum response values over the visual angle scale remain constant; thus, the neurons show *angle* but not *size* constancy. *Vertical stripes:* In contrast to the prey-catching behavior, the neuronal response to 2° wide stripes of different length, arranged perpendicularly to the movement direction, changes in a manner similar to that for corresponding large squares (Fig. 61 A–C,b). This would mean that information about the object's length, particularly across the movement direction, would be transmitted by a ganglion cell as a modulation of its firing rate. *Horizontal stripes:* Experiments with appropriately long stripes extended in the direction of the movement seem to confirm this, since lengthening beyond 2° does not induce, on the average, an obvious change in firing frequency (Fig. 61 A–C,a).

Detailed studies suggest that R3 neurons in toads may possibly be divided into two sub-classes. The response characteristics of type 3a are as described (Fig. 61 B). Type 3 b is distinguished from 3 a by the fact that stripe lengthening *in* the horizontal movement direction – within limits (up to 8°) – also produces a slight increase in the firing rate. But the responses to vertical stripes extended perpendicularly to the movement direction are on the whole somewhat higher than the responses to corresponding stripes oriented in the movement direction (in contrast to tectal class T5 (1) neurons, see later p. 107 Fig. 63 B). Types 3 a and b are also different in their *on-off* characteristic: the *off*-response is more clearly marked in 3 a and the *on*-response in 3 b.

The Angular Velocity Parameter. If angular size and contrast are held constant, the discharge rate for the neurons of all three classes increases with increasing angular velocity within the range studied (1° to 30°/s). These results are similar in principle to that observed in the behavioral experiments. The minimum velocity for the activation of a neuron is very small in class R2, clearly higher in R3 and especially high in R4. The class R2 neurons have a further response property: if a stimulus is moved in the ERF and then suddenly stopped in the field center, R2 neurons continue to discharge for a couple of seconds in response to the stationary stimulus [45]. The delayed response of these neurons may form a basis for the behavioral "after effects" mentioned above (p. 88).

The Stimulus-Background Contrast. In agreement with the behavioral experiments, the neuronal response to a moving pattern increases with

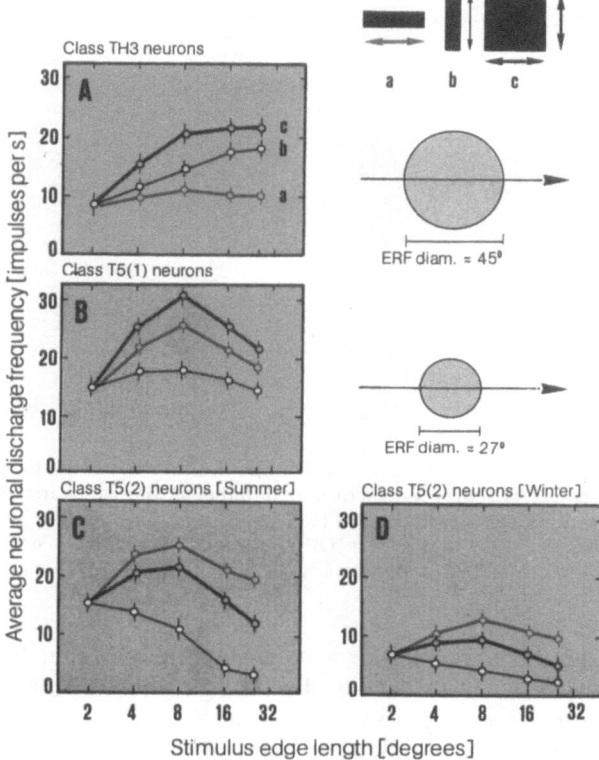

Fig. 63 A–D. Responses of different classes of neurons [T5(1), T5(2), TH3] from the central visual system of the common toad to moving configurational stimuli, *a–c,* studied in behavioral experiments. Visual angular velocity: 7.6°/s; horizontal movement. The width of the stripe pattern was always 2°. Curves of the means from *n* = 20. [Modified from Ewert JP, Wietersheim A v (1974) J Comp Physiol 92: 131–148]

increasing contrast against the background [45]. The discharge rate of class R2 neurons is a function of the direction of the stimulus background contrast and exhibits a surprising parallelism with corresponding data obtained in prey-catching experiments (p. 88). In particular the neuronal white-black, or black-white preference is dependent on: (1) the location of the receptive field in the visual field, (2) the visual angular size of the stimulus pattern, and (3) the season. This is a neurophysiological indication that the visual world for the toad is constrained by the system characteristics of the signal processing nervous networks. They may also be subject to seasonal variations [46].

Let us return to the original question. A comparison between Figures 48 A and 61 A-C shows that the dependency of the behavioral activity on

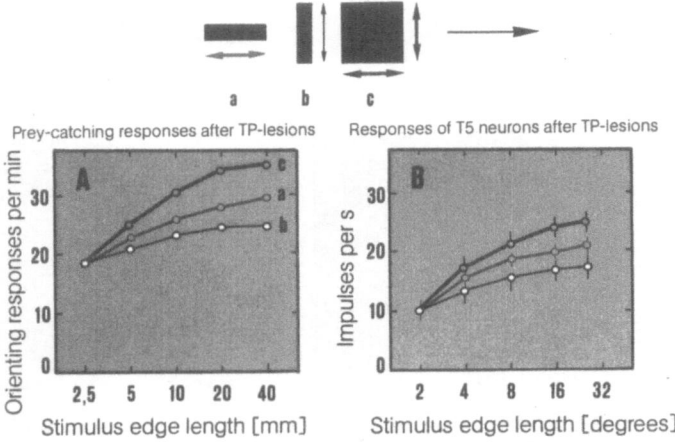

Fig. 64A and B. The effect of thalamic-pretectal lesions on the prey-catching orienting activity **(A)** of the common toad for different moving configurational stimuli *(a–c)* and on the discharge rate **(B)** of tectal T5 (1,2) neurons from layers *d, e* (in Fig. 10). Visual angular velocity: 20°/s **(A)**, 7.6°/s **(B)**; horizontal movement. Curves of the means from *n* = 20. [Modified from Ewert JP, Wietersheim A v (1974) J Comp Physiol 92: 149–160]

Gestalt parameters of a moving pattern cannot be described by the response characteristics of retinal ganglion cell classes. *Thus, there are neither special prey nor enemy detectors in the toad eye.*
Let us now follow the optic tracts into the brain and study in a similar way the responses of the neurons connected "downstream" from the retina.

4.3.2 Processing of Visual Input in the Thalamic-Pretectal Region

There are in the TP-region a large number of different neuron types activated in more or less specific stimulus situations [48]. Once again we record the impulses of a single neuron extracellularly with the aid of a microelectrode in the awake, immobilized animal. Some neurons respond well to large stationary objects, others only to moving visual stimuli. There are also neurons which fire mainly when an object is moved in the z-axis toward the toad's eye. Many thalamic neurons respond to stimulus situations that would normally elicit avoidance behavior of the animal (to use a general expression): retreat, or turning away from an enemy; avoidance of a stationary obstacle.
We carry out quantitative studies on a monocularly driven neuron type frequently found in the TP-region, called class TH3 neurons (TH, for thalamus). The average diameter of their excitatory receptive fields is about 45°. Systematic mappings of the fields show that all retinal areas of the contralateral eye are mapped in the TP-region (Fig. 62, *red*) in a

relatively rough scanning pattern. Figure 63 A shows the effect of the various configurational parameters of a moving object on the discharge rate of these neurons. Accordingly, stepwise lengthening of a stripe in the movement direction does not lead to a clear change of the neuronal discharge frequency. Conversely, the firing rate increases with lengthening of the stripes perpendicular to the movement direction. This effect is even more dramatic in response to squares of increasing size.

Class TH3 neurons are sensitive to the entire area of a moving pattern and, with respect to configurational cues, predominantly to its extension perpendicular to the direction of movement. The behavioral results therefore, cannot be interpreted solely on the basis of this stimulus transformation (cf. Fig. 48 A).

4.3.3 Processing of Visual Input in the Optic Tectum

In the optic tectum, too, there are numerous different neuronal classes responding to visual moving stimuli; most of them are clearly distinguishable from TP-neurons according to their response characteristics. They are localized in different layers. As a rule the deeper the layer, the larger the receptive field – a phenomenon based on the principle of converging connections (Figs. 10 B and 74). Some neurons receive visual information from both eyes [47].

Since toads identify prey objects monocularly, once again we restrict the quantitative experiments to monocularly excitable neurons. In the central tectal layers (Fig. 10 B, d, e) extracellular responses are regularly recorded from single neurons with an ERF-diameter of approximately 27°. We call them class T5 neurons (T, for optic tectum) [48]. In this tectal layer, too, a visual field projection is present (Fig. 62, *black*) that corresponds roughly to that of the more superficial retinal ganglion cell axon terminals (cf. Fig. 60). In relation to the sagittal axis the retinal projection diagram in the optic tectum is like a mirror image of the projection diagram in the TP-region (cf. Fig. 62, *black* and *red*). Thus, displacing the recording electrode from rostral to caudal in the optic tectum would produce a shift of the receptive field positions from dorsal, anterior to ventral, posterior and in the TP-region from ventral, posterior to dorsal, anterior [49].

First we test the activity of T5 neurons in response to different moving configurational stimuli: a $2° \times 2°$ square and a $2° \times 20°$ antiwormlike stripe, each of which is moved through the ERF center horizontally at the same constant angular velocity. The ratio of the neuronal discharge frequencies to the square ($R_{2°}$) and to the antiworm ($R_{20°}$) – $R_{20°}/R_{2°}$ – is shown in Figure 65 A for a sample of 38 different T5 neurons. From the sample distribution it is evident that T5 neurons are not homogeneous. There are obviously two different types, that can be distinguished by

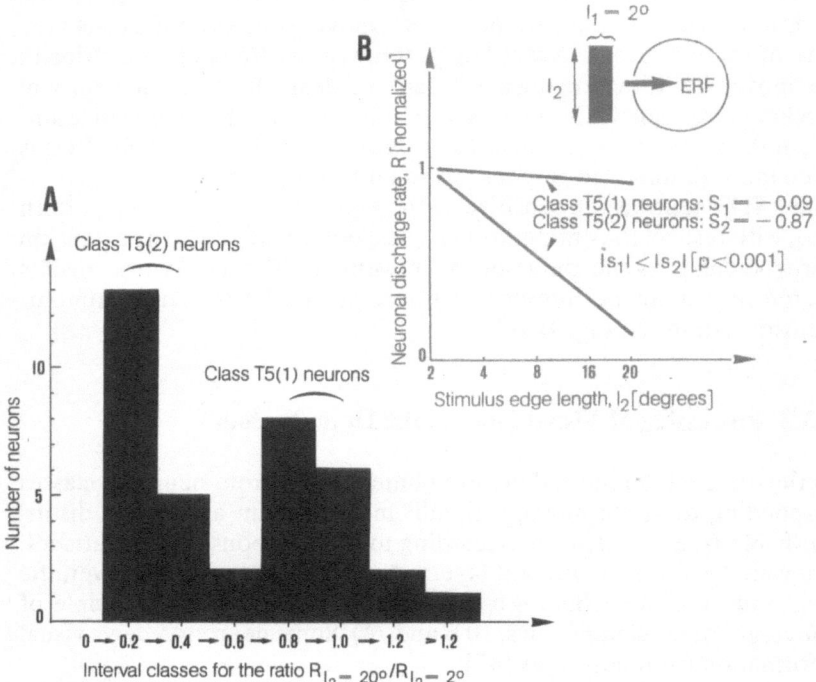

Fig. 65 A and B. Statistical analysis of two neuronal types, T5(1) and T5(2), studied quantitatively in the optic tectum of the common toad. **A** The response criterion chosen is the ratio of the discharge rates with a 2° × 2° square to that with a 2° × 20° stripe, oriented perpendicular to the direction of movement. For all neurons of layers d and e (Fig. 10) we obtain a clear two-peak distribution whose maxima coincide with the two response types. **B** The dependence of the neuronal discharge rate (R) on the logarithm of the stripe length (l_2) may be described by a linear function: $R = -s \log l_2 + k$. After calculating the linear regression of R on $\log l_2$ the slopes s for tectal T5(1) and T5(2) neurons are significantly different in the t-test with p < 0.001. [Modified from Wietersheim A v, Ewert JP (1978) J Comp Physiol 126: 35–42]

their response selectivity to moving configurational stimuli (cf. also Fig. 65 B).

In T5(1) neurons (Fig. 63 B) stepwise stripe lengthening *in* the movement direction up to 8° leads to an increase and beyond 8°, to a decline of the discharge frequency. This effect is even more pronounced for correspondingly wide squares. The firing rate is only slightly changed in response to stripes of varying lengths extended perpendicularly to the movement direction. These results suggest that T5(1) neurons are sensitive to the area of the moving pattern and, with respect to configurational cues, predominantly to the extension in the horizontal direction of movement. The corresponding behavioral results, there-

fore, cannot be interpreted by this stimulus transformation alone, either
(Fig. 48 A).

T5 (2) neurons are distinguished from T5 (1) neurons by the fact that
their firing rate is relatively strongly diminished by lengthening of
components of a pattern perpendicular to the direction of movement
(Fig. 63 B and C). Thus, the response of these neurons depends on *both*
Gestalt components of a moving pattern, and it would, at first
approximation, describe the probability that the stimulus fits the prey
category.

4.3.4 Quantitative Measure of Neuronal Response Selectivity

A quantitative measure for selection between wormlike (W) and
antiwormlike stripes (A) of corresponding sizes is given by the
"form-contrast" formula

$$D_{W,A} = \{\overline{R}_W - \overline{R}_A\} \ \{\overline{R}_W + \overline{R}_A\}^{-1},$$

where $\overline{R}_W$ is the average response to a wormlike moving stripe and $\overline{R}_A$
the response to the same stripe moving as an antiworm (the stripe is
moved horizontally in both trials). $D_{W,A}$ is plotted against the Gestalt
parameters xl_1^* (length in direction of movement) and xl_2^* (length per-

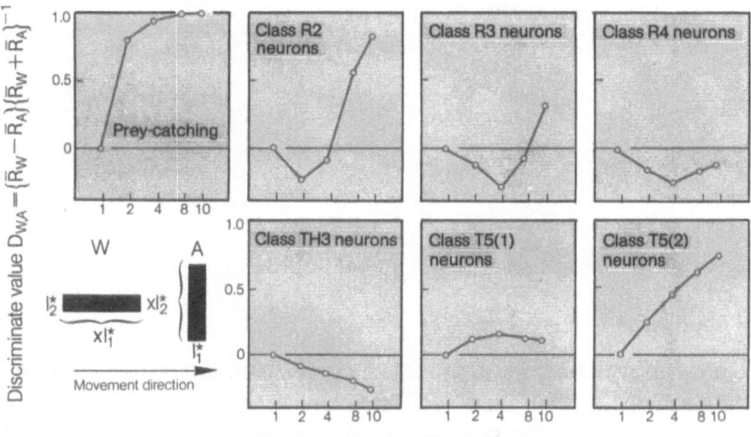

Fig. 66. Discriminate value $D_{W,A}$ as a measure of selection between wormlike *(W)* and
antiwormlike *(A)* moving stripe stimuli of different length xl_1^* or xl_2^* respectively; $l_1^* = l_2^*$
= 2.5 mm (in prey-catching behavior experiments), or 2° visual angle (in neurophysiological experiments). The values were obtained for prey-catching orienting activity and for the
activation of neurons belonging to different classes in the visual pathway of the common
toad. $\overline{R}_W$ is the average response to the worm and $\overline{R}_A$ to the antiworm stimulus. [Modified
from Ewert JP, Borchers HW, Wietersheim A v (1978) J Comp Physiol 126: 43–47]

pendicular to the direction of movement). $D_{W,A}$ is a type of correlation coefficient and its values can vary between $+1$ and -1. When wormlike stripes are preferred to the same ones moving as antiworm, $D_{W,A}$ is positive. When $D_{W,A} = +1$ or -1, both stimuli are distinguished by a clear yes/no decision. The results are shown in Figure 66.

In addition to this data evaluation, the stimulus response relationships from behavioral experiments

$$\bar{R}_b = f\,(xl^*_{1;2}) \quad [\text{orienting responses} \times \text{min}^{-1}]$$

and the corresponding neurophysiological response relationships

$$\bar{R}_n = g\,(xl^*_{1;2}) \quad [\text{impulses} \times \text{s}^{-1}]$$

are compared (n = R2, R3, R4, TH3, T5(1), T5(2) – the neuronal classes investigated). The Pearson correlation coefficient r allows one to compare quantitatively the similarity of two waveforms. Continuity within each curve is approximated. The routine yields a value r between $+1$ and -1 where something close to $+1$ indicates a positive correlation, -1 a negative correlation and $r = 0$ a zero-correlation. Thus, r_W is calculated for responses to wormlike (parameter xl^*_1) and r_A to antiwormlike stripes (parameter xl^*_2). The question of correlation between behavioral and neuronal responses has to be analyzed for *both* parameters, xl^*_1 as well as xl^*_2. In the case of perfect positive correlation a combination $\{r_A; r_W\} = \{1; 1\}$ should be expected. The results are shown in Table 5.

Table 5. Correlation coefficient pair $\{r_A; r_W\}$ for comparison between behavioral and neurophysiological responses to wormlike and antiwormlike stripes [cf. Borchers HW, Ewert JP (1979) Behav Processes 4: 99–106]

Neuronal class	$\{r_A; r_W\}$
R2	0.6; 0.5
R3	−0.7; 0.2
R4	−0.9; 0.6
T5(1)	0.0; 0.8
T5(2)	0.9; 0.7
TH3	−0.9; 0.9

What can we conclude from these correlation analyses? Among retinal ganglion cells the class R2 neurons show best but not optimal correlations with the behavioral results on prey-catching according to the configurational parameters xl^*_1 and xl^*_2. No positive correlation pair is found in R3 and R4 neurons nor among neurons from central retinal projection fields, class TH3 and class T5(1). The best positive correlation of all investigated neurons from the retina and the retinal projection fields is obtained in tectal T5(2) neurons. Thus, we suggest that the initial steps of information processing concerning evaluation of the stimulus configuration are performed at the retinal level. This information is further processed in neuronal populations beyond the retinal level. TH3 and T5 (1) neurons show different kinds of *sensitivity* to moving configurational stimuli. T5 (2) neurons exhibit *selective* responsiveness to Gestalt parameters tested. However, these neurons show no worm-*specificity!* They may have additional properties which have not yet been quantitatively investigated. The activity of a class T5 (2) neuron in response to moving configurational stimuli reflects approximately the probability that a stimulus fits the prey category (cf. Fig. 66).

No quantitative data are presently available to answer the question of how the toad's visual system transforms visual angular sizes into absolute dimensions to produce the effect of "size constancy". The questions of how and where size constancy phenomena enter the evaluation systems, may be answered by recording from freely moving animals during prey-catching behavior. It should be emphasized that size constancy mechanisms basically are not involved in the discrimination among stimuli having equal (absolute) size but different configurations, as in the case of a stripe moved either as *worm* or as *antiworm*.

4.3.5 Question of Directional Invariants

Using $D_{W,A}$ as an index of the property of T5 neurons to distinguish between a stripe moving as worm or as antiworm, we may investigate questions of directional sensitivities. The configurational *selectivity* of T5 (2) neurons is largely independent of the direction of stimulus movement in the x–y coordinates. T5 (1) neurons, however, may show various degrees of configurational *sensitivities* at different movement directions. Furthermore, long-term recordings of T5 neurons indicate that the configurational sensitivity in *some* T5 (1) cells can change with time. This variability appears to be restricted mainly to the antiworm configuration. T5 (2) neurons differ from these T5 (1) neurons in that they are strongly selective *and* very stable [51].

4.3.6 Motivation-Determining Factors

Time of the Day. Common toads are active during twilight. From qualitative observations it is known that the discharge rate of T5 (2) neurons in response to adequate stimuli is higher in the morning and afternoon, rather than at noon time. (The background luminance in the perimeter was constant).

Seasonal Effects. We have mentioned previously that the response characteristics of R2 retinal neurons with respect to the contrast direction of a moving stimulus are subject to seasonal changes which parallel changes in prey-catching behavior. Seasonal variations in the overall amount of activity are also shown by tectal T5 (2) neurons. If one compares the curves for the three experimental stimulus series carried out in the summer (Fig. 63 C) with those carried out in the winter (Fig. 63 D), it appears that the mean firing rates of the latter are, in general, strongly decreased in the winter. In spring during the spawning season no responses or only weak activity could be recorded from T5 (2) neurons in the optic tectum of the common toad (cf. p. 89).

Neurons were recently recorded from the optic tectum of *Triturus vulgaris* whose sensitivity for the color contrast on the body of the sexual partner showed a positive correlation restricted to the mating season [52].

Hunger State. The excitability of T5 (2) neurons seems to be correlated to the state of hunger of the animal. Toads fed to satiation do not react with prey-catching, even to an optimal prey object; only feeble responses or no responses at all of T5 (2) neurons may be recorded from their tectum, whereas activities of retinal ganglion cells, class TH3 and T5 (1) neurons appear to be unchanged in comparison with a satiated animal [53].

Immobilization and Motivation. The excitability of T5 (2) neurons appears to be state-dependent. It can be assumed therefore that the stimulus responses (discharge frequencies) of T5 (2) neurons, recorded from the immobilized animal, are below the threshold for possible involvement in triggering a behavioral response (prey-catching orienting). In the future, comparative recording experiments in freely moving toads will therefore constitute another important step for the understanding of the neuronal basis of the toad's prey-catching behavior (cf. p. 276).

In connection with the question of neuronal coding of prey and enemy key-stimuli in common toads we may summarize some essential points:

1. At retinal level the ganglion cell classes R2, R3, and R4 already perform the first important preliminary operations on the visual input where prey and enemies must be distinguished. These properties consist of a modulation of firing rate as a function of angular velocity, angular size, and contrast. In the retina, there are, however, neither specific prey nor enemy detectors. R2 and R3 neurons are activated mainly by prey stimuli, R3 and R4 neurons predominantly by enemy objects.

2. The visual field is mapped in different areas of the toad's brain. The optic tectum receives inputs from class R2, R3, and R4 neurons; the thalamic-pretectal (TP) region receives inputs from classes (R1), R3,

and R4. There are also other projection fields of optic nerve terminals in the brain.

3. Neurons with *sensitivity* to moving configurational stimuli (cf. Fig. 66) are already found in the retina. Configurational sensitivity is further elaborated in neurons of the retinal projection fields in the optic tectum (class T5 (1), sensitive to wormlike stimuli) and the thalamic-pretectal region (class TH3, sensitive to antiwormlike stimuli).

4. T5 (2) neurons show *selective* responses to configurational stimuli, corresponding to the prey-catching behavior. They respond well to a stripe moving as a worm in direction of its axis but they show only weak activity if the axis of the same stripe is oriented perpendicular to the direction of movement (antiworm). The activity of T5 (2) neurons reflects approximately the probability that a stimulus configuration fits the prey category. No neurons were found with *specific* response to a stimulus of a certain configuration.

5. The worm/antiworm preference of T5 (2) neurons is within limits invariant for changes of other stimulus parameters, such as (a) the direction of the stimulus movement within the x–y coordinates of the visual field, (b) the contrast direction with the background, (c) the background structure (visual noise), (d) the movement velocity, (e) the distance from the eye of the animal.

6. The general degree of activation of T5 (2) neurons appears to be dependent on motivation-determining factors in a way similar to its influence on the prey-catching activity.

7. At present it is not known whether the neurophysiological results of this chapter can be generalized for other amphibian species. Presumably, here too, as in behavior, variations are to be expected [51].

4.4 A Working Hypothesis on Neuronal Decision-Making

Consider the diagrams of Figures 67–69. In the central visual system of the common toad there are apparently neuronal *"Gestalt filter"* systems which evaluate the extension of the area of a moving pattern mainly along (T5 (1)neurons) or perpendicular to (TH3 neurons) the direction of movement (cf. Te_1 and TP_1 in Fig. 68). Comparing the neurophysiological results with the corresponding behavioral findings, it appears that the toad brain is unable to differentiate prey objects from irrelevant stimuli or enemies solely on the basis of one of these Gestalt filters.

Theoretically, it could fulfill this task if there were interactions between the two filter systems producing *"Gestalt decoder"* systems (cf. TP_2 and Te_2 in Figs. 68 and 69).

The firing rate of tectal T5 (2)neurons is dependent on both configurational parameters of a moving stimulus pattern. Their selective response characteristic might result from excitatory inputs of T5 (1) and inhibitory inputs from the TH3 neurons. The T5 (2) *Gestalt decoder* neurons may belong to a system that triggers or commands (cf. pp. 213 and 214) the prey-catching orienting movement after a certain response threshold has been reached (cf. Te_2 in Fig. 68). The response spectrum would include all objects which could be classified into the prey category according to their configuration.

How would a subtractive interaction between the two Gestalt filter systems operate? We can illustrate this with an example (cf. Te_2 in Fig. 68). The T5 (2) neurons are strongly activated by worm-shaped objects, as a result of excitatory inputs from T5 (1) neurons (Te_1 in Fig. 68). If, however, such a worm were oriented with its longitudinal axis across the movement direction, the T5 (2) neurons might be inhibited by TH3 neurons (TP_1 in Fig. 68). The object is thereby classified as unsuitable for food; the threshold for eliciting the orienting turn will not be reached. It is also conceivable that by an additive interaction between the two Gestalt filter systems – T5 (1) and TH3 neurons – further neuronal systems of the TP-region are activated (cf. TP_2 in Fig. 68), eliciting avoidance responses if the stimulus is highly elongated in and perpendicular to its movement direction.

According to this hypothesis the fundamental steps of configurational prey-enemy recognition are based on *subtractive* and *additive* interactions between tectal and thalamic-pretectal Gestalt filter systems. There are numerous experimental indications that support this idea. Some of them will be briefly discussed.

1. The response of T5 (2) neurons to a moving visual stimulus may be inhibited by an immediately preceding electrical stimulation of a point in the TP-region. This is a physiological indication of an inhibitory influence of the TP-region upon the optic tectum (Fig. 67B). There is also anatomical evidence for pretecto-tectal projections [54].

2. When stimulation and recording positions are interchanged (Fig. 67A), wide-field neurons of the TP-region can be activated by large moving visual stimuli as well as by focal electrical stimulation of the optic tectum. Furthermore, after prolonged electrical stimulation, TP neurons display after discharges, indicating the possibility of positive feedback via interneurons [55].

3. Electrical point stimulation of the optic tectum in freely moving toads elicits prey-catching orienting movements, whereas point stimulation of the TP-region invariably elicits avoidance responses.

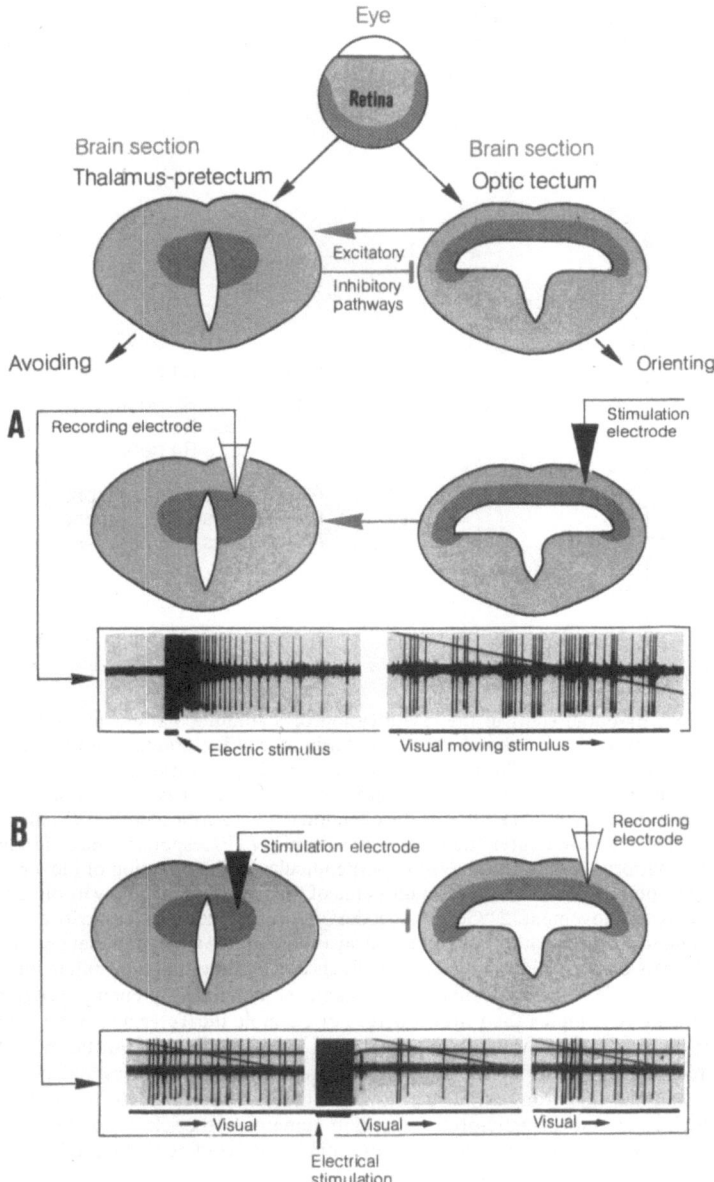

Fig. 67 A and B. Electrophysiological clues for excitatory **(A)** and inhibitory **(B)** connections between the optic tectum and the thalamic-pretectal area of the common toad. See text for details. [Modified from Ewert JP (1974) Sci Am 230: 34–42. Ewert JP, Hock FJ, Wietersheim A v (1974) J Comp Physiol 92: 343–356]

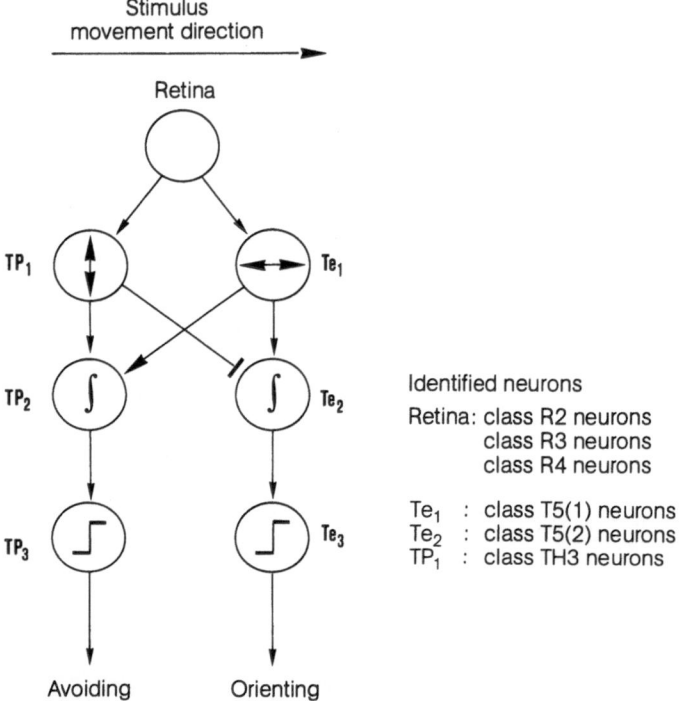

Fig. 68. Essential steps in the classification of a stimulus pattern under the categories prey/nonprey/enemy are taken on the basis of configurational features linked to the movement direction. The diagram illustrates a simple working hypothesis concerning neurophysiological elements that could form the basis of configurational prey/enemy recognition in the visual system of the common toad. *Te* optic tectum, *TP* thalamic-pretectal region. *1* "Gestalt filter" neurons, some of which *(TP)* respond to the total stimulus area of the pattern, mainly its dimension perpendicular to the direction of movement; others *(Te)* respond also to the total stimulus area of the pattern, but mainly its dimension in the direction of movement. *2* "Gestalt decoder" neurons, *3* neurons with critical threshold for turning toward (orienting) or turning away (avoiding). *Arrows* designate excitatory, *lines with cross bars* inhibitory connections. Each *circle* in the diagram stands for a neuronal population. The whole information-processing system for prey/enemy recognition of the toad is more complex than shown here. For example the Te_1-neurons in common toads possibly also receive (weak) inhibitory inputs from TP-neurons. Furthermore Te and TP systems appear to be under the control of telencephalic and hypothalamic influences. "Gestalt filter" neurons may also be involved in resolving other detection problems. In the optic tectum and thalamic-pretectal region numerous other classes of neurons exist [48] responsible for functions other than orienting and avoiding. [Modified from Ewert JP (1973) Fortschr Zool 21: 307–333]

4. After TP-lesions (or application of curare to the optic tectum) the configurational prey/enemy recognition system of the toad is no longer functional (Fig. 64 A). In addition the T5 neurons lose any configurational response selectivity (Fig. 64 B; cf. also Fig. 70).

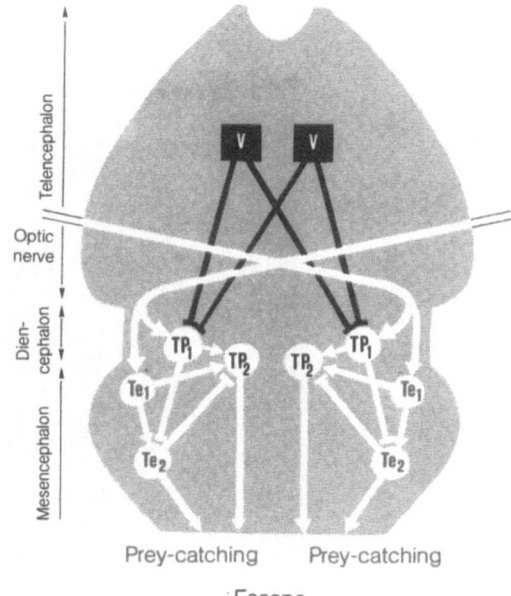

Fig. 69. Postulated central nervous interactions for the control of prey-catching and escape behavior of the common toad. This diagram is a synthesis of data obtained in brain lesion and brain stimulation studies and results from single cell recordings (cf. Fig. 68). $Te_{1,2}$ neuronal systems in the optic tectum, $TP_{1,2}$ neuronal systems in the thalamic-pretectal region. *Arrows* indicate excitatory, *lines with cross bars* inhibitory connections. [Modified from Ewert JP (1967) Z Vergl Physiol 57: 263–298]

A modification – or rather an extension – of this hypothesis should be discussed:
After TP-lesions, the ERF's of T5 neurons in the central layers of the ipsilateral optic tectum may be changed. All of the T5 neurons show "disinhibited" activity in response to moving configurational stimuli. Furthermore, T5 neurons lose their habituation properties [56]. It seems likely that these response characteristics (Fig. 64B) are general properties of tectal T5 neurons and hence we may designate them as T5 (0). Therefore, it is possible that, in the intact brain, different neuronal populations in the optic tectum are produced by differentially weighted inputs to these neurons from the thalamic-pretectal (TP) region. One population would have the response characteristics of class T5 (1) neurons (Fig. 63B), and a second group, characteristics of class T5 (2) neurons (Fig. 63C). There, of course might be others which remain to be discovered. It is important to consider the results of recent studies which indicate that the response characteristics of T5 (2) neurons to different moving configurational stimuli is relatively constant. Their general level of activity appears to be "state"-dependent. However, some of the T5 (1) neurons may alter their response characteristic to configurational stimuli depending upon the circumstances ("temporal surround") which are as yet unknown. Specific interactions between neuronal populations of the TP-region and the optic tectum may result in "hardwired" and "softwired" systems, necessary for the analysis of behaviorally relevant visual signals and their classifications into innate and acquired categories of meaning.
Consequently, telencephalic and hypothalamic influences [57, 58] must also be considered in the framework of Figs. 68 and 69. There is one loop (1) concerned with the *striatum* of

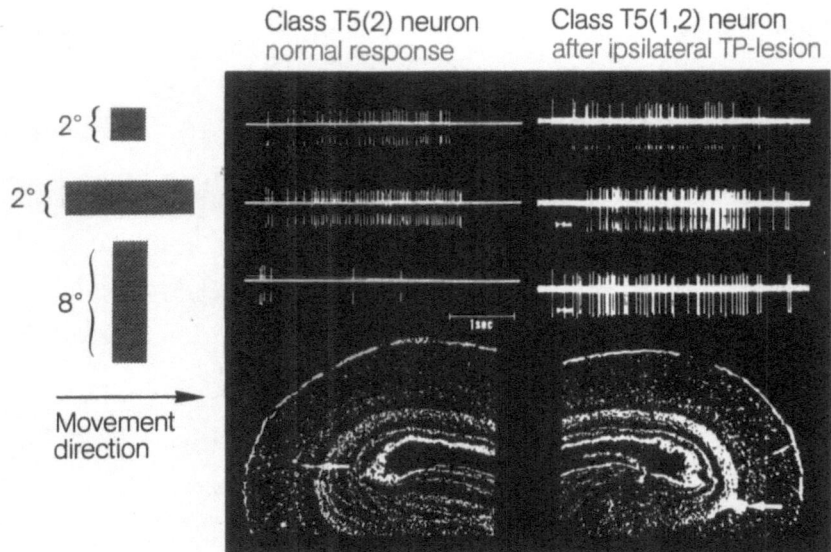

Fig. 70. *Left,* Original record of a T5(2) tectal neuron in response to three different configurational black stimuli moved at 7.6°/s against a white background. *Right,* Representative record from a neuron in the same tectal layers after ipsilateral lesions of the thalamic-pretectal region, showing loss of selectivity properties in response to the same stimuli. The recording positions in the central layers of the optic tectum are indicated by a *white arrow.* (Cf. also Figs. 63 B, C and 64 B) [Modified from Ewert JP (1978) In: Die Psychologie des 20. Jahrhunderts. Stamm, RA Zeier, H (eds) Lorenz und die Folgen, vol VI. Kindler AG, Zürich]

the lateral telencephalon and mediated by the lateral forebrain bundle: Both retina and optic tectum are projecting to postero-lateral thalamus and from there to striatum, which in turn feeds back to optic tectum, indirectly, via postero-lateral thalamic or tegmental cell groups. Another loop (2) is concerned with the *medial pallium* (medial telencephalon, medial forebrain bundle): Retina and optic tectum project to anterior thalamus and from there to medial pallium, which in turn feeds back to preoptic area and hypothalamus, both nuclei send axons to the deeper layers of the optic tectum. Loop (2) is associated with some nuclei, the homologs of which belong in mammals to the "limbic system" (p. 219). With respect to training experiments (p. 89) olfactory and visual cues could be "combined" in loop (2). The resulting information arriving in the optic tectum might also be fed into loop (1), thus changing properties of thalamic-pretecal cells and thereby those of tectal neurons. Briefly, according to levels and states of motivation (hunger, season) the *general activity* of class T5(2) neurons might be turned on or off by loop (2). The *response characteristics* of these neurons could be modified (in relation to experience) by the involvement of loop (1). These concepts, however, are speculative at present.

To summarize, what are the basic points of the proposed mechanism for recognizing moving configurational stimuli?

In the present context configurational pattern recognition is defined as the classification of two-dimensional space and time-dependent distributions of brightness from the environment into innate and learned classes of functional significance: *prey, nonprey, enemy*. This operation proceeds in two steps:

1. The extraction of behaviorally relevant Gestalt features from the pattern to be perceived, by neuronal Gestalt filter systems.

2. The decoding and decision-making process based on subtractive interaction between filter systems and weighted neuronal threshold-value operation. The intensity of the corresponding behavioral reaction (number of orienting movements) would then depend on the number of threshold crossings.

We may try now to simulate the neuronal responses by means of neuronal models (two-dimensional networks) on the basis of specific spatial and temporal parameters. The principle consists of "working back" to the underlying system from the response to the optimal stimulus. The results of these studies have yielded important indications of the system properties of the neuronal Gestalt "filters" and "decoders" [59].

4.5 Relevance of the Supposed Gestalt Recognition System

4.5.1 Orientation Detectors and Gestalt Decoders

Toads as well as monkeys are able to distinguish a horizontal stripe from a vertical one. The underlying filter mechanisms are, however, different so far as *orientation detectors* and *Gestalt decoders* are concerned.

The *orientation detectors* (simple cells) of the visual cortex of mammals, on the one hand, respond to the orientation of a stripe pattern in relation to the orientation of the main axis of their longitudinal (anisotropic) receptive field; the response is dependent on the stimulation of excitatory and inhibitory field sections (Fig. 71 A, a and b). On the other hand, the *Gestalt decoders* studied so far [T5(2) neurons] in the common toad have ellipsoidal, on the average, approximately radially symmetrical receptive fields. They evaluate the orientation of a stripe pattern relative to its movement direction. Such a processing principle is largely independent of the absolute direction of movement of the stripe pattern in the x–y coordinates (Fig. 71 $B_{1,2}$, a and b) [51]. Furthermore,

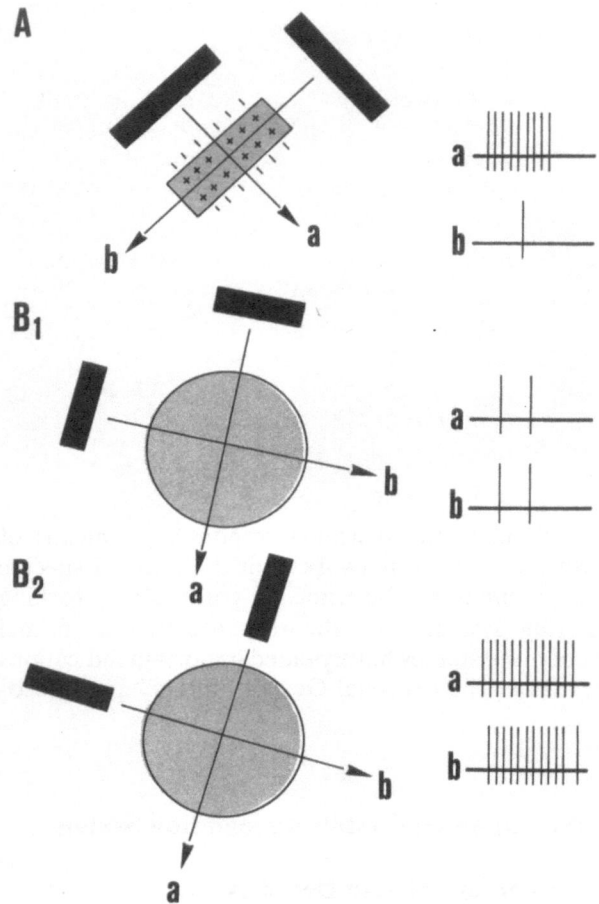

Fig. 71 A and B. Perception of black bars by: **A** "Orientation detector" from the visual cortex of cat. The elongated receptive field has excitatory (*off* zones " + ") and inhibitory (*on* zones " − ") subdivisions arranged in parallel; **B** "Gestalt decoder" from the optic tectum [T5(2) neuron] of the common toad. See text for explanation. [Modified from Hubel DH, Wiesel TN (1962) J Physiol (London) 160: 106–154. Ewert JP, Borchers HW, Wietersheim A v (1979) J Comp Physiol 132: 191–201]

the simple cells of the visual cortex extract the background structure from a stimulus if both, i.e., stimulus and background, are moving simultaneously (cf. p. 140). For Gestalt decoders in toads, the stimulus is almost completely masked in such a situation. However, if the stimulus moves over a structured background, the selectivity of prey Gestalt decoders is maintained, whereas simple cells lose their characteristics as orientation detectors. Thus, Gestalt decoders of toads appear to have a greater resemblance to some functional properties of "complex" or "hypercomplex cells" in the mammalian visual cortex (cf. p. 139) [60]. The particular response properties of Gestalt decoders may be related to the absence of involuntary eye movements in toads.

What might sensitize a Gestalt filter system to one or the other configurational components of a moving stimulus pattern? To understand this we should clarify first which stimulus features are changed in one or the other case. (1) In the case of area expansion perpendicular to the movement direction only spatial components of the stimulus are changed. (2) If an area is enlarged in the movement direction, temporal as well as spatial stimulus components are varied: at the same angular velocity, a longer horizontal bar darkens a larger area (spatial component) and the shadowing remains for a longer time (temporal component).

In the former case area components perpendicular to the movement direction may be evaluated by summation of spatial excitation in the ERF.

In the latter case area extension in the movement direction might be evaluated by summation of spatial excitation in connection with a relatively large delay between excitatory and inhibitory processes in the receptive field.

By differential weighting of operations (1) and (2), neurons might develop the capacity to evaluate mainly one or the other extension component of a moving pattern as a consequence of the asymmetry in the time domain [59].

Ontogenetic and species-dependent variations in prey selection might be the result of the various interactions produced by different weighting of the two operations in space and time domains.

4.5.2 What is the Performance of Such a Recognition System and Where Are Its Limits?

The hypothesis represents only a gross simplification of what possibly occurs in the toad brain during prey-enemy recognition, since not only configurational cues, but also other stimulus parameters, such as movement and contrast directions, movement patterns and stimulus distance enter into the total evaluation process. Moreover, individual experiences partly combined with motivational determinants and seasonal factors also influence the evaluation process. The model therefore can only indicate fundamental steps for classifying the configuration of moving visual stimuli as either prey, nonprey or enemy. For detailed differentiations within a corresponding category as yet unknown identification systems must be assumed.

Of course, various other neuronal systems exist in the same brain areas studied and they have to fulfill many sensory and motor functions, besides the release of orienting and avoidance movements. It is also conceivable that the Gestalt filter neurons described [class TH3 and T5(1)] are involved in other detection problems. In conclusion, the question of the neurophysiological basis for prey-enemy recognition in the toad remains incompletely answered, but we have presumably come a step closer to its solution. Of course, the "worm-antiworm" phenomenon is a useful tool for the experimenter to test the pattern recognition system of the toad with simple, behaviorally meaningful stimuli (key stimuli). Clearly, this should not imply the assumption that the CNS of the toad is simple as illustrated in Figure 68; but this wiring may resemble an important feature of the configurational prey recognition system.

4.5.3 Some Comments on Innate and Acquired Pattern Recognition

Have two different mechanisms been developed in the visual system for the identification of patterns?

One of these mechanisms could serve for the recognition of key-stimuli. The basic operations fit the concept of an *innate releasing mechanism,* IRM. They consist of an extraction and separation of feature vectors by discrimination functions that are based on the system properties of fixed or preprogrammed neuronal circuits. The efficacy of a key-stimulus could then be represented by the response characteristics of neuronal trigger or command systems (cf. pp. 213, 214). For prey-orienting and enemy-avoidance responses of the toad the $Te_{2,3}$ or $TP_{2,3}$ neurons may be involved (Fig. 68). These command elements with recognition properties release upon appropriate activation the corresponding motor program which – as a fixed action pattern (FAP) – is also prewired in the CNS [61]. Starting from such a "hardwired" recognition system, modifications are possible to produce particular adaptations. The interaction of neuronal filter mechanisms with inputs from systems that control motivational states may determine the properties in the "softwired" aspects of a releasing mechanism (cf. also IRME).

Further systems may exist that associate stimulus patterns with acquired classes of meaning. Anatomical considerations alone compel us to reject the idea that a corresponding number of "wired in" complex feature detectors – redundant ones at that – have been developed for all possible patterns which are stored in an individual lifetime and which remain recognizable by means of an ARM. That is, for example, we may hardly expect to find in the brain of man specific "grandmother detectors" [74].

Of course, for specific individual recognition phenomena, too, we may postulate different neuronal classes, where each class responds to a certain partial stimulus characteristic. A stimulus pattern is thereby assumed to be represented by a corresponding pattern of excitation and inhibition in large neuronal assemblies. Different patterns of excitation and inhibition in those neuronal assemblies will represent other stimulus patterns. However, so far unanswered is the question of *how* and by *whom* such a complex code is read in the brain, how decisions are made and how they are transformed into efferent commands.

4.5.4 Sensori-Motor Interfacing

Sensory signals – after they have been identified – may elicit an appropriate motor response depending on states and levels of motivation of the animal. What is the substrate of sensori-motor interfacing? The results of electrical brain stimulation experiments in freely moving toads suggest that motor programs for the different components of the visually guided prey-catching sequence are apparently triggered in the optic tectum. During recordings in awake immobilized toads and frogs various types of neurons have been identified in the optic tectum [48]. According to their stimulus response characteristics, some neurons may participate in the release of the prey-catching sequence, as shown in Fig. 72 and Table 6, cf. also Figs. 68 and 74. Thus, neurons in the optic tectum might have to exert not only *sensory* analytical, but also *motor* command functions.

The "command neuron" concept was originally developed in invertebrate neuroethology mainly on the basis of results obtained in experiments in which stimulation of single cells elicited particular *motor* patterns. In the present instance we are attempting to apply this concept to an experimental situation where single cell activity is correlated with various behaviorally relevant *sensory* patterns. With this distinction in mind the following scheme is offered as a tentative interpretation of single unit activity related to prey-catching behavior in frogs and toads.

In connection with the concept of "command systems" – summarized on pp. 213/214 – we may assume that the output of several particular tectal neuron types, each forming a *command element* ( ), together constitute a *command system* [ ] which itself activates a specific motor pattern. The entire sequence of motor patterns during prey-catching is released by different sequentially activated command systems that collectively belong to a *multiple action system* { }. In this sequence each executed command provides the stimulus situation of the next following command, as shown in Fig. 41. Operation of each command system requires simultaneous activation of all of its command elements, thus constituting an "AND"-gate mechanism. Command systems differ from each other by a distinct combination of command elements which can be shared by the different command systems, for example

$$\left\{ \begin{array}{llll} [(T4)(T5\cdot 2)] & \to & [(T2\cdot 2)(T5\cdot 2)^*(T1\cdot 1)] & \to & [(T5\cdot 2)^*(T1\cdot 2)] & \to & [(T5\cdot 2)^{**}(T1\cdot 3)(T3)] \\ \text{ORIENT!} & & \text{APPROACH!} & & \text{FIXATE!} & & \text{FIXATE and SNAP!} \end{array} \right\}$$

If a predator crosses the visual field of the toad, then
$$[(T4)(T5\cdot 1)(TH3)]$$
AVOID!

The same command ORIENT for different goals, e.g., mate, may be issued by different combination of command elements, e.g. $[(T4)(T5\cdot 1)]$. Of course, additional or alternative command elements

Table 6

Stimulus	Class of neurons in the optic tectum	Motor pattern activated
Prey object traversing a part of the lateral visual field	*Class T5(2) neurons*, monocularly driven. They have excitatory receptive fields of 15°–30° diam. T5(2) neurons of the corresponding visual field projection are best activated by moving configurational stimuli that fit the prey category. Prey recognition is feasible at all positions of the visual field and this process precedes the orienting turn!	T5(2) neurons are involved in a system that commands the orienting movement toward prey (once a particular level of neuronal activity has been reached)
Movement of the prey object from temporal to nasal visual field positions	*Class T2 neurons*, monocularly driven. They have relatively large ERFs extending more than 90° into the visual field including the frontal binocular part. Some of the T2 neurons are best activated if the stimulus moves in the visual field from temporal to nasal. (If eyelids are closed during turning the image of the object may be "re-adjusted" at the end of the turning movement by re-elevation of the eyes)	Control of the amount of the turning movement (however, frogs and toads obviously ignore any visual feedback obtained *during* the turn!)
Prey object moves in the binocular visual field	*Class T1 neurons*, binocularly driven. Their ERFs extend 15°–30° through the frontal visual field. Some of the T1 neurons are activated if angular size and position of the visual stimuli are adequate for the ERFs in both retinae. This information might be used to estimate distance	Control of approaching the prey (and snapping)
Image of the stimulus increases on the retina as the animal approaches the prey during fixation	*Class T3 neurons*, monocularly driven. Their ERFs extend over 20°–30° of nasal regions of the visual field. T3 neurons are strongly activated if a visual object moves in the z-axis toward the animal. The relative increase in size of the object's image is larger, the closer the object. Information might be used to estimate distance	Control of snapping

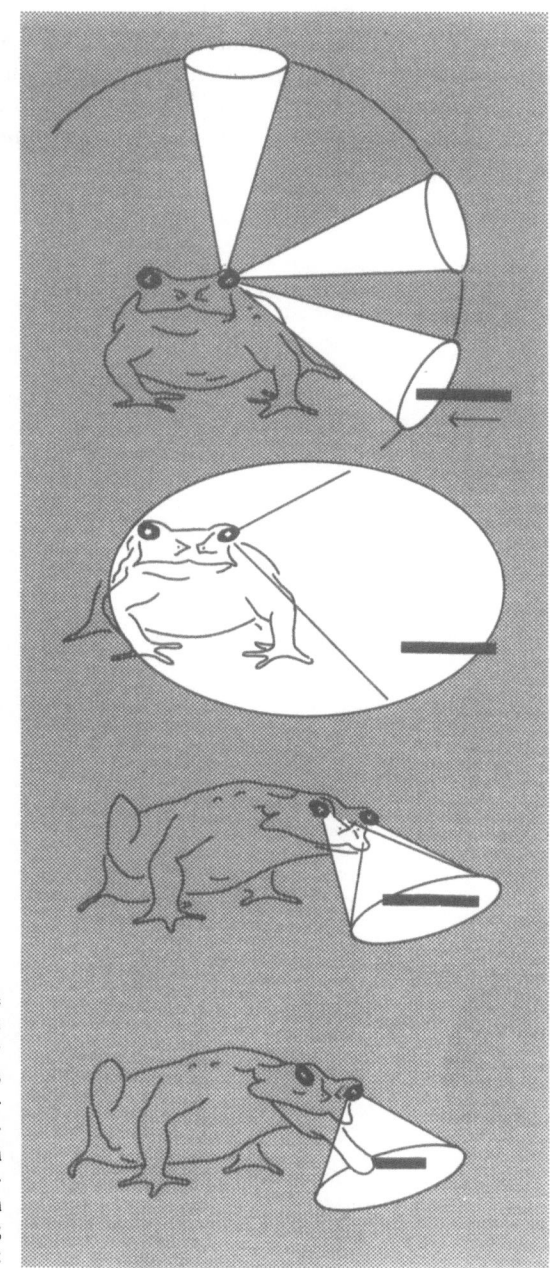

Fig. 72. Basic steps of the release and control of the prey-catching sequence by activation of different classes of tectal neurons, cf. also Figure 41 and Table 6. [Modified and combined from Grüsser OJ, Grüsser-Cornehls U (1970) Verh Dtsch Zool Ges 64: 201–218; Ewert, JP (1974) Sci Am 230: 34–42]

of the above systems remain to be discovered. Whether or not these neurons correspond to "command elements" within such a proposed logical structure of command systems serving specific behavioral goals is as yet unknown [50]. But this question can be answered by recording from these neurons in the *behaving animal* (cf. Methodological Appendix p. 276).

5 Functional Areas in the Visual System of Vertebrates: Comparative Aspects

The studies on toads provided our first insights into neuronal processes that might be responsible for the release and control of motor behavior. In this connection a whole series of basic questions and problems arose, concerning not only toads in particular, but also other vertebrates and invertebrates. The same questions also apply, finally, to other sensorimotor systems.

Let us now consider the visual system of mammals. Here, too, there must also be three principal functions, which may be designated by the terms *localization, space constancy* and *identification.* We shall closely examine the comparative structural and functional correlations.

5.1 Substrates for Stimulus Localization

5.1.1 Basic Functions

The localization processes required for the direction of an orienting movement are guided mainly by the optic tectum or the superior colliculus (homolog of mammals) in all vertebrates studied. We are dealing here with a phylogenetically old system whose neurophysiologically and neuroethologically measurable functions show in part an astonishing constancy through the whole range of vertebrates from fish to monkey. Thus, it is possible, for example, to trigger retinotopically adjusted turning reactions by electrical stimulation of a point in the optic tectum in freely moving fish as well as in the cat (Fig. 73). An interesting correlation of orientation movements of two different sense organ systems is shown by such stimulation experiments in the rabbit [62]. First the ear is turned toward a corresponding area in the visual field; head and eye follow afterwards in the same direction.

The optic tectum (superior colliculus) shows common properties of basic neuronal organization in all vertebrate species which have been studied. The visual receptive fields of the neurons are enlarged with increasing depth of the cells in the tectum (Fig. 10B), a phenomenon due to the convergence of visual input (Fig. 74). In the deep layers adjoining the ventricle the visual wide-field neurons may also receive additional inputs from other sensory systems (Fig. 74, subtectum). Some of them originate

Optic tectum (dorsal surface)

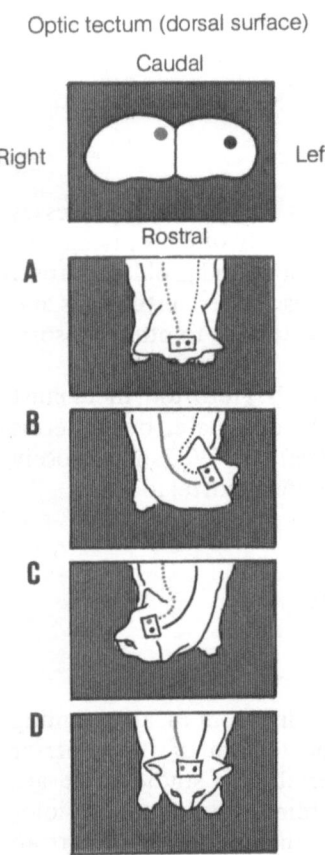

Fig. 73 A–D. Electrical point stimulation of cat optic tectum. **A** Starting position, unstimulated. **B** Stimulation of right posterior optic tectum elicits head turning toward the opposite side. **C** Analogous response after stimulating the left tectum. **D** Simultaneous stimulation of both tecta leads to a resultant head position. [Modified from Hopf et al (1970) in Schäfer KP (1970) Brain Behav Evol 3: 222–240]

in the ear, others in the skin. In the rattlesnake the outputs from the infrared receptors of the head region also project topographically onto the optic tectum, but maps of the two modalities are not precisely congruous. In certain layers infrared afferents and visual information may converge [63].

The optic tectum is thus a brain region for multimodal sensory integration, which, like the ancient image of the Janus head, looks in two directions: toward sensory and motor functions. The midbrain tectum represents a correlation center for afferent messages and efferent motor commands. It is sometimes regarded as a kind of microcosm of the whole brain.

5.1.2 Plasticity

How does a frog behave if its eyes innervate the wrong sides of the brain? That is, what happens if each eye projects to the ipsilateral instead of the contralateral optic tectum? This condition can be produced experimentally by sectioning the optic nerves of both eyes in front of the chiasm and preventing the regenerating axons from decussating. After restoration of the topographical order of the axonal endings in the tectal layers on the wrong side, the frog's vision is restored, but it now behaves as if visual

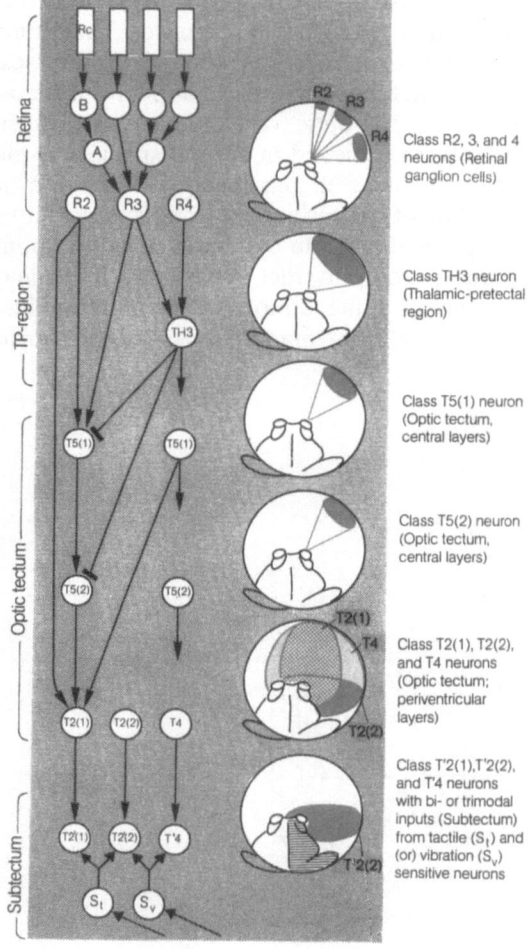

Fig. 74. Principles of visual convergence, divergence and multisensory integration in the toad's visual system: retinal, thalamic-pretectal, tectal and subtectal levels. *Arrows* indicate excitatory, and *lines with cross bars* inhibitory connections. Only a few neuron classes, representing populations, are depicted in this diagram, cf. also Figures 10, 61 and 68. Note that bimodal class T′2(2) neurons in subtectal regions *(bottom)* have both a contralateral excitatory *visual* receptive field and a contralateral excitatory *cutaneous* receptive field. [Modified from Ewert JP, Borchers HW (1971) Z Vergl Physiol 71: 165–189. Ewert JP (1971) Z Vergl Physiol 74: 81–102]

objects appear to be located in positions of the visual field which are mirror images of their actual positions, reflected across the midline. Incidentally, the frogs behave in a similar way if an optic nerve is induced by means of a different experimental maneuver to innervate the ipsilateral tectum (Fig. 13 F). In these experiments, the optic tectum on one side is ablated, and the regenerating optic tract fibers become re-routed to the opposite tectum. The frogs never learn to correct the errors in their visual motor responses. That humans can learn to correct these kinds of errors has been shown by experiments with prism spectacles that make the world appear upside down.

The re-routing of optic tract axons to the wrong side of the midbrain has recently been achieved in several species of small mammals. Of these, the Syrian hamster has been studied the most extensively. The re-directing of the retinal projections occurs after unilateral ablation of the visual layers of the tectum in newborn animals. After this surgical procedure, the optic tract axons grow into the damaged tectum where they develop some abnormal termination in the remaining deeper layers. In addition, many axons grow, abnormally, across the tectal

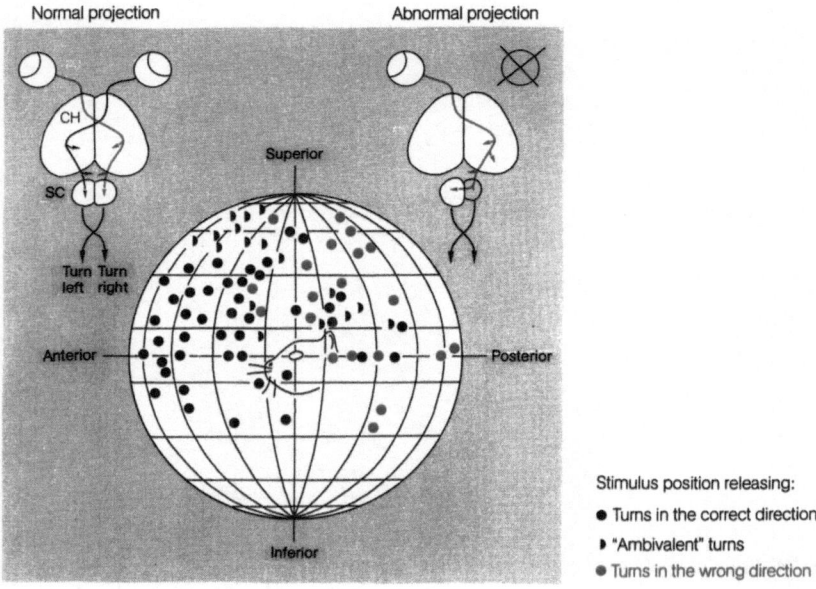

Fig. 75. Orienting responses of Syrian hamster mediated by anomalous retino-tectal pathways to the wrong side of the brain. Results of videotape analysis of the left visual field. The right eye and the superficial layers of the right superior colliculus were ablated at birth. Grid lines are 20° apart, with extra lines at horizontal eye level and at 90° from the straight ahead. *CH* cerebral hemisphere, *SC* superior colliculus. See text for explanation. [Modified from Schneider GE (1973) Brain Behav Evol 8: 73–109]

midline into the undamaged tectum; this connection is greatly increased if the eye on the side of the tectal lesion is also removed (Fig. 75, top right). The re-crossing axons, which reach the tectum after crossing the midline twice (once at the optic chiasm and once at the tectal midline), terminate in positions which, as in the case of the frog, are mirror images of their normal positions. However, the topographic ordering is abnormal with many small deviations from a perfect map. The topographic order of the projection in the correct side, in the residual tectum, is even more disorderly.

The visually elicited behavior of the hamsters with these abnormal connections of the optic tract is not the same as that of frogs with a retinal projection to the wrong tectum. The hamsters do turn in the wrong direction, but for stimuli in only part of the visual field of the remaining eye (Fig. 75). They turn in the correct direction for stimuli in the lower nasal field, and in the wrong direction for stimuli in the upper temporal field. For stimuli in parts of the field transitional between these two areas, the hamsters often begin a turn in one direction and then quickly change to the opposite direction. The area of wrong-direction turns corresponds to the medial part of the optic tectum, where the termination of re-crossing optic tract axons is most dense. It is also interesting to note that the hamsters, when they make a wrong-direction turn, fail to turn to the mirror image position as described for frogs. They tend to stop their turn short of this position, as if they notice when the food object disappears from their field of vision. Frogs, which complete their turns in about 1/3 the time taken by the hamsters, appear to ignore any visual feedback information obtained during the turn. The analysis of head and body orienting in mammals is complicated by the possibility of modulating feedback effects.

Adult hamsters with ablations of the visual layers of the optic tectum completely lose their ability to turn toward seeds presented visually (see later p. 140). The hamsters with such lesions suffered soon after birth do not show this complete behavioral defect because of the continued ingrowth of the developing optic-tract axons in the neonate. This does not occur in the mature animal after the lesion. It is tempting to propose that this kind of neuroanatomical plasticity can explain differences in early and late brain lesions in humans, as well as other mammals. It is commonly believed that "it is better to have your brain lesion early – that is, if you're going to have one!", in the words of the late neuropsychologist Hans-Lukas Teuber. However, the early unilateral tectal lesions in hamsters still resulted in some functional defects. Not only was maladaptive wrong-direction turning observed, as described above; but also, the correct-direction turning to visual stimuli in the lower-nasal field was often slow and inaccurate [64]. Comparable findings of incomplete sparing of function and alterations in function after early lesions in man have been reported by Hans-Lukas Teuber [65].

5.2 Space Constancy and Optokinetic Nystagmus

If we look at a passing train while seated in another train our eyes unconsciously carry out stepwise pursuit movements; for a short while they keep a fixed relation to the moving surroundings (passing coaches), then they are left behind somewhat, only to jump back into a new position; the same process is repeated over and over again. The function of these stepwise movements (optokinetic nystagmus) is to keep the structured visual world in the visual field constant for a brief moment – a prerequisite for its identification.

Optokinetic nystagmus may be studied experimentally, especially in those animals which show head-nystagmus. The releasing stimuli are richly structured visual patterns with large surfaces, for example, a screen of vertically oriented stripes rotating around the experimental animal in the horizontal direction. The essential, effective stimuli are image movements on the retina in a naso-temporal direction (Fig. 76 A). Accordingly, a frog blind on the right side (Fig. 76 B) would, for example, respond only to fields moving clockwise, and vice-versa.

Brain structures which are especially old phylogenetically, participate in the control of optomotor eye or head movements. In mammals it is the

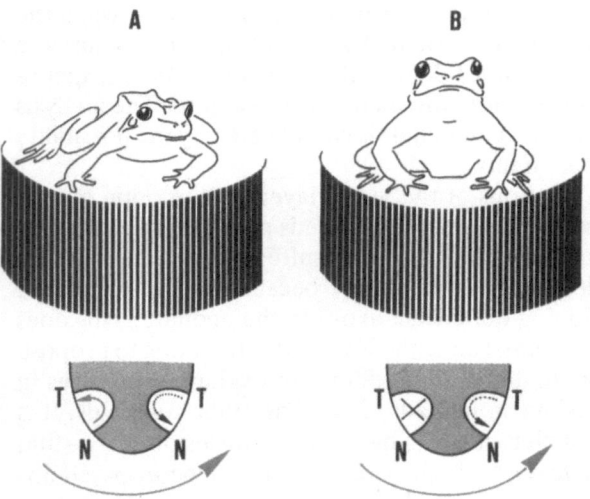

Fig. 76 A and B. Optomotor responses of the frog *Rana temporaria* in a moving stripe drum. **A** Behavior of intact animal. The *large red arrow* indicates the rotation of the drum, the *small red arrow* shows effective naso-temporal image displacement on the retina. **B** After extirpation of right eye, the optomotor reaction is extinguished in the same experimental conditions; it would, however, reappear as soon as the stripe cylinder rotates in the opposite direction and images in the retina of the seeing eye are shifted in naso-temporal direction. [Modified from Birukow G (1938) Z Vergl Physiol 25: 92–142]

nucleus of the optic tract, NOT (pretectum); in anuran amphibians it is the nucleus of the basal optic tract, NBOT (ventral mesencephalon). Experiments in cats and rabbits show that the corresponding neuronal systems receive their main inputs directly from the retina of the contralateral eye. Neurons have receptive fields of 40°–150° and are chiefly activated by richly structured patterns such as, for example, "visual noise". For neuronal excitation as well, the most effective stimulus patterns are those whose images on the retina of the contralateral eye are shifted in naso-temporal direction. Ipsilateral inputs are also present in cats: they are, however, weaker, and are extinguished after ablation of the visual cortex. The monocular optokinetic nystagmus becomes "symmetrical" with the aid of these ipsilateral inputs.

There is evidence that the *horizontal* optokinetic nystagmus in cats is controlled by cells of the NOT, and the *oblique* optokinetic nystagmus by the medial terminal nucleus (MTN) of the accessory optic system. The corresponding cells of the NOT may be part of the accessory optic system [66].

Electrical point stimulation of the nucleus of the optic tract can activate optokinetic nystagmus in the freely moving animal in the absence of visual patterns. After elimination of the corresponding brain structures, the function disappears in the frog as well as in the cat.

Recently it has been found that the pretectal "optomotor neurons" of the cat convey their impulses, among other places, directly to the inferior olivary nucleus and from there via climbing fibers (Fig. 169) to the Purkinje cells of the cerebellar cortex [67].

5.3 Brain Structures for Stimulus Identification

5.3.1 Retinal Level

Comparative neurophysiological studies of the vertebrate retina have identified distinct classes of neurons which exhibit particular response characteristics to visual stimuli. Studies of the retinal output of various vertebrates have demonstrated the following properties:

1. Enhancement of contrast of light-dark borders (by lateral inhibition in the neuronal networks).
2. Measurement of *angular size* of a stimulus pattern (by particular spatial distributions and relative strengths of excitatory and inhibitory processes in the receptive fields; for example, in *on* and *off* center neurons which are concentrically organized).
3. Coding of the stimulus parameter *movement* (proposed to be based on local adaptation of the subsynaptic membrane between bipolar and ganglion cells: "movement-specific neurons").

4. Coding of the *orientation* of a light-dark border (on the basis of specific orientation of longitudinally arranged excitatory and inhibitory receptive fields: "local edge" or "orientation-sensitive neurons").

5. Evaluation of the *direction* in which a stimulus pattern moves within the x–y coordinates (by unilateral inhibition: "directional sensitive neurons").

6. Coding of *wavelengths* of a light stimulus (by "color sensitive neurons").

Although the vertebrate retina consists uniformly of six neuron types – receptor, horizontal, amacrine, interplexiform, bipolar, and ganglion cells [70] – the degree of information processing in the individual species may be very different. There is possibly a correlation between retinal specialization and the development of the cerebral cortex (neopallium).

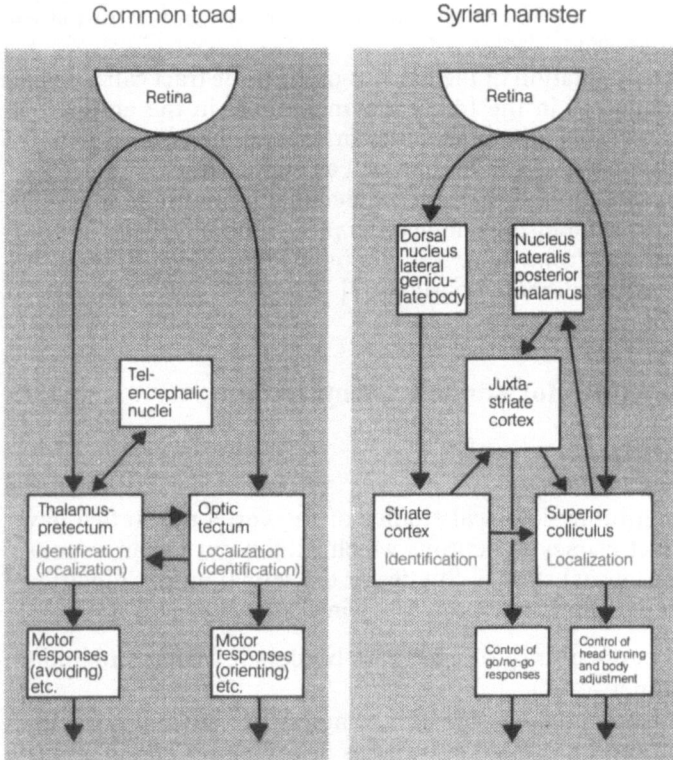

Fig. 77. Brain structures for localization, identification and motor control in the visual systems of toad and hamster. See text for details. [Modified from Schneider, GE (1975) In: Ingle D, Sprague JM (eds) Sensorimotor function of the midbrain tectum, Neurosci Res Prog vol XIII. MIT Press, Cambridge Mass; Ewert JP (1973) In: Lindauer M (ed) Orientierung der Tiere im Raum I. G Fischer, Stuttgart]

In amphibians and birds visual signals are already filtered to varying degrees in the eye. Ground squirrels and rabbits occupy an "intermediate" position [68]; although they do have a neocortex, it is functionally 'not as highly developed as in higher mammals (cats, monkeys) whose retina appears to have a comparatively simple organization.

We know, however, from recent studies that even the concentrically organized *on* and *off* center neurons of the cat retina consist of different cell types, called X-, Y- and W-cells [69]. The X-cells are brisk conducting "sustained neurons" belonging anatomically to the β-type (soma diam. ≈ 25 μm); their fibers project via particular classes of neurons of the lateral geniculate body (diencephalon) predominantly to the visual cortex. The Y-cells are brisk conducting "transient neurons" belonging to the α-type (soma diam. ≈ 35 μm) and project to visual cortex *and* superior colliculus. W-cells are sluggish conducting, sustained or transient, and they belong anatomically to the γ-type (soma diam. ≈ 12 μm). The concentrically organized W-cells are coding color, direction, and local edges of a stimulus. The W-fibers project mainly to the superior colliculus. With regard to the projection relationship, the visual cortex appears to "look" mainly to the visual field center with the relatively small receptive fields of the X-cells, whereas the superior colliculus "looks" into the periphery with the Y-and W-cells.

On the basis of electron microscopy studies, together with intracellular recordings from various vertebrate retinae, we have been able to obtain a more profound insight into their synaptic organization [70]. An increased amount of information processing in the retina, is paralleled by an increasing incidence of amacrine cells functioning as interneurons between bipolar axon terminals and ganglion cell dendrites. Presumably they are primarily responsible for the complex response characteristics (1)–(5) of retinal ganglion cells.

According to response properties and central projection patterns, the appropriate output of retinal ganglion cells is involved in the formation of neuronal systems for different functions.

5.3.2 Diencephalic and Mesencephalic Levels

In lower vertebrates the evaluation and recognition of visual signals takes place in the networks responsible for localization (optic tectum) by connection with other subcortical systems (thalamic-pretectal, TP, areas). Thus, localization and identification systems are relatively tightly enmeshed in frogs and toads (Fig. 77, left). After tectal ablation both the ability to localize and to recognize moving stimuli are abolished; however, stationary barriers may still be localized with the remaining retino-pretectal (thalamic) projection. After bilateral lesion of the TP-region, however, tectal neurons may lose their selective response properties and the animal fails to recognize moving configurational visual signals.

In mammals, too, it is known that tectal (collicular) neurons can be driven by neurons located in other brain areas. For example, the directional selectivity of tectal neurons disappears when the connecting tracts to the visual cortex are sectioned [71].

In mammals the phylogenetically old, subcortical identification systems in tectum/thalamus/pretectum are also important to regain the ability to learn simple pattern discrimination tasks after cortical lesions [72]. Such a recovery of function after ablation of visual cortex appears to occur predominantly in the neonates of some species. If, in newborn hamsters, the striate cortex is destroyed, certain pattern and form distinctions may be carried out later by means of the mesencephalic-diencephalic neural networks and their connections with remaining parts of the neocortex (Fig. 77, right; cf. also p. 140).

5.3.3 The Cortical Level

In mammals information from the eye is conducted to the cerebral cortex via particular neuronal populations in the diencephalon (lateral geniculate body). This information is filtered in the visual cortex by

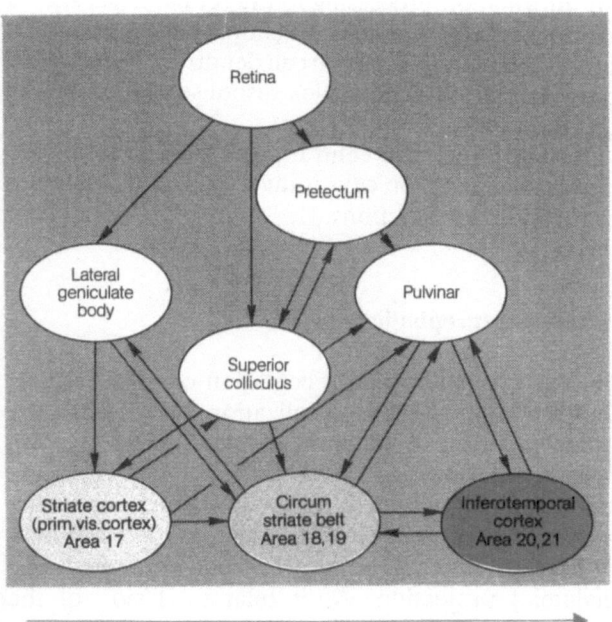

Complexity of visual information processing

Fig. 78. Simplified representation of subcortical *(black)* and cortical structures *(red)* participating in visual pattern recognition in cat. (Projections from dorsal geniculate body to circum striate belt were not found in monkeys). Increase of complexity of visual information processing at cortical level is indicated by the *red arrow (bottom)*

specific intrinsic neuronal circuits as well as by connections with thalamic nuclei. Such processes are believed to form the basis of pattern recognition. In areas 17, 18, and 19, three main types of neurons have been identified which, according to their response characteristics, are called "simple", "complex", and "hypercomplex" cells [73]. In cats, for instance, simple cells receive inputs from retinal X-fibers via particular neurons of the dorsal lateral geniculate body whereas Y-fibers project via other cells of the dorsolateral geniculate to the complex cells.

The diagram of Figure 78 shows a highly simplified representation of brain areas that are involved in visual data processing in cats. The retina is mapped in all of these cortical and subcortical nuclei – partly with different field magnification according to particular neural functions. At the cortical level the extent of configurational pattern evaluation ranges from (1) relatively "simple" coding of the orientation and length of light-dark borders *(striate cortex)* via (2) "hypercomplex" transformation of angles formed by borders moving in definite directions *(circum striate belt)* to (3) association of visual signals to learned (acquired) classes of functional significance *(inferior temporal lobe)*. Neurons of the inferior temporal lobe of monkeys show highly complex response properties. They appear to extract particular configurational features of behaviorally relevant stimuli, such as those, for example, associated with a monkey's hand or a face, respectively [74]. Thus, various characteristics of a visual stimulus are possibly "represented" by the activity in different classes of cortical neurons. However, nothing is known of how this activity is "read" and presumably transformed into efferent "commands". There is evidence that efferent fibers of the visual cortex project to various subcortical motor systems.

Recently the response characteristics of simple and complex cells of area 17 have been tested in situations where a structured background (two-dimensional visual noise) was superimposed on the visual object (signal, e.g., a bright bar): (1) The object was moved together with the background simultaneously in the same direction at constant angular velocity (8°/s), (2) the object was moved over a stationary background. The results are summarized in Table 7. The three cell types investigated in cat's visual cortex may function as follows: *Simple cells* extract the form of an object from its background after onset of a "tracking" eye movement; *complex cells (type I)* determine the velocity of the background including the object after measuring the spectral energy of the whole stimulus; *complex cells (type II)* perceive a moving object against a stationary background. Of course, other areas of the visual cortex are involved in the complete analysis of the form. The pattern recognition system must also satisfy certain invariance conditions concerning stimulus intensity, velocity, and position. In cats neurons of the Clare-Bishop area (close to area 19), may participate in these highly complex operations.

Lesion experiments in combination with quantitative behavioral tests reveal further insights into the functions of visual cortical neurons. After bilateral ablation of area 17 cats show deficits in pattern discrimination tasks, which may be partially compensated by relearning in the course of time. However, compared with controls, the operated cats show long-lasting (permanent) deficits in test situations which require

Table 7. Influence of a structured background on the response of simple and complex cells in the visual cortex of cats [K.-P. Hoffmann, W. v. Seelen: Biol. Cybernetics *31*, 175–185 (1978); W. v. Seelen, H. Heitländer, Tutzing, 1979, pers. comm.]

Cortical neuron in area 17 (cats)	Object and background are moving simultaneously	Object moves on a stationary background
Simple cells	Response to object, no response to background	No response at all
Complex cells (Type I)	Response to background (including the object)	Response to object; activity drops with increasing noise level of the background
Complex cells (Type II)	Response to object and to background	Response to object, no response to background

identification of an object against a structured background, i.e., a signal that is masked by noise. If the lesion extends also to areas 18 and 19 (and Clare-Bishop area in cats), visual pattern discrimination is abolished. Bilateral ablations of the inferior temporal lobe (sparing areas 17, 18 and 19) lead to "psychic blindness". Animals suffer from a deficit in visual pattern discrimination. However, they fail to link visual stimuli to corresponding categories of meaning [75].

5.3.4 The Concept of "Two Visual Systems"

In some mammalian species the cortical identification system may operate, as it were, "independent" of the subcortical localization system (Fig. 77, right). After elimination of the superior colliculus the Syrian hamster, for instance, is no longer able to localize visual stimuli in space by turning toward them; but it still can distinguish them from each other. If, conversely, the visual cortex is eliminated and the superior colliculus is left intact, the hamster behaves in a manner reminiscent of the thalamic/pretectal lesioned toad; although the signal is localized in space, it is no longer recognized.

The results from ablation studies of the hamster led to the concept of "two visual systems" [76] which is often taken to mean that completely separate neural pathways are involved in discerning *where* and *what* a visual stimulus is. Anatomical and physiological studies, however, indicate many interconnections between the neural structures involved in these functions. Also, it seems clear that certain aspects of stimulus localization are affected not only by lesion of the midbrain tectum, but also by neocortical lesions in hamsters and Mongolian gerbils as well as in cats, monkeys, and man. The two-systems dissociation is clearest when we consider the output side of the visual system: control of turning movements of the head may be quite separate from control of starting and stopping of simple approach movements (Fig. 77). Rats and gerbils [77] with complete bilateral lesions of the superior colliculus can learn to locomote directly toward visual targets. However, such animals will not show oriented head turns toward novel visual stimuli presented in the peripheral visual field. Therefore, the orientation behavior mediated by the superior colliculus appears to be mainly one of head turns toward peripheral stimuli rather than the visual control of locomotion toward visual targets. But these are only two aspects of visual-motor control; other responses may also depend on distinct neural subsystems.

5.4 Formation and Deterioration of the Visual System: Environmental Effects

5.4.1 Early Experience

Experimental observations in young cats and monkeys suggest that the complex connections for visual orientation tasks already exist at birth as "basic wiring". The development of the visual cortex is, however, incomplete.

To what extent does the environment affect the morphology and physiology of the brain?

The effects of visual isolation may be studied in deprivation experiments. Thus, in mammals and birds deprived of visual experiences during development, definite morphological changes are found in neurons of the cerebral cortex (visual cortex) as well as in neurons of the lateral geniculate body. Cells are smaller than normal; thickness and vascularization of the cortex, as well as the development of dendritic spines, may also be reduced. Modifications induced by visual deprivation may affect the neurophysiological response characteristics as well as visually guided behavior. Here are some examples.

5.4.2 Monocular Deprivation

In kittens the time of 4–5 weeks after birth is a very sensitive phase of "visual development"; at an earlier time the eyes are still closed. A comparable sensitive phase in monkeys starts after birth and lasts 8 weeks. When in kittens one eye is closed by suturing the lid immediately after birth (or within the sensitive phase), reopened about 3 months later and, after suturing the lid of the other eye, the animals are now compelled to manage in the visual world with the "untrained" eye, strong and partly irreversible deficits are manifested in their visually

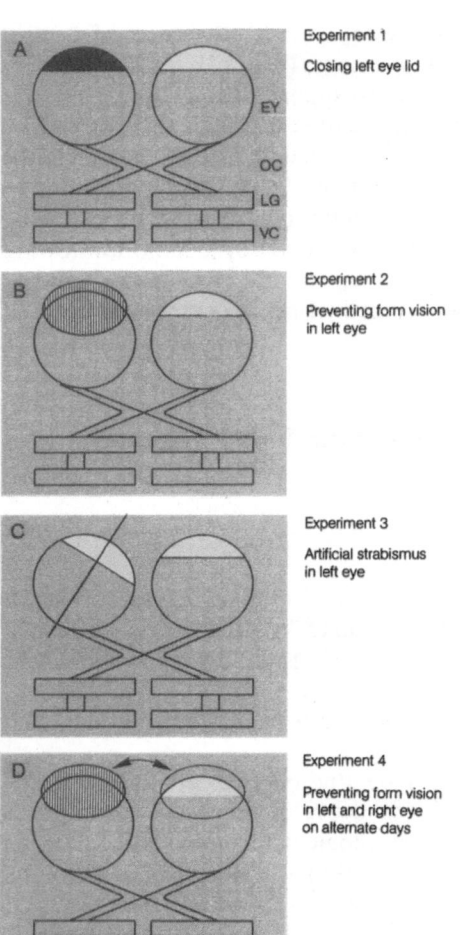

Experiment 1

Closing left eye lid

Experiment 2

Preventing form vision
in left eye

Experiment 3

Artificial strabismus
in left eye

Experiment 4

Preventing form vision
in left and right eye
on alternate days

Fig. 79 A–D. Influence of various kinds of "monocular experience" on the development of neurons in the cat visual cortex. *EY* eye, *OC* optic chiasm, *LG* dorsolateral geniculate body, *VC* visual cortex. Explanations in text. [Results of four different experiments carried out by Wiesel TN, Hubel DH (1963) J Neurophysiol 26: 1003–1017 (1, 2). Hubel DH, Wiesel TN (1965) J Neurophysiol 28: 1041–1059 (3, 4)]

guided behavior (Fig. 79 A). Cats can still locate objects with their monocularly deprived eye in the monocular region of the visual field. However, they show a marked deficit toward the midline and almost total failure in the binocular region contralateral to the deprived eye. The animals learn quickly with the deprived eye to distinguish visual objects, such as horizontally and vertically oriented stripes. However, if the patterns have to be detected on a structured background, or say, the signal is masked by noise, the performance is three times better with the normal eye than with the deprived eye [78].

If tasks require specific visuomotor coordination, the cats no longer perform adequately. They collide with objects and fall over the edges of tables.

Changes are also shown in the response characteristics of neurons in the superior colliculus. Whereas numerous neurons of the superior colliculus normally respond selectively to stimuli moving in certain directions, this directional specificity is now lost. Moreover, only the neurons which had been innervated from the previously closed eye exhibit this deficit. These cells behave now in a way similar to what is seen after lesions of the ipsilateral visual cortex.

Whereas the responses of the retinal ganglion cells appear to be normal (with respect to the stimulus parameters that are tested here) visual acuity in lamina A_1, of the lateral geniculate body, is reduced to half the normal value. There are strong alterations in the responses of cells of the visual cortex. Only about 6% of the neurons can be activated by stimulation of the deprived eye and almost all show abnormal receptive fields. The ability to extract the stimulus from a structured background is abolished or strongly reduced. During the sensitive phase deprivation of 3–4 days may seriously decrease the *binocularity* of cortical cells. In an adult cat, however, even eye closure over one year does not lead to a comparable syndrome.

We might suggest that some basic information transfer from the eye directly to the midbrain is innate, while pathways leading from the eye to the visual cortex via lateral geniculate neurons have to be acquired with regard to certain processing operations. These latter circuits are stunted when the opportunity of modification by experience is withdrawn.

What can be "learned" here?

5.4.3 Binocular Coincidence

Is the crucial aspect of the deprivation for instance, the brightness? Kittens raised with a plastic occluder (instead of suturing the eye, Fig. 79 B) exhibited the same deficiencies. Obviously the *binocular interaction* of form vision is essential. In another experiment (Fig. 79 C) artificial squint was produced in kittens during the period of highest susceptibility by cutting the extraocular muscles. Under these conditions the images on the two retinae failed to superimpose. Result: cells of the visual cortex showed relatively normal receptive fields, however, only a few of the neurons could be activated by both eyes. Thus, some of the cells exhibit a strong "monocular dominance" for the contralateral and others for inputs from the ipsilateral eye. In a further experiment (Fig. 79 D) the eyes of the kittens were provided with a plastic occluder which was switched from one eye to the other on alternate days. Here *both* eyes received the same information – but not simultaneously. The neurophysiological results were similar to those of the squint experiments. We may conclude that precise coincidence of adequate visual input in space and time domains during certain periods of early life is essential for the development of normal synaptic interactions.

There is some evidence that the locus coeruleus of the brain stem and its noradrenergic projections make an important contribution to the high level of plasticity of the visual cortex in the critical period [79]. Interestingly, the number of neurons in this nucleus declines with age, an uncommon phenomenon in brain stem nuclei. Furthermore, cortical plasticity might be turned off by interaction of the serotonergic projections from the raphe nuclei.

5.4.4 Influence of the Environment

What happens, if normal kittens are raised in different monotonously structured environments? We have already mentioned the cortical simple cells (Fig. 71 A). They respond to a stripe pattern, whose longitudinal axis is oriented parallel to the main axis of their longitudinal receptive fields. Under normal conditions "orientation detectors" with horizontally and vertically oriented receptive field axes are equally abundant. However, this equal ratio may depend upon the visual environment in which the kittens have been raised. If the environment consists simply of vertically striped walls (Fig. 80 A) the cortical simple cells are mainly activated by vertical bars. Correspondingly, simple cells are predominantly activated by horizontal stimuli if kittens are raised in horizontally striped environment (Fig. 80 B).

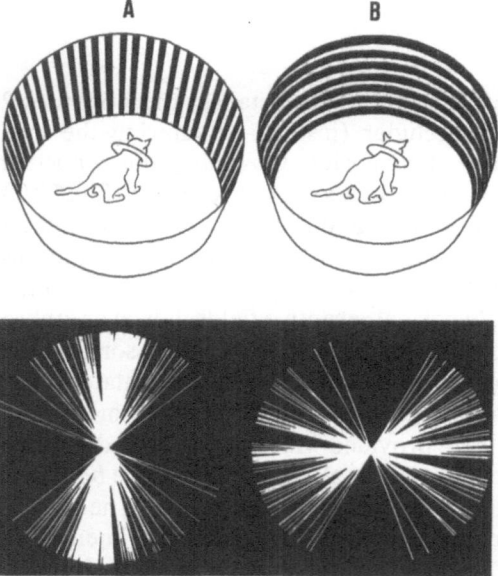

Fig. 80 A and B. Influence of striped environments *(above)* on the main axis orientation of the receptive fields *(below)* of "orientation detectors" during development of cat visual cortex. *Left,* vertically striped environment (72 neurons tested); *right,* horizontally striped environment (52 neurons tested). Further details in text. [Modified from Blakemore C, Cooper GF (1970) Nature (London) 228: 477–478]

In another kind of experiment kittens were fitted with a pair of particular goggles [80]: one eye saw three horizontal bars and the other, vertical bars. Animals spent one part of the day wearing the goggles and the other part in the dark. In the course of time cortical neurons lost their binocularity; furthermore, simple cells could be activated from one eye mainly by those bar stimuli that had an orientation corresponding to the stripes of the goggle.

How can these phenomena be explained? Various possibilities might be considered. (1) Neurons have changed their orientational sensitivity. (2) Cells with "open coding capacities" become committed to the features of the environment. (3) Only those "orientation detectors" which received the *appropriate* information from the environment – during early life – can later be activated by corresponding stimuli. (4) Alternatively, cells with other orientation preferences may not have developed or are silent. Thus, experience may represent a kind of "functional validation". However, there is a great deal of controversy about the plasticity of the visual system and the mechanisms by which the environment shapes the visual system. Further studies are required to answer these questions [82].

5.4.5 Sensori-Motor Interactions

We have briefly considered sensory deprivation, but it must be emphasized that adaptation to the environment may develop only by close feedback interactions between sensory and motor functions. Cats raised in darkness, for instance, do not learn to orient themselves later in a structured environment, if they are merely carried about. They learn to see only when they move about by themselves [81].

The *interaction* of sensory and motor functions is important here. We know of comparable results from experimental human psychology; if a human subject is transported, as it were, passively into a structured environment while wearing prism glasses, he does not learn to compensate for the visual distortion produced by the prism. Only when he stands and moves on his own feet can he adapt quickly to the changed perceptive conditions [81].

In summary we may say that the sensori-motor interfacing with the environment including social interactions, as well as training and learning processes, participate in the development of the cortex in the growing mammal. Rats raised in a "rich environment" (with toys, running wheels, ladders, social partners, etc.) compared with those raised in a "poor environment" (animal alone in small dim cage) showed statistically significant differences in (1) cortical weight, (2) cerebral capillary supply, (3) number of glial cells, (4) metabolic rate, (5)

ACh-esterase activity, and (6) development of dendritic structures (spine synapses). The values were higher in the animals from the "rich" environment [83]. The earlier the deprivation occurs, the less reparable it is. The interesting fact is that the phylogenetically older subcortical brain structures – with the exception of the hippocampus – remain, as far as have been tested, practically unchanged by this kind of sensory deprivation.

Let us recapitulate:

1. There are parts of the visual system of vertebrates whose specific behavioral functions (localization, space constancy) have developed at an early stage of evolutionary history.

2. Recognition systems for visual signals obviously developed first at subcortical levels in networks connected with structures relevant for signal localization: the optic tectum. This brain structure thus has sensory as well as motor functions.

3. Given the importance of signal evaluation for the evolution of behavior, particular structures have been developed in mammals for processing of sensory information, such as the visual cortex [84].

4. Certain brain areas become functionally differentiated only during sensori-motor interaction with the environment. There are, in the visual system of vertebrates, information-processing pathways which are innate ("hardwired" systems), but also evidently others which must be acquired ("softwired" systems).

Here, especially, certain "sensitive phases" during ontogeny (early life) may play a role, perhaps comparable in some way to the phenomenon of imprinting investigated by Konrad Lorenz in greylag geese [84].

6 Examples for Release and Control of Behavior Patterns by Other Sensory Systems

6.1 Olfactory Sense: Scent Coding in Insects

In invertebrates, as well as in vertebrates, there are innate behavior programs and corresponding innate recognition systems for behaviorally important signals (key-stimuli). Animals have evolved different ways of dealing with environmental conditions. As a generalization, "hard-wired" systems may work very well, especially in simple animals that live in relatively stable and predictable environments. For a species to exist in a variety of environments or in very unpredictable ones neuronal systems capable of responding to a wider variety of sensory stimuli may be necessary. We find especially clear examples of animals with *narrowband* sensory filters ("specialists") and *wideband* sensory filters ("generalists") in the olfactory domain of insects.

The ability to correlate a scent with certain innate and acquired classes of meaning is widespread in the animal kingdom. Its neurophysiological basis has been particularly well studied in insects for several reasons: (1) We know of numerous scents affecting behavior in insects. They may originate from food (grass, carrion), from a partner (e.g., sex-attractant substances), or they may coordinate the social hierarchy in the community as pheromones do in socially organized insects: the laying down of track marks (scent marks), calling for assembly, alarming against enemies, signalling casualties, etc. (2) The chemical structures of many behaviorally relevant odorous substances are known. They can be synthesized in original or modified configurations and may be used in the laboratory as dummies for the investigation of releasing cues. (3) The receptors are morphologically well studied and are relatively easily accessible for electrophysiological recordings.

The olfactory receptors of insects are located on the antennae. The basic unit is the sensillum; it consists of several primary sensory cells (olfactory receptors) whose dendrites are located inside hair-like formations of the cuticle (scent hairs). There are several types of sensilla (e.g., Fig. 81 A and B). Their common characteristic is a "pore-tubular" system in the cuticle that connects the liquor filled hair lumen, containing the antennal receptor dendrites, with the exterior.

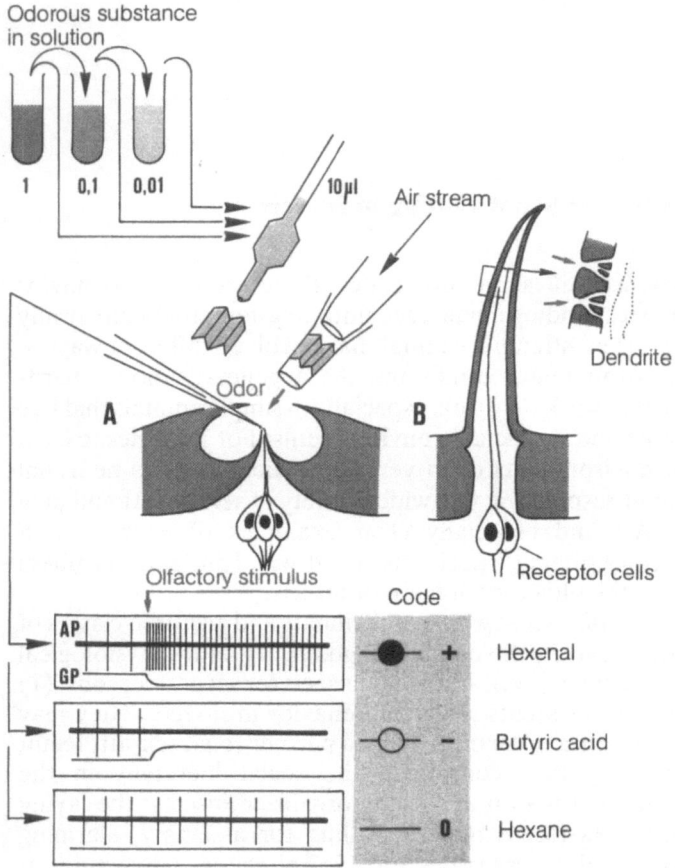

Fig. 81 A and B. Neurophysiological procedure for studying scent coding in insects. **A** Sensillum coelonicum of *Locusta*. **B** Sensillum basiconicum of *Calliphora*. The enlarged section *above right* shows the pores in the cuticle of the sensillum with their branched channels by which the scent molecules reach the dendritic structures of the primary sensory cells. *Below,* Examples of the "ternary scent code". *AP* Action potentials, *GP* generator potential. Further details in text. [Modified and combined from Schneider D (1967) Naturwiss Rundsch 20: 319–325. Boeckh J (1967) Z Vergl Physiol 55: 378–406]

6.1.1 The "Scent Code"

The mechanism by which the dendritic membrane of an olfactory antennal receptor is excited by an odor molecule is still unknown. The first neurophysiologically measurable response to a stimulus is the generator potential which is coded in a sequence of action potentials at the trigger zone of the receptor cell (Fig. 81 A). In an "neutral-odor" environment a weak permanent activity of one impulse per 10–20 s is recorded from an antennal receptor. If an odor is now blown over the hair sensillum, one of three different effects may be observed: (1) the discharge rate of the receptor can increase, (2) it can decrease, (3) it can remain unaffected. The outcome depends on the odor molecule.

Let us therefore keep in mind that the receptor cells of the olfactory sensilla are able to encipher a chemical compound according to a plus/minus/null-code.

6.1.2 Scent "Specialists" and "Generalists"

If we collate the responses of individual receptor cells to various odors we arrive at an important result (Fig. 82). In the migratory locust two

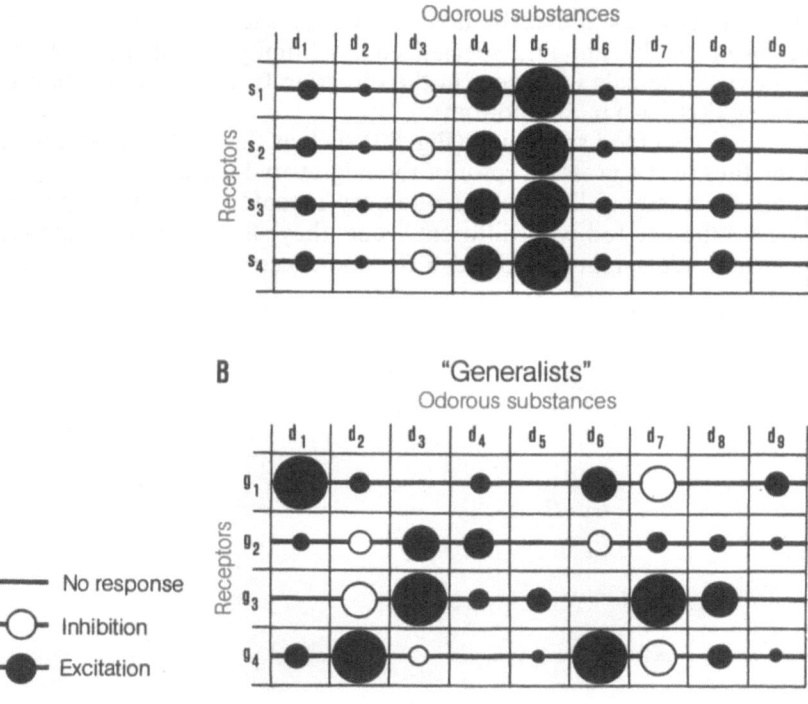

Fig. 82 A and B. Principles of scent coding by two different sensory cell types – "specialists" (s_{1-4}) and "generalists" (g_{1-4}) from sensillum coelonicum of the locust antenna. The strength of a receptor response (excitation or inhibition) to different odorous substances (d_{1-9}) is illustrated by the *size of the circle*. Further details in text. [Schematized from Schneider D (1967) Naturwiss Rundsch 20: 319–325; cf. also Kaissling KE (1971) In: Beidler LM (ed) Handbook of sensory physiol, vol IV/1. Springer Berlin Heidelberg New York]

prominent different reaction types appear to exist among the olfactory receptors. Representatives of one group show nearly the same response modes to a selection of various scents; the response threshold to a specific, behaviorally significant odor is particularly low (Fig. 82 A, d_5). We might call such receptors with *narrowband* response spectra "specialists". But we must keep in mind that these cells are *not exclusively* responsive to one specific odorous substance. On the antenna of the male silk moth *Bombyx mori* about 70% of the receptors are specialized in the way described for the sex-attractant bombycol, which is an unsaturated fatty alcohol (trans-10-cis-12-hexadecadien-1-ol). The sensitivity of the receptor cell is so great that a single attractant molecule is sufficient to induce an action potential in the corresponding axon; two impulses are generated when two molecules impinge on a receptor. Of course the transduction of molecule hits into an impulse *train* is more complex. The demonstration of such a "single hit reaction" has been obtained by means of tritium (^3H) labeled bombycol. About 200 simultaneous molecule hits at 200 of the more than 25,000 receptors on the antenna specialized for bombycol tell the male of *Bombyx:* "a female is present in the direction from which the wind is blowing" (cf. also p. 206 bottom).

There are also food specialists (Fig. 83) such as the carrion receptor of many flies or the grass receptor of the migratory locust.

Apart from "specialists" other olfactory receptor cells are regularly found. When tested with different scents they show a broad response spectrum. We might call these *wideband* receptors or "generalists".

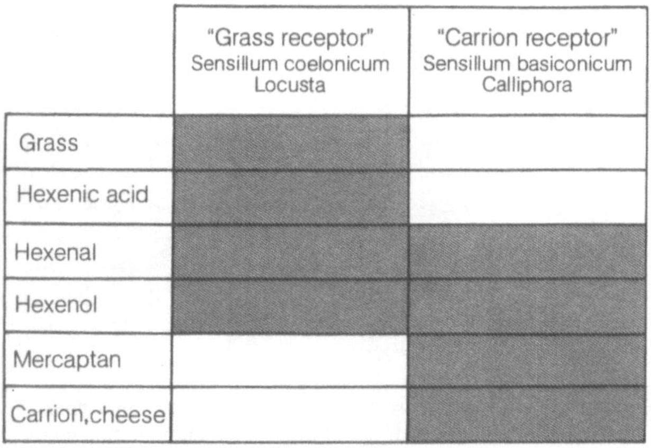

	"Grass receptor" Sensillum coelonicum Locusta	"Carrion receptor" Sensillum basiconicum Calliphora
Grass	▓	
Hexenic acid	▓	
Hexenal	▓	▓
Hexenol	▓	▓
Mercaptan		▓
Carrion,cheese		▓

Fig. 83. "Food specialists" on the antenna of *Locusta* and *Calliphora.* Odorous substances responded to are shown by *grey panels.* The responses of the two specialist types may overlap (see hexenal and hexenol). [From Schneider D (1967) Naturwiss Rundsch 20: 319–325]

Among these receptors, too, there are manifestly different types with constant response characteristics; some of them are illustrated in Figure 82 B. The response spectra of representatives of the different types strongly overlap. There are also mechano- and temperature-sensitive receptors. Certain postsynaptic neurons obviously receive the convergent inputs related with different sense modalities (multimodal sensory convergence).

6.1.3 Effective Components of Scent Molecules

The search for the active components of a chemical molecule may be illustrated with the example of the grass receptor of the migratory locust. The natural key-stimulus is the smell of fresh grass. The chemically active component is hexenol and hexenal contained therein. The chemical specificity of such a molecule can be pinpointed by various modifications. Their efficacy depends on different components (Fig. 84): (1) The chain length of the aliphatic hydrocarbon molecule: C_6-compounds are optimal, aromatic compounds are not effective. (2) Unsaturated bonds in the C_6-chain: hexan and hexene do not elicit a receptor response; hexyne-1, however, is effective. (3) Functional groups: introduction of an hydroxy or aldehyde group into the hexene molecule "activates" the substance. In saturated compounds these

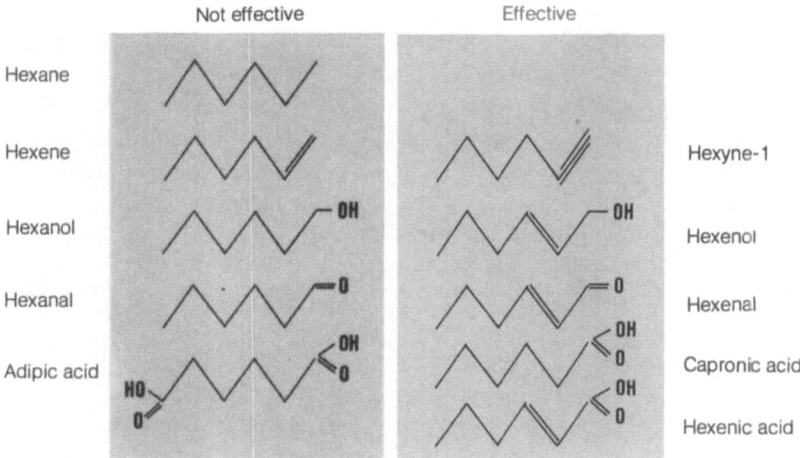

Fig. 84. The effect of unsaturated bonds and functional groups in hydrocarbon molecules on the activation of grass scent receptors (sensillum coelonicum) in the locust. The H-atoms at the C-atoms *(represented by an edge)* are not depicted in this diagram. Further details in text. [Modified from Boeckh J (1967) Z Vergl Physiol 55: 378–406; cf. also Kafka W A (1970) Z Vergl Physiol 70: 105–143]

groups have no effects, but a terminal carboxyl group activates hexane. (4) Asymmetries: substitutions at both terminal C-atoms (e.g., dicarboxylic acids) make the molecula ineffective.

In summary we may say that none of the factors mentioned is decisive by itself. Evidently the interaction of several factors is required. Other experiments suggest that the dendritic membrane has excitatory as well as inhibitory acceptor areas, but it is not clear at present how the odor molecule fits the acceptor structure of the dendritic membrane. In analogy with enzyme reactions an odor molecule may perhaps be classified on the basis of its configuration (substrate specificity). The molecule may be classified *and* recognized within the functional groups on the basis of its electron distribution and electron mobility (action specificity). The acceptor, functioning as gating mechanism, may then – according to its binding properties with the odor molecule – trigger a process that changes the ionic conductivity of the dendritic membrane in the direction of depolarization or hyperpolarization. Following its action the scent molecule becomes inactivated.

Finally let us keep some important points in mind:

1. Already at the olfactory receptor level, insects possess systems which are activated predominantly by behaviorally relevant scent signals. They might be called "specialists". The food specialists, together with the postsynaptically connected neuronal populations, may form the IRM for feeding. Thus, important components of an IRM appear to be determined here by certain properties of the dendritic receptor membrane.

2. Specialists are not optimally suited for the versatile coding of numerous different odors. For this purpose the "generalists" and the downstream neuron populations are available and form presumably a substrate for an IRME or an ARM (cf. p. 58). With a few receptors of this type coding of thousands of different odors would be possible by the plus/minus/null-code. Here, again, it is by no means clear how such a code is read in the brain.

3. We encounter similar problems of recognition-specialization and generalization in various neural and corresponding behavioral levels of integration, in vertebrates as well as in invertebrates. Even in the insects the concept of odor specialists and generalists at receptor level does not hold for all species. In cockroaches *Periplaneta americana*, for example, generalist properties occur in neurons that are postsynaptic to the receptors. Their response characteristic might be determined by "quality-convergence" of narrowband olfactory receptor cells.

6.2 Vibration Sense: Prey-Catching of the Back Swimmer

Light-reflecting water surfaces may act as a trap for some flying insects by causing them to lose their orientation and fall into the water. There are preying insects specialized for this ecological niche.

The back swimmer *Notonecta glauca* hangs by its extremities on the underside of the water surface and waits to ambush prey animals falling into the water. Its sensory system, just like that of the toad, must perform two important operations during prey-catching that is released by signals

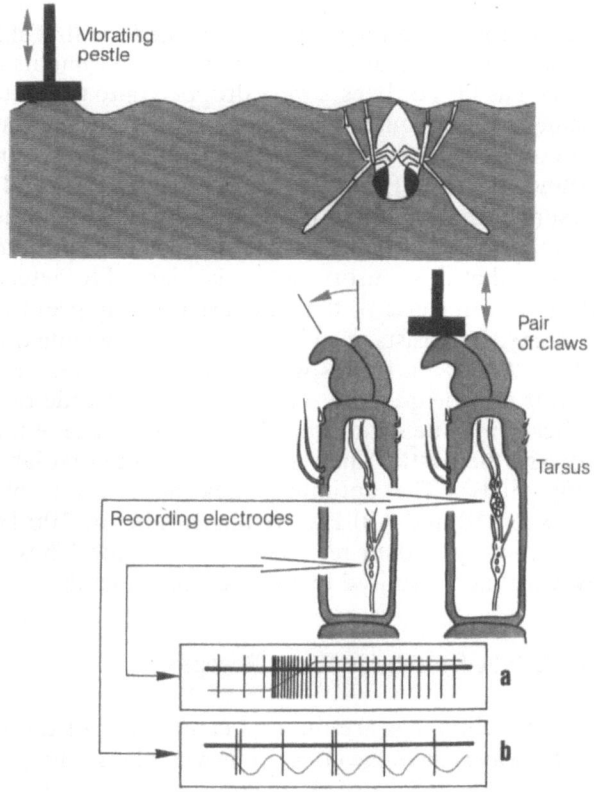

Fig. 85. Prey-catching of the back swimmer *Notonecta glauca. Above,* principle of experimental set-up for generating waves on the water surface. *Below, a* phasic-tonic receptor responses of the proximal scoloparium after bending of the pair of claws (see arrow). *b* Phasic receptor responses of the distal scoloparium to vibration stimuli after direct coupling of the vibrating piston with a claw. [Modified and combined from Markl H, Wiese K (1969) Z Vergl Physiol 62: 413–420. Wiese K (1972) J Comp Physiol 78: 83–102. Wiese K, Schmidt K (1974) Z Morphol Tiere 79: 47–63]

originating from the stimulating object: it must be able to localize and identify them.

The back swimmer uses for localization and identification not visual stimuli but a signal contained in the circular waves produced on the water surface by the struggling prey. The initial components of prey catching consist of the orienting reaction toward the origin of the waves (prey); next follows the swimming approach. From the whole functional process we extract three questions: (1) Which components of the surface wave signify prey for the back swimmer? (2) Where is the signal received and how is it evaluated? (3) How is the prey signal localized?

6.2.1 Key-Stimuli

Laboratory experiments have furnished the first characterization of the behaviorally relevant signals. In these experiments natural prey animals – such as flies or bees – were dropped onto the water surface and at the same time a frequency spectrum of the resulting waves was recorded by photoelectric methods (Fig. 86 A). Maximal wave amplitudes are always found in a low frequency range between 10 and 100 Hz. It can be assumed that the relationship between the amplitude and frequency of water waves is a significant feature of the information about the prey stimuli that this sensory system is adapted to detect.

The natural cause of the waves (struggling prey) is now replaced by a dummy. It consists of a vibrating pestle touching the water surface from above (Fig. 85). In this way the frequency and amplitude of the water waves may be quantitatively varied over a wide range and the stimulus efficacy may be determined by the percentage of the animals swimming toward the pestle. The results confirm the first laboratory experiments (Fig. 86 B). The optimum releasing effect is observed with waves between 10 and 250 Hz, frequencies below 100 Hz having a clear cut effect owing to their relatively low damping. These are just the kinds of waves that are caused by natural prey animals.

6.2.2 The Search for Wave Receptors

The paddling legs are the first candidates for the site of receptors: the water waves have the effect of moving the animal, together with its tarsi, toward the surface of the water by the updraft acting on the insect's body. The prey-catching behavior is in fact strongly impaired if a fine insulating tube is pulled over claw, tarsus, and tibia of the fore and middle legs. A mechano-receptor called scolopidial organ is localized in the tarsi (Fig. 85).

Scolopidial organs contain receptors that respond to mechanical stretching. They are widespread among insects. In principle they may

Fig. 86 A–C. Relationship between key-stimulus **(A)**, behavioral threshold **(B)** and receptor sensitivity **(C)** during prey-catching in the back swimmer *Notonecta glauca*. [Modified from Wiese K (1972) J Comp Physiol 78: 83–102]

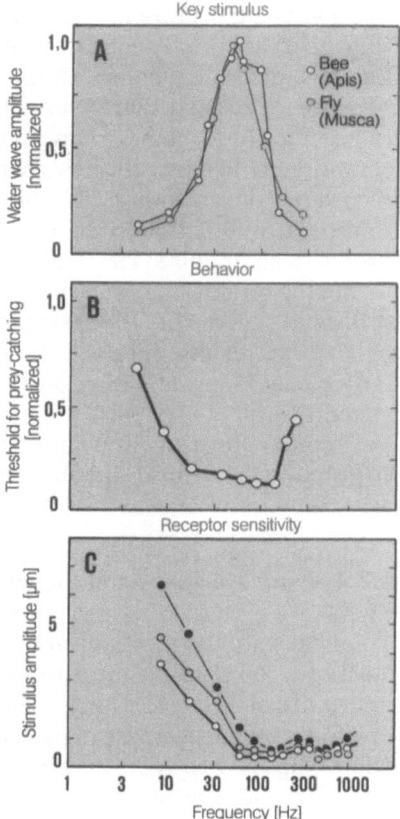

carry out different functions, e.g., positional and posture reception, but they may also serve for the perception of vibration or sound.

6.2.3 Signal Filtering

It might be assumed that the water surface waves result in a bending of the claw at the tarsus, and thus excite the stretch receptors in the scoloparium. This assumption may be tested electrophysiologically in a "dry experiment". In a *Notonecta* leg removed from the animal, a microelectrode is carefully inserted through a hole near a receptor cell (Fig. 85). The claw can now be moved in relation to the fixed tarsus with the aid of an angle-measuring instrument, and the receptor responses may be recorded on an oscilloscope.

Evidently there are different types of receptor: some (Fig. 85 a) indicate the angular velocity of the movement as well as the position of the pairs 'of claws and therefore have phasic-tonic response characteristics; others (Fig. 85 b) are briefly activated only if a single claw is moved and that by flexion (phasic response characteristic). As is well known, phasic receptors are particularly suitable for the detection of surface waves (e.g., vibration signals). This may be demonstrated by connecting the claw directly to the artificial wave producer – i.e., a vibrating pestle – and recording the response of the phasic receptors (Fig. 85 b). With sinusoidal stimulation in the range of 60 Hz each vibration (sine-wave) produces an impulse. The sensitivity range of the receptors therefore lies within the range of the wave frequencies produced by prey objects. If the threshold curve of the receptor response is compared with the behavioral results (Fig. 86 B and C), differences appear at wave frequencies below 60 Hz, suggesting the involvement of further receptor systems.

In another investigation the experimental arrangement has been adjusted to the natural stimulus conditions. Here, too, a clear correlation was found between water wave frequency and receptor discharge rate.

6.2.4 Basic Mechanism of Signal Localization

To investigate the mechanism by which the source of the water waves is localized, comparative measurements between the left and the right pair of legs are of interest. What is measured thereby? Two possibilities come to mind: (1) The decrease of the wave amplitude from the left to the right pair of legs or vice versa may be measured by the claw deviation in each case. The differences are, however, too small (below 0.1 μm). (2) The time difference $t_i - t_j$ between the waves reaching the left and the right leg may serve as a measure of the angle by which the body must be turned so as to swim toward the wave producer.

Before testing the second possibility, we shall first determine the accuracy of the orienting reactions. For this purpose the turning angle α_t is measured for various initial body positions α by photographs: with wave frequencies between 20 and 60 Hz a deviation $|\alpha_t - \alpha|$ of 2° to 18° is found in the range $0° \leqq \alpha \leqq 180°$. The localization is therefore relatively accurate even in the first turning. Possible localization errors may be compensated for by a subsequent correction.

Since the turning reaction of the back swimmer is very rapid and is carried out "open-loop", without peripheral control, it may be assumed that, for different time differences $t_i - t_j$, corresponding programs are present in the central nervous system as "movement commands" for the appropriate turns.

6.2.5 Prey Localization in Toad and Back Swimmer

We may recall that the *telotactic* turning reaction of the toad is triggered according to the *position difference* x−w of a pattern w projected on the retina outside the fixation point x. In its brain as well, corresponding motor programs for the turning movements must be represented. After an errorless turn, x−w = 0. Such position differences may be simulated experimentally by electrical stimulation of corresponding retinal projection fields in the brain (optic tectum).

The *tropotactic* localization mechanism of the back swimmer requires paired sensory organs. Here the phase relationship between sense organs of a multireceptor system has an important function for prey localization. The orienting turn in the stimulus field is obviously controlled by the *time difference* $t_i−t_j$ in the two sensory organs. After an errorless turn, $t_i−t_j = 0$.

In a simulation experiment we may now ask if the back swimmer shows corresponding orientation responses when the tarsal claws of the two

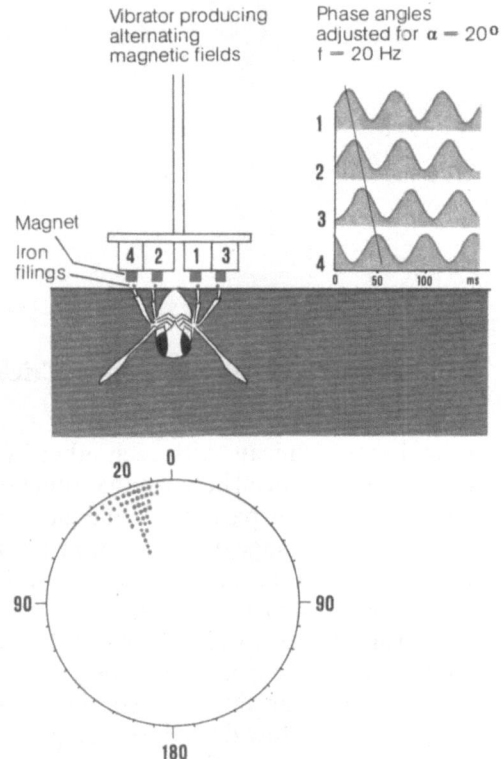

Fig. 87. Search for the principle of prey localization in the back swimmer *Notonecta*. Experimental set-up for the simulation of water wave stimuli by local alternating magnetic fields *(1–4)*. The four magnets are positioned immediately above the four claws of the animals fore- and middle-legs; each claw is covered with glued-on iron filings *(red)*. Example: A phase delay of $t_1−t_2 = 8$ ms set between stimulators above receptors *1* and *2*, $t_1−t_3 = 7.5$ ms between receptors *1* and *3*, and $t_3−t_4 = 16$ ms between receptors *3* and *4* is followed by a turning reaction of the animal with a turning angle of 19.2° ± 5.5° *(below)*. The setting of the phase shifters was adjusted to the time pattern corresponding to a wave approaching at an angle $\alpha = 20°$. The wave frequency was 20 Hz. The turning angles (*red dots* in circle diagram) were determined from photographs taken before and after a reaction. [Modified from Wiese K (1974) J Comp Physiol 92: 317–325]

pairs of legs are diverted one after the other by varying time differences. Iron filings are glued to the claws. By means of magnets fixed above the water surface the claws can be moved, first one, then the other, according to the alternation of magnetic fields (Fig. 87, above). For appropriate delay periods the animal indeed shows corresponding orientation reactions whose margin of error is entirely within the range of responses elicited in natural stimulus situations (Fig. 87, below).

We may summarize briefly:

1. The prey-catching behavior of the back swimmer *Notonecta* provides a clear analytical model for neurobiological studies. In comparison with the prey-catching behavior of the toad, both insect and vertebrate are confronted by basically similar demands. They are, however, fulfilled by completely different mechanisms in the two sensory systems.

2. The prey recognition system of the back swimmer has a much simpler organization. The localization system does not function in response to topical position differences, but responds to time differences. This localization system represents the most primitive precursor of a mechanism which is found in principle in the auditory system of even the highest vertebrates where it has been refined to a greater extent.

3. A common postulate for the prey-catching behavior of the back swimmer and the toad is fixed motor programs for the turning reaction which can be called into play by appropriate stimulus situations. We shall return later to the subject of *fixed action patterns* (FAP) and *command systems* (p. 203).

6.3 Acoustic Communication in Crickets and Frogs

Animals communicate with each other by signals. Communication as a characteristic of social behavior has functions in various contexts, such as attracting a sexual partner, demarcating a territory, and defense. The basis for the exchange of information is a signal generating system which has a corresponding receiver system in the partner that evaluates and recognizes the signal and releases an appropriate behavioral reaction. The term species-specific releaser has already been mentioned. It includes various sensory modalities.

Three essential questions are especially relevant to neuroethology in this connection: (1) How does the signal generating system work? (2) What makes receptors specific for certain modalities and which neuronal processes serve to evaluate the signal? How is the resulting neuronal activity processed for recognition and localization? (3) Which processing steps finally lead to the behavioral reaction? Complete answers to all three questions have not been obtained so far in any sensory communication system. It has, however, been possible to obtain important insight into a series of fundamental processes. Two examples from acoustic communication in terrestrial animals, a domain where the signals are transmitted through air, will indicate some experimental approaches to neuroethological questions.

6.3.1 Cricket Song

When a male cricket or grasshopper is motivated to mate, it produces a species-specific calling song. Receptive female grasshoppers answer the song, thus producing a duet. The sexual partners find each other phonotactically by mutual localization of the sound producer. That the acoustical signal is decisive, even within visual range, is shown by the following experiment. To the left of a female cricket we present a singing male cricket – acoustically insulated under a bell-glass; to her right, a loudspeaker emits a tape recording of the calling song. Result: the female approaches the loudspeaker.

a) Motor Control of the "Transmitter"

How is the Song Produced? The requisite "instruments" consist of a stridulating edge pulled over a serrated stridulating ridge (file). Crickets chirp, i.e., stridulate, in this manner by rubbing their forewings together

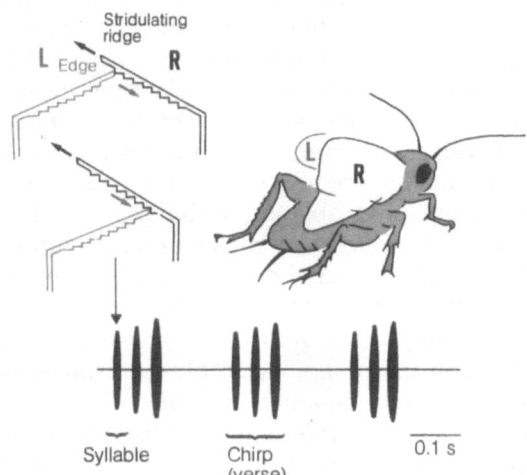

Fig. 88. Cricket song: a wing edge stroke over the stridulating ridge produces a syllable; several syllables give a chirp (verse), exemplified by the calling song

(Fig. 88); acridid grasshoppers rub their hindlegs up and down on their forewings. Whatever the exact means of production, with each stroke over the stridulating ridge, specific wing structures resonate, thus enhancing the sound by complicated resonance effects. The sound produced during the cricket's mating call has a frequency of 4–5 kHz: it constitutes a syllable. Several syllables – in the case of the calling song there are three syllables – constitute a verse or chirp (Fig. 88). How do we know that one stroke of the edge over the stridulating ridge corresponds to a syllable? Experimental proof was obtained by simultaneous recordings of the sound and the movement process (For special recording techniques see Chap. 9, Methodological Appendix, Fig. 165).

Neuronal Correlates of Sound-Production. Whereas in the analysis of the prey-catching behavior of the toad and the back swimmer we tried to shed light into the "black box" – i.e., the underlying system character- istics of the CNS – from its sensory input, here we proceed in the opposite direction; we begin from its *motor output* and first of all ask what correlation exists between the activation of the motor neurons and the coordinated muscle contractions. For this purpose thin recording wires are implanted into the "song muscles" and the afferent motor nerves in the freely moving animal. Result: an action potential originating in the motor neuron produces a muscle potential in the corresponding motor unit. The muscle contraction initiates the stroke over the stridulating ridge (Fig. 89) [85].

Neural Control of the Song. How is the cricket song organized in the CNS? Two possibilities may be considered: (1) we are dealing with an autonomous function of the CNS, (2) the sound production is orga- nized and controlled by feedback via sensory inputs (perhaps according to the principle "How can I know what I mean before I hear what I say"). The second possibility may be examined by a simple experiment. Recordings from the song muscles show the typical neuronal stridulation pattern even after elimination of the organs of stridulation (wings). The behavior, therefore, is centrally organized or, in other words, it is generated by a central program. By a combination of brain stimulation, recording, and ablation techniques we have acquired far-reaching knowledge concerning the central structures and complex interactions within and between thoracic ganglia on which singing behavior is based. The different songs obviously exist as fixed programs; they are presumably called into play by "command fibers" in the brain. The choice of the program possibly corresponds to an excitation of varying intensity in the command system.

The task of the brain in the auditory behavior of the cricket consists of spatio-temporal coordination of corresponding motor sequences and thereby determining when and what shall be sung. Thus, it should also be

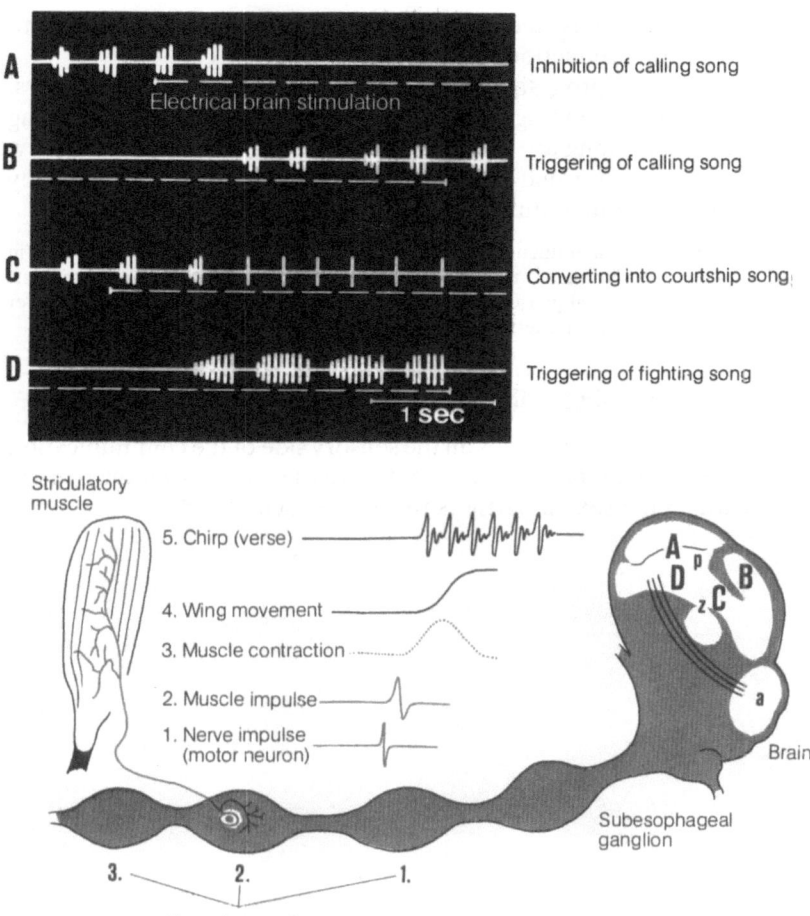

Fig. 89 A–D. Activation and inhibition of different song types (**A–D**) in the cricket *Gryllus campestris* by electrical point stimulation of specific brain sites; *a* antennal lobe, *p* corpus pedunculatum, *z* central body. *1–5* Neurobiological events between action potential of a motor neuron from the 2nd thoracic ganglion and production of sound. [*Above,* Modified from Huber F (1970) Rheinisch-Westf Acad Wiss, N 205, 41–91. Westdeutscher Verlag, Opladen. *Below,* modified from Bently D, Hoy RR (1974) Sci Am 231: 34–44]

possible to elicit songs in the freely moving animal by electrical point stimulation of the brain via an implanted electrode. This idea was elegantly proved by the now classical experiments of Franz Huber (Seewiesen). Various song types seem to be coordinated with specific brain regions [86] (Fig. 89 B–D). Moreover, there are regions whose stimulation may inhibit a song (Fig. 89 A). Accordingly, the song may be "disinhibited" by targeted ablation of brain structures, leading to continuous song production.

The thoracic song programs may also be influenced by other factors. After copulation (stripping off of spermatophore) both the calling song and the courtship song are extinguished in the male. As long as no new spermatophore is formed, the songs may not even be induced by electrical brain stimulation.

An example of peripheral influences on central programming was recently discovered in acridid grasshoppers. Normally, the stridulation patterns produced at the right and left sides of the body show slight phase delays. Following leg extirpation (about one week after surgery) the corresponding patterns are synchronized [87].

b) Sensory Processing in the "Receiver"

Let us now focus our attention on the sensory side of the communication system, that is to the *receiver side.* What are the auditory features in the song by which the sexual partners recognize each other?

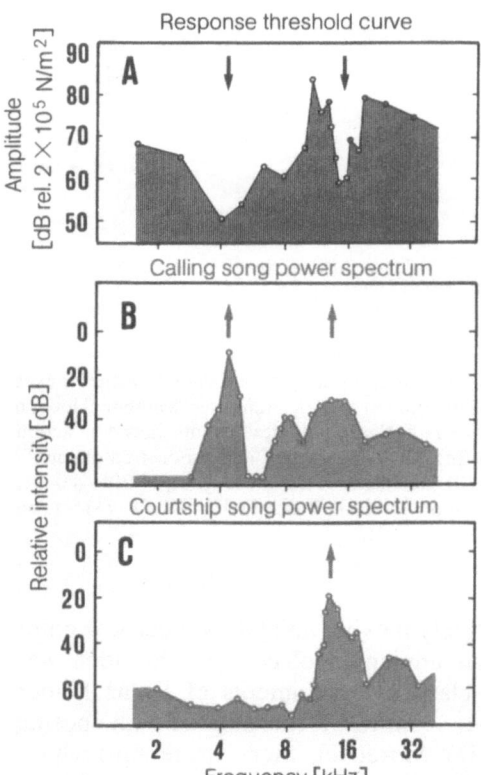

Fig. 90 A–C. Auditory threshold curve of the tympanic organ **(A)**, frequency spectrum of the calling song **(B)**, and of the courtship song **(C)** in *Gryllus campestris.* [Modified from Nocke H (1972) J Comp Physiol 80: 141–162]

Key-Stimuli. Experiments using artificial calling songs synthesized in the laboratory by means of function generators gave us our first insights into the sound recognition scheme and showed that the female cricket apparently recognizes the specific sound of the male cricket predominantly by the repetition-rate of the syllables; the fundamental frequency and overtones as well as the verse duration are also taken into account. It appears to be important that the pattern of syllables contains short as well as long intervals.

We may now search for possible neuronal filters capable of evaluating the acoustic signals.

The Auditory Organ. The ears of the cricket, the tympanal organs, are located in the tibiae of the forelegs. This organ may be exposed to tones of different frequencies; by recording the responses of the tympanal nerves to a wide range of frequencies, a tuning curve can be established (Fig. 90 A). It appears that the cricket is particularly sensitive around 4 kHz and 14 kHz. This agrees well with the frequency spectra of the calling song and the courtship song. The power spectra of these songs (transmitter) fit the optimum sensitivity range of the auditory organ (receiver). Both songs therefore may already be pre-filtered by virtue of the different frequency ranges of the receptor systems.

Central Auditory Neurons. Let us now follow the pathway of the tympanal nerve into the CNS: protocerebrum, subesophageal ganglion, thoracic ganglia. By comparing the tuning curves of single neurons from such areas with that of the auditory organ, in the acridid grasshopper, it

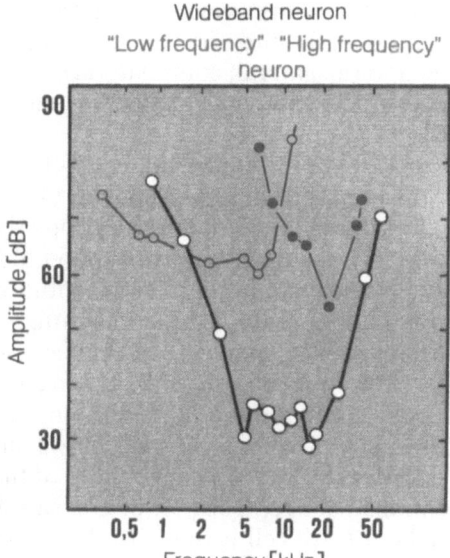

Fig. 91. Tuning curves of three different neuron types from the central nervous system of the grasshopper. [Modified from Adam LJ, Schwartzkopff J (1967) Z Vergl Physiol 54: 255–264]

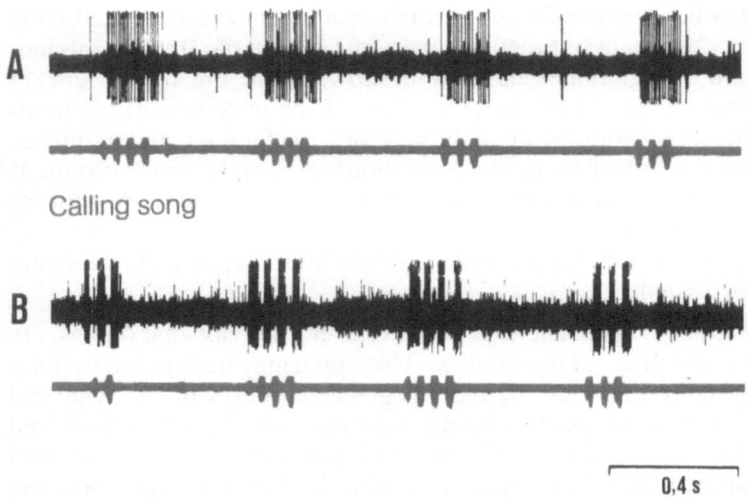

Calling song

Fig. 92 A and B. Records of two auditory central neurons from the ventral connective between subesophageal ganglion and prothoracic ganglion of *Gryllus campestris. Red,* play-back of the calling song. *Black,* neuron that copies the whole verse and the verse periods **(A)** and another neuron that copies single syllables and syllable periods **(B).** [Slightly modified from Stout JF, Huber F (1972) Z Vergl Physiol 76: 302–313]

was found that the frequency sensitivity is unaltered. There are low- and high-frequency sensitive neurons as well as wideband neurons (Fig. 91). Moreover, numerous other neuron types are found which code different sound parameters, such as: (a) the onset, (b) the duration, (c) the amplitude, (d) intensity changes, (e) frequency increases, (f) sound direction (ipsilateral, contralateral), and (g) rate of presentation. There are also neurons capable of coding more than one of these parameters [88].

Of special interest is the discovery of particular auditory neurons in the ventral cord (prothoracic ganglion) of the cricket (Fig. 92). Representatives of one type predominantly copy the verse in the calling song as a whole (A), others, however, respond to the individual syllable (B). We may be dealing here already with important components of the recognition system for the calling song.

Most recently, combined intracellular recording and staining techniques have been applied to the study of two types of cells in the cricket auditory pathway. Both are considered first-order interneurons (Fig. 93 A). One of them, called *omega cell,* shows intensity and pattern response properties that are similar to those of the ascending pulse-coding units mentioned above. One pair of omega cells is located in the prothoracic ganglion. Each cell is tuned to the carrier frequency of the conspecific song, and these cells merely seem to relay receptor signals to higher

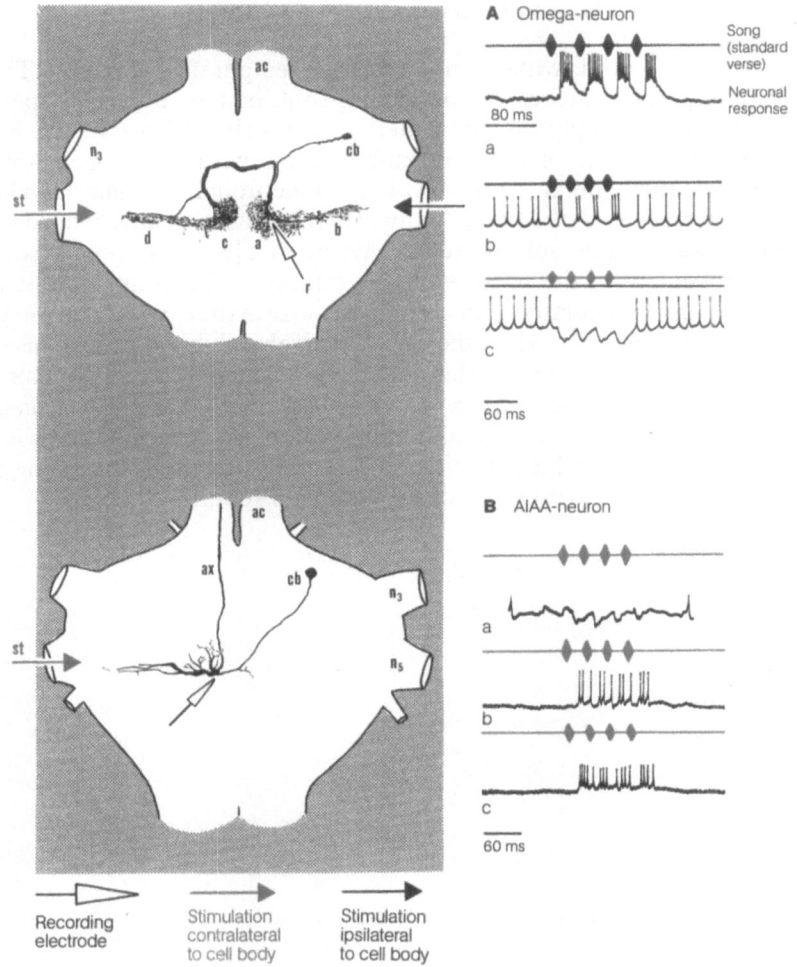

Fig. 93 A and B. Reconstruction of two different auditory interneurons in the prothoracic ganglion of *Gryllus*. Neurons have been stained intracellularly with cobalt nitrate (see Chap. 9, Methodological Appendix). **A** Intracellular recording of synaptic and spike activity of an *Ω*-neuron in *G. bimaculatus*. The acoustically isolated ear is stimulated with the standard temporal pattern of an artificial calling song; syllable period 50 ms, standard verse. *a* and *b* Stimulation of the side ipsilateral to the cell body at 4 kHz (75 dB) or 10 kHz (100 dB) elicits summated EPSP's and superimposed spikes, correlated with each syllable of the verse (cf. *black color*). *c* Stimulation of the side contralateral to the cell body at 10 kHz (100 dB) induces a clear inhibition, correlated with each syllable of the verse (cf. *red color*). **B** Intracellular recording of synaptic and spike activity of an AIAA-neuron in *G. campestris*. The ear is stimulated contralateral to the cell body, using an artificial calling song with the standard temporal pattern (*cf. red color*). *a* At 4.5 kHz (95 dB) spontaneous activity is inhibited by syllable-coded summated IPSP's. *b* At 14 kHz (90 dB) and *c* at 20 kHz (90 dB) the cell being activated responds to each syllable of the verse with a group of spikes. *a, b, c, d* dendritic areas, *ac* anterior connectives, *ax* axon, *cb* cell body, *n3* nerve 3, *n5* nerve 5 (leg nerve), *r* recording electrode, *st* stimulation site. [Modified from Wohlers W, Huber F (1978) J Comp Physiol 127: 11–28]

order interneurons without modifying the temporal information. Thus, the temporal structure of the calling song of the respective species appears to be simply copied by these neurons (Fig. 93 A a, b). We have evidence that both omega cells inhibit each other (Fig. 93 A c). Acting together they might be involved in a neuronal mechanism which determines *sound direction*. Another type of auditory interneuron with an ascending axon, abbreviated *AIAA neuron* [89], presumably serves as an element in the process of *recognition of courtship song*. If the tympanum is stimulated with a standard verse at the carrier frequency of the calling song (4.5 kHz) the unit is actively inhibited in response to each syllable (Fig. 93 B a). This neuron is activated, however, by sounds above 10 kHz (Fig. 93 B b, c). It is assumed that the frequency dependence of excitation and inhibition in these cells is linked with a neuronal process which modifies the response to the calling song (of other males) in the presence of the courtship song.

In summary:

1. The example of the cricket song demonstrates that transmitter and receiver systems are neurobiologically adapted to each other to a high degree.

2. The conspecific songs already exist as complete motor programs in the CNS. They may be called into play (or inhibited) from specific brain areas. Singing can be "disinhibited" by particular localized brain lesions.

3. In the auditory system of the female cricket there are neurons weighting certain parameters of the songs important for recognition, such as the fundamental and overtones, the verse duration, and the repetition rate of the syllables. But at present it is not known how the information conducted along the central auditory pathway is read out by the ultimate receiver, in the actual process of recognition.

4. It seems likely that the acoustic information is first processed for the purpose of sound identification. The subsequent localization of the recognized signal seems to be an independent mechanism. However, in numerous relay points those neurons, meant to respond to the question *where*, must – in a way similar to the prey decoding neurons of the toad – also recognize *what* they respond to.

6.3.2 Frog Chorus

In spring when adult male frogs want to attract females to the pond, or attain and maintain their territory against conspecific males, they break into a kind of canon, stimulating each other to call. Mating calls are only

Fig. 94 A–C. Sound analysis of the mating call of the male American bull-frog *Rana catesbeiana*. **A** Mating call consisting of 6 croaks. **B** A croak expanded in time showing the 100 Hz fundamental period, which is marked by *x*. Time scale cf. top of C. **C** Frequency analysis of a croak. [Slightly modified from Capranica RR (1966) J Acoust Soc Am 40: 1131–1139]

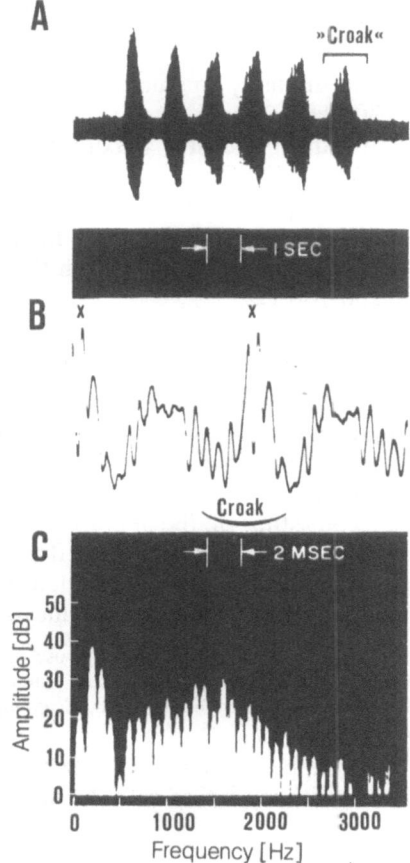

produced by males, but they may serve as species-specific signals for both sexes. We shall ask what components of a frog mating call are behaviorally relevant in this context and how are they recognized by the conspecifics? Let us choose the American bullfrog *Rana catesbeiana* as an example [96].

a) Sensory Processing of Mating Calls

Call Analysis. The natural mating call is composed of 4–15 similar successive croaks (Fig. 94 A). In the field they can be recorded on tape and may then be played back to male bullfrogs in the laboratory from which they evoke a vocal response. To analyze the key-stimulus it is important to obtain information on the frequency power spectrum and temporal structure of the croak (Fig. 94 C). Certain features attract our attention:

1. An intensity maximum in a low-frequency region around 200–300 Hz.
2. Another intensity maximum in a high-frequency region around 1400–1500 Hz.
3. The amplitude of the low-frequency components is typically about 10 dB greater than that of the high-frequency components.
4. A relative amplitude dip in the mid-frequency region between 500 and 700 Hz.
5. A fundamental periodicity of the complex temporal waveform of about 100/s (Fig. 94 B "x").

Synthetic Calls. It might be assumed that these five components are important features of the key-stimulus in the mating call. This can be tested in the laboratory by mixing components (1) to (5) with the aid of function generators and playing them back to the males [90]. A synthetic call consisting of the features (1)–(3) and (5) in fact elicits a vocal response. It is, however, less effective when the fundamental periodicity (5) is either lowered or raised, also when the low-frequency (1) or the high-frequency tone (2) is reduced in relative amplitude.

At present we know that bullfrogs use mating calls to attain and maintain territories (male-male interactions.) Phonotaxis of females to loudspeakers playing back mating calls has been demonstrated in *Hyla, Scaphiopus, Bufo, Pseudacris, Acris, Littoria* and *Gastrophryne*.

What happens when a 600 Hz-tone (4) of sufficient energy is added to the synthetic mating call? In this case the efficacy of the artificial call

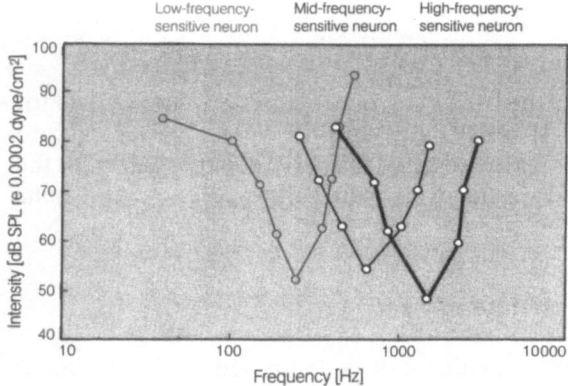

Fig. 95. Excitatory tuning curves of low-frequency, mid-frequency and high-frequency sensitive fibers from the auditory nerve of the American bullfrog. Further details in text. [Slightly modified from Feng AS, Narins PM, Capranica RR (1975) J Comp Physiol 100: 221–229]

decreases. Indeed, the bullfrog fails to respond when this component is 10 to 24 dB above the intensity of the 200 Hz excitatory tone. The fact that an acoustic key-stimulus may be extinguished by adding a tone of specific frequency and relative intensity is at first surprising. The possible biological significance of this phenomenon will be discussed later.

Question of Peripheral Signal Filters. The inner ear of amphibia contains two receptor systems (Fig. 96, 1 and 2): The amphibian papilla (AP) with a sensitivity range from below 100 up to 1000 Hz and the basilar papilla (BP) excited by sounds of higher frequencies (from about 1000 Hz upward). The overall frequency range of the bullfrog's inner ear extends from below 100 Hz to about 3000 Hz.

We shall now discuss neurophysiological correlates of the acoustic key-stimuli. When the activity from single auditory nerve fibers is recorded by microelectrodes in response to tones of different frequencies, three types of unit can be identified: (a) *High-frequency-sensitive* fibers with maximum sensitivity in a range between 1000 and 1700 Hz (Fig. 95), (b) *mid-frequency-sensitive* fibers with best responsivity between 500–900 Hz, and (c) *low-frequency-sensitive* fibers optimally tuned below 500 Hz (Fig. 95). Representatives of this last type are marked by a peculiarity: their response to a 200 Hz low frequency tone, for example, may be inhibited by the simultaneous addition of a second tone in mid-frequency range, 600 Hz for example. In a way similar to the animal's behavior, the response may even be totally suppressed if the inhibitory tone exceeds the intensity of the excitatory tone by 10 to 24 dB. We call this phenomenon "tone-on-tone suppression".

The inhibitable low-frequency-sensitive fibers (c) and the noninhibitable mid-frequency-sensitive fibers (b) originate at the amphibian papilla; the basilar papilla gives rise to the noninhibitable high-frequency-sensitive neurons (a) [91].

It would be interesting to know what kind of inhibitory mechanism is responsible for the response characteristics of the inhibitable neurons. Several possibilities may be considered here: (1) a system of lateral interaction might be present in the plexus of nerve branches terminating on the hair cells in the amphibian papilla; (2) it might be possible that neurons with maximum responsivity around 600 Hz have an inhibitory effect on those with maximum sensitivity around 200 Hz; (3) the inhibition could have a peripheral, mechanical origin rather than a neuronal one; (4) an efferent mechanism might also be considered. However, the response characteristics of the inhibitable fibers remain unchanged after section of the auditory nerve. (5) Finally biochemical mechanisms at the membranes of the cilia of the hair cells should also be considered. Indeed particular tuning properties of primary afferent nerve fibers in frogs were found to be a function of temperature [93]. Further studies are needed to provide answers to these questions.

The Search for Mating Call Detector Systems. As already mentioned, the mating call as a key-stimulus for the conspecific has, among other things, an important characteristic; it combines a particular low-frequency and a high-frequency tone; in the bullfrog *R.catesbeiana:* 200 Hz and 1400 Hz; in the leopard frog *R.pipiens:* 300 Hz and 1700 Hz.

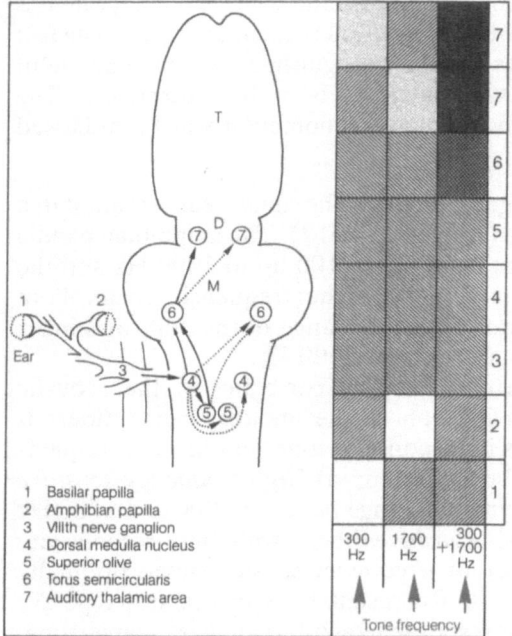

1 Basilar papilla
2 Amphibian papilla
3 VIIth nerve ganglion
4 Dorsal medulla nucleus
5 Superior olive
6 Torus semicircularis
7 Auditory thalamic area

Fig. 96. Responses of neuronal populations at different levels of the auditory system evoked by low- and high-frequency tones, presented alone and in combination. Simplified representation for the leopard frog *Rana pipiens*. Note that in the midbrain torus semicircularis *(6)* the magnitude of the evoked potential to the combination tone is equal to the algebraic sum of the responses to the individual tones presented one at a time. In the postero-central grey of the auditory thalamic area *(7)* the magnitude of the evoked potential to the combination tone is much greater than the algebraic sum of the responses to the component tones individually. [Mudry KM, Constantine-Paton M, Capranica RR (1977) J Comp Physiol 114: 1–13]

Are there in the brain of the leopard frog neuronal systems responding particularly well to the combination of both tones? For this purpose the frog's ear is exposed first sequentially to single tones of 300 Hz and 1700 Hz and then to a combination of both these tones. The neuronal responses are recorded as "evoked potentials" (p. 267) from different ascending parts of the auditory system. Figure 96 shows a preliminary result of such studies. In the auditory pathway, beginning at the level of the inner ear up to the superior olive, different neuronal populations exist; some showing maximum responses to 300 Hz, others to 1700 Hz. In the midbrain torus semicircularis units were identified having "W"-shaped tuning curves with the two best frequencies in the range of 300 Hz and 1700 Hz. Thus, the same neuron is activated, when the stimulus contains energy in either one (300 Hz) *or* the other (1700 Hz) of their two ranges of sensitivity. We might term these neurons

"OR"-gates. The response of the OR-gates to the combination of a 300 Hz and a 1700 Hz tone corresponds approximately to the algebraic sum of responses to the single tones offered individually (Fig. 96;6). In the postero-central thalamus of the diencephalon neuronal systems exist showing selective responsiveness to specific tone combinations, e.g., the combination of frequencies in the mating call. In some of these neuronal populations the response to the 300 *and* 1700 Hz combination tone is much greater than the algebraic sum of the responses evoked by the presentation of either of the component frequencies (Fig. 96;7). It seems likely that individual neurons – which have recently been recorded extracellularly – are functioning as "AND"-gates. These neurons are perhaps important components of a detecting system for the species-specific mating call. Quite comparable relationships were found in the bullfrog and tree frog auditory thalamic nuclei.

Temperature effects: solving an "experimental controversy". In the auditory thalamic nucleus (postero-central grey) of the green tree frog *Hyla cinerea,* the magnitude of the neuronal response to a combination of certain high- and low-frequency tones is also far greater than the algebraic sum of the responses to the same tones presented one at a time. However, in these experiments a discrepancy was noted between behavioral and neurophysiological results concerning the optimal frequency of the low-frequency component, which was 500–600 Hz in the neurophysiological and 800–1100 Hz in the behavioral studies [92]. What caused these significant differences? For technical reasons the neurophysiological experiments had to be done when the body temperature of the animals was 18°–22°C, whereas in the behavioral studies animals were tested at 24°–28°C. Thus, it is reasonable to suggest that temperature might affect frequency preferences of the animal's auditory system. Indeed, if females were given a choice between complete synthetic calls with low-frequency components of 600 Hz or 900 Hz they always chose the 900 Hz call at a body temperature between 24° and 26°C; but they always chose the 600 Hz call when the temperature was between 18° and 20°C. This, by the way, was an unexpected result because the low-frequency peak in the calls of *male* green tree frogs changes only slightly with temperature. We do not know why the female's receiver is temperature-dependent whereas the male's vocalization system appears to be independent under the same experimental conditions in this particular species. Normally, in the acoustic communication of frogs it is found that transmitter *and* receiver are both temperature-dependent and change with temperature in a roughly parallel fashion [93]. The grey tree frog *Hyla versicolor,* for instance, mates over a temperature range of 9°C. If gravid females are tested at different temperatures, they usually prefer synthetic mating calls with temperature-dependent temporal properties similar to those produced by the males at about the same temperature as their own. – From this story we learn that sensory processing of auditory information is very complex, and that results obtained in a particular species cannot be simply "transferred" to another species.

b) Significance of Certain Frequency Bands in Different Species

Selective Meaning of Low- and High-Frequency Bands in Long-Distance Communication in the Tree Frog. The mating call of green tree frogs *Hyla cinerea* has a low-frequency peak around 900 Hz and a high-frequency peak around 3000 Hz. In a *simple* phonotaxis experiment the female will also respond to synthetic signals consisting only of either the low- or the high-frequency component if sound pressure levels are adequate. It will respond to a stimulus consisting of 900 Hz at 50 dB, but to a call of 3000 Hz only at 90 dB. Although this result might suggest that the high-frequency component has little effect at moderate sound levels, in combination with a low-frequency tone it influences the female's choice in a *selective* phonotaxis situation. More specifically, at sound pressure levels of 55 dB females choose a combination sound of 900 Hz + 3000 Hz over a pure tone of 900 Hz. We might suggest that the low-frequency component of the male's call attracts the female at *distance.* As the female approaches the chorus, high-frequency components also become important for *species recognition* [94].

Significance of the High-Frequency Band for the Distinction of Local "Dialects" in the Cricket Frog. Like toads, frogs are typically loners for most of the year with the exception of the mating period. As we have seen, filter systems exist in their auditory system optimizing communication with the conspecifics, and excluding to some extent communication with other species. The female cricket frog *Acris crepitans* apparently goes a step further in signal differentiation: it prefers calls of sexually mature males of the proper geographical race. Thus, a female from Georgia shows selective phonotaxis to the calls of males from Georgia in a loudspeaker test (Fig. 97). The calls of potential partners from Alabama, Texas, New Jersey, or South Dakota are ignored. Obviously the auditory system of the female cricket frog is selectively tuned to the characteristics of the local "dialect". The power spectrum of mating calls of Georgian males, for instance, has maximal energy in the high-fre-

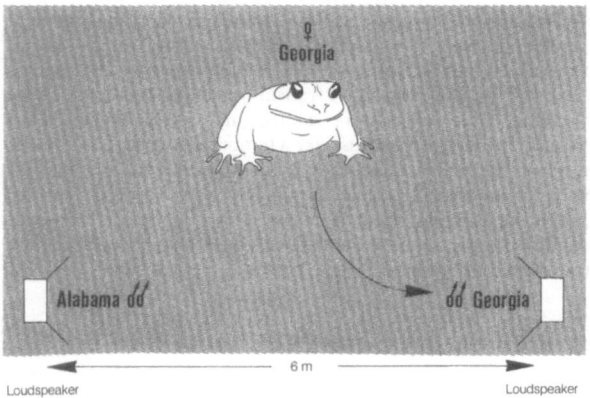

Fig. 97. Discrimination of "dialects" in the mating calls from different geographical races of the cricket frog *Acris crepitans.* In selective phonotaxis test a female from Georgia is attracted mainly by the mating calls of males from Georgia. The mating calls of males from Alabama were ignored. [Based on results obtained by Capranica RR, Frishkopf LS, Nevo E (1973) Science 182: 1272–1275]

quency range at about 4100 Hz, that of Texas males around 3000 Hz. Females from New Jersey even distinguish between males of their own local population (high-frequency peak: 3550 Hz) and those from South Dakota (high-frequency peak: 2900 Hz). Neurophysiological recordings from the dorsal medullary nucleus indicated that units had "best frequencies" that matched the dominant high-frequency components of the mating call of the local population. It is interesting to note that the high-frequency maxima in the call spectra of cricket frogs from geographically intermediate populations may lie between those of the two dialects. The degree of selectivity of intermediates ("intergrades") is so far an unanswered question.

It is not known whether this matched sensitivity within these species is completely under genetic control. But there is also no evidence that cricket frogs learn their "dialects". Probably frequency tuning is mechanical and correlated with body size which, in turn, may be correlated with environmental factors, e.g., moisture.

The "600 Hz-Inhibition" in Auditory Behavior of Bullfrogs. To begin with, it is important to find out the natural occurrences of 600 Hz sounds or corresponding frequency components. They appear in the mating call of the adult male only with relatively low energy (cf. Fig. 94 C). But when we analyze the calls of the young frogs, which do produce mating calls, even when sexually immature, we find not only the usual intensity maximum around 1400 Hz but also a low-frequency spectral peak around 500–700 Hz, much higher than the 200 Hz peak in the adult's calls. Therefore, the intensity of the "600 Hz peak" in the calls of the young frogs, when combined with the adult mating call, has an inhibitory effect on the vocal response of the adult (cf. p. 169). Thus, there are two reasons why the calls of young bullfrogs should be less effective than those of adults in the same behavioral context.

How does the voice of the young male bullfrogs change with aging? A complicated central control might be the first thing that comes to mind. The answer, however, appears to be astonishingly simple: the higher-pitched peak in the low-frequency range can obviously be attributed to the smaller vocal cavities of the young frog. During growth their volume increases as well, and the low-frequency spectral peak in its mating call shifts downward from the "inhibitory" mid-frequency region into the "excitatory" low-frequency band; at the time of sexual maturity it is about 200 Hz [95]. The change of the voice of young male bullfrogs is apparently due to a feature of the peripheral sound-producing system.

To summarize:

1. In a manner reminiscent of the situation in crickets we find in the auditory communication of frogs that sound producing and receiving systems are closely attuned to each other. An important function of

the conspecific mating calls produced by the males is bringing together of sexual partners.

2. Certain low- and high-frequency bands in the mating call spectra may have specific meanings. In tree frogs *Hyla cinerea* the low-frequency component attracts the female at long distance. The combination of low- and high-frequency components is important for species recognition. The female cricket frog *Acris crepitans* distinguishes the local "dialects" of mating calls on the basis of high-frequency bands which vary with geographic origin of the males. In bullfrogs *Rana catesbeiana* young males are probably excluded from mating by the 500–600 Hz band in their calls, which is not only ineffective but may also be inhibiting with regard to intraspecific communication.

3. A first stage of processing of these acoustic signals may already be the specific tuning properties of the peripheral auditory system. In higher auditory centers there are neuronal populations which participate in the further analysis and detection of behaviorally relevant auditory signals (e.g., *specific combinations of features* in mating calls). In this connection it is of interest to note that we are discussing an area in the postero-central thalamus. Certain neuronal populations of the adjacent postero-lateral thalamic region obviously participate in the evaluation and discrimination of behaviorally relevant visual signals (e.g., *configurational features* of prey and enemy).

6.4 Active Exploration of the Environment: Echo- and Electrolocation

The visual system is characterized by a relatively high capacity for spatial frequency analysis. As an important orientation system it is widely distributed in the animal kingdom. What are the possible mechanisms of orientation of animals that live in almost complete darkness or in an optically opaque medium?

It could be assumed, on the one hand, that their *visual system* is sharpened by particular neuronal circuits of lateral inhibition. On the other hand, animals could orientate themselves by touch, using mechanoreceptive inputs from the *somatosensory system*. A disadvantage in the latter case would be its almost total reliance on information that is available in the immediate environment and gathered by continuous bodily contact. Special mechanoreceptive organs (e.g., lateral line organs) have been developed in aquatic animals for detecting water streams which are caused or modified by objects in the immediate

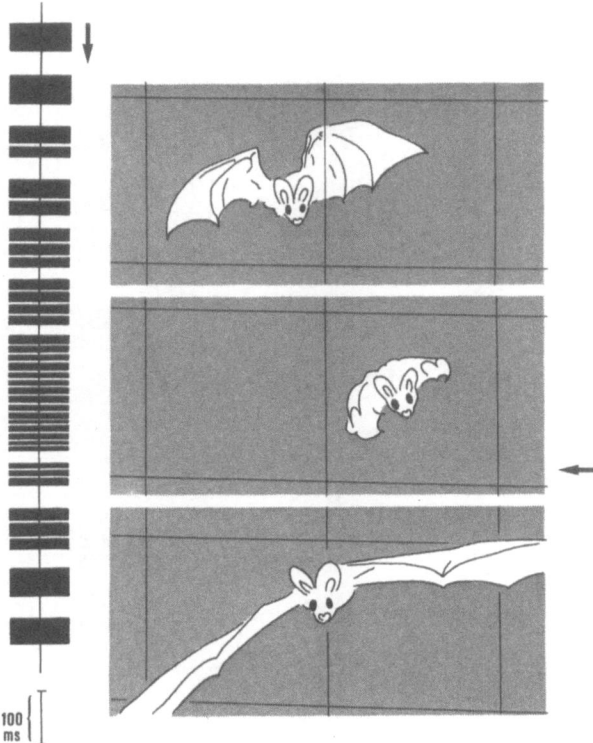

Fig. 98. *Right,* Bat, directed by its echo-locating system, flies through a 14 cm mesh-grid obstacle consisting of 80 μm nylon threads. (From photographs of Möhres FP, Neuweiler G (1966) Z Vergl Physiol 53: 195–227). *Left,* duration and frequency of orientation sounds emitted during approach, fly-through *(horizontal arrow),* and after passing the obstacle. [According to measurements of the large horseshoe bat, *Rhinolophus ferrume-quinum,* by Schnitzler HU (1973) In: Lindauer M (ed) Orientierung der Tiere im Raum I. G Fischer, Stuttgart.]

environment. The *olfactory system* appears to be of limited suitability for tasks of general orientation owing to its restricted capacity for spatial and temporal resolution. Moreover, it transmits information of special signals that are often associated with particular objects or environmental conditions.

6.4.1 Echolocation

What About the "Auditory" System? If used "passively" the auditory system is hardly suitable for general orientation purposes because many behaviorally important objects are silent. They may, however, become audible when exposed to sound; that is their echo can be perceived and

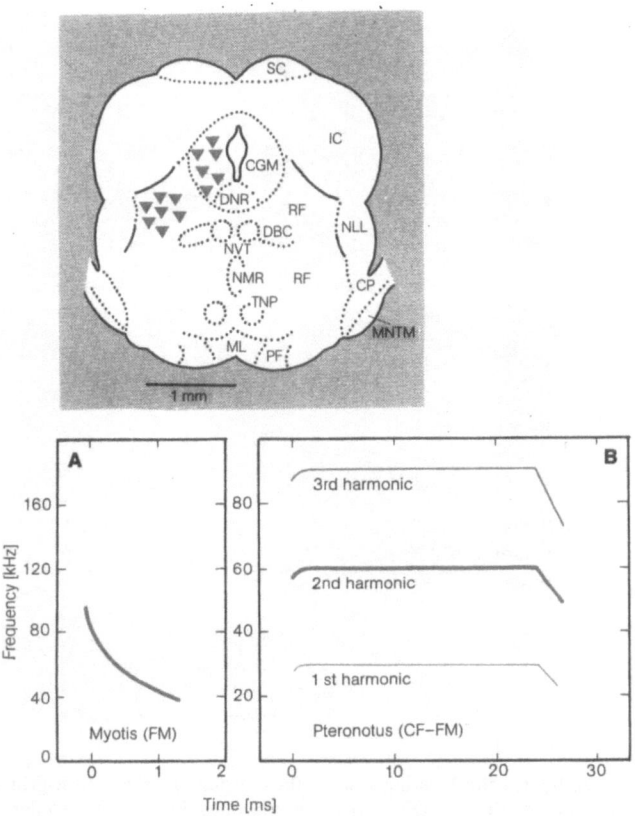

Fig. 99 A and B. Species-specific orientation sounds evoked from bats by electrical point stimulation of specific sites of the brain below the inferior colliculus in *Myotis* (**A**) and *Pteronotus* (**B**). *Above,* Cross section through midbrain. Orientation sound eliciting stimulation points are indicated with *red* symbols. Important abbreviations: *CGM* central grey matter, *DBC* decussation of brachium conjunctivum, *DNR* dorsal nucleus of raphe, *IC* inferior colliculus, *NLL* nucleus of lateral lemniscus, *NVT* nucleus of ventral tegmentum, *RF* reticular formation, *SC* superior colliculus. [Modified from Suga N, Schlegel P, Shimozawa T, Simmons J (1973) J Acoust Soc Am 54: 793–797]

they can be located – provided the animal continuously emits orientation sounds. In this way, for example, many blind persons are able to detect large obstacles in their path. They listen to the echo of the sounds produced by the snapping of their fingers or the tapping of their cane. Although the human ear is not particularly effective in this respect, there are certain groups of animals that specialize in echolocation, depending on their biotope and their life style [97]. The performance of such biosonar systems can even surpass that of our modern electronic instruments. Echolocating animals include toothed whales (e.g., dolphins), most bats and certain species of cave-nesting birds (oil birds). These animals analyze the reflections of their own vocalizations as so-called biosonar signals.

The echolocation system of the bat has been particularly well studied [97, 100]. With its biosonar system the bat can (1) estimate its own flying speed, (2) determine the distance and the velocity of target objects, (3) localize the direction in which an object is situated, and (4) recognize patterns according to material, configuration, and surface structure. Let us examine some neurobiological prerequisites for these capabilities. We study the echolocation system in the large horseshoe bat *Rhinolophus ferrumequinum* and the Panamanian moustache bat *Pteronotus parnellii rubiginosus*.

Behavioral Experiments with Obstacles. Bats may be trained to fly through a grid of thin threads in order to reach a food reward. Their locating ability can be tested by an appropriate choice of thread thickness and grid width. The results are astonishing (Fig. 98): a bat with a 40-cm wing span flies in total darkness through the grid with folded wings even when the grid is 14×14 cm^2 and the thickness of the nylon threads only 80 µm. Collision rarely occur. Presumably the bats can still hear echos from 60µm thick threads. In this case the reflected sound energy was so small that it could not be measured with presently available instruments.

Moving echos may be an even more effective cue for echolocation [98]. When barriers consisting of vertical nylon lines spaced 30 cm are made to oscillate in horizontal direction, animals *(Myotis lucifugus)* avoid the moving obstacles more successfully than the stationary ones. The just noticeable movement speed of the obstacles perceived by the echolocating bats is 4.2 ± 2.7 cm/s.

How Does This Highly Sensitive Localization System Function?

Orientation Sounds. Bats use ultrasound for orientation. Some species emit it through the open mouth, others through the nose. In the little brown bat *Myotis lucifugus* each sound is frequency-modulated (FM), i.e., it starts at 100 kHz and sweeps down to 40 kHz within 2 ms (Fig. 99 A). The moustache bat *Pteronotus parnellii* (Fig. 99 B) produces different orientation sounds: they consist in the 2nd harmonic (which is the most intense component) of a 5–30 ms constant frequency 61 kHz tone (CF component) which declines to 50 kHz within 2–4 ms (FM component). The horseshoe bat *Rhinolophus ferrumequinum* produces a 83.3 kHz CF tone for 8–65 ms, which sweeps down in the FM component by 20 kHz within 5 ms.

Thus, there are two main types of orientation sound: FM and CF-FM sounds. There are also neural correlates: in response to electrical point stimulation of the vocalization centers in the midbrain (Fig. 99) members of the genus *Myotis* emit only frequency-modulated (FM)

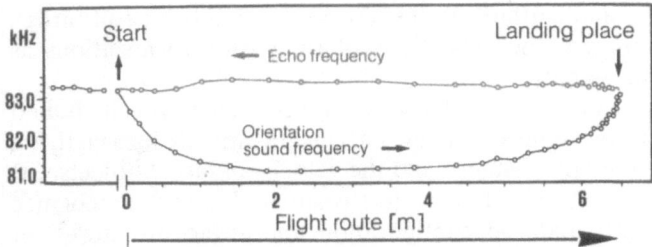

Fig. 100. Dopplershift compensation during the flight of the large horseshoe bat from take-off to landing. Relationship between frequency of orientation sounds and echo frequency reflected from target. The latter is held constant within an extremely narrow frequency band. Further details in text. [Modified from Schnitzler HU (1968) Z Vergl Physiol 57: 376–408]

sounds, typical for them, while members of the genus *Pteronotus* emit their CF-FM orientation sounds.

One sound or a group of sounds is produced per respiration or wingbeat. When a horseshoe bat in the obstacle experiment approaches a nylon thread, the duration of the CF component of each sound is typically abbreviated and the sound rate is increased (Fig. 98, left). Rapid and repeated scanning of the environment increases the flow of information.

The Doppler Shift Compensation. Let us now observe the acoustic behavior of the large horseshoe bat from take-off to target, e.g., a landing spot (Fig. 100). Before take-off it emits at first sounds with a frequency of 83.3 kHz in the CF component. The echo reflected by the target has the same frequency. What happens when the bat flies in the direction of the target? Owing to the Doppler shift the frequency of the returning echo is always higher than the emitted frequency and depends on the flying speed.

The Doppler shift principle can be explained with the help of an example. If we stand still on the surf we are hit by a certain number of waves within a certain time. When we swim against the waves the number of the waves we encounter in the same time increases, depending on the speed with which we swim against them.

Two Doppler effects are coupled to the flight of the bat, one to the sound emission and the other to the reception of the echo. The bat is, however, aware of the Doppler shift, and compensates: during the flight it reduces the frequency in the CF portion of its orientation sounds in such a way that the CF of the Doppler-shifted echo is stabilized in a narrow 200-Hz-frequency band between 83.05 and 83.25 kHz. The band serves as a standard "set point" (Fig. 100). Within the framework of this feedback system the acoustic information received by the ear is processed in the auditory centers and transformed into an appropriate vocalization command [99]. Moustache bats also perform Doppler shift

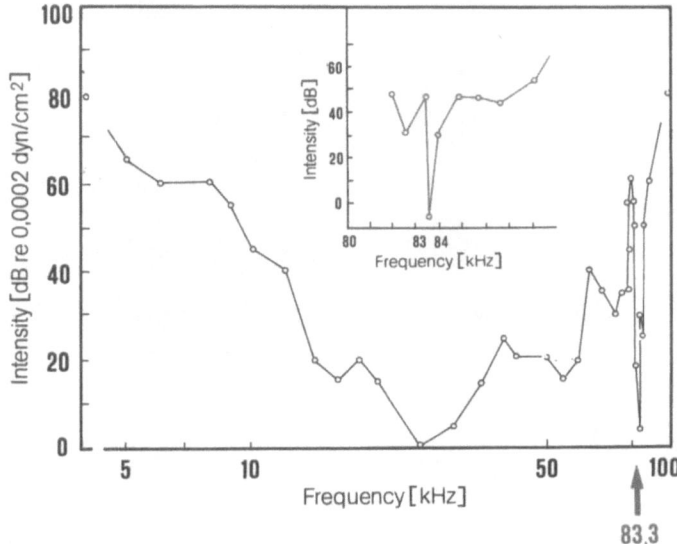

Fig. 101. Tuning curve from the inferior colliculus of the large horseshoe bat. Note the extremely narrowband filter at 83.3 kHz. The *black curve* was constructed on the basis of averaged evoked potentials recorded from the inferior colliculus, the *red curve* on the basis of microelectrode recordings from a single neuron of this brain area. [Modified and combined from Neuweiler G (1972) Verh Dtsch Zool Ges 66: 168–176]

compensation [100]: they stabilize the second harmonic CF of the Doppler-shifted echo at about 61 kHz.

By the Doppler shift compensation the bat can increase the accuracy of its estimate of its own flying speed, the relative speed between it and a flying prey insect and also the direction of the latter.

Search for a Filter for Doppler-Shifted Echoes. The fact that horseshoe and moustache bats always tune their sending frequency to obtain an echo in an extremely narrowband suggests the existence of a corresponding filter in their auditory system. Ablation experiments provided the first evidence of such a filter. Even if the auditory cortex is destroyed, the echo localization capacity remains intact; it is lost, however, when phylogenetically older auditory centers, such as the inferior colliculus of the midbrain, are eliminated. This broadly delineates the sites of the neural structures in question. For detailed analysis (Fig. 101) the bat's ear is exposed to CF sounds of different frequencies and the evoked potentials are recorded from the inferior colliculus, which is in comparison to the visual centers (superior colliculus) very strongly developed in bats (Fig. 99, above). Indeed, the tuning curve of this brain area in the horseshoe bat indicates an

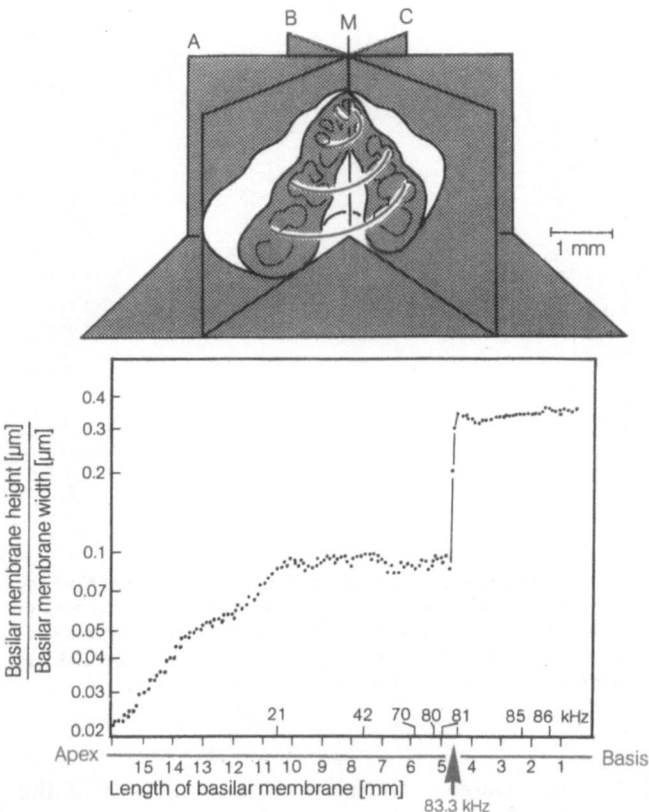

Fig. 102. *Top*, Right cochlea of the large horseshoe bat in a 60° plane diagram *(A, B, C)*. A 60° section is removed showing the basilar membrane *(red)*. *Bottom*, Relation between height and width of the basilar membrane (measured from base to apex). Note the abrupt change of the stiffness of the basilar membrane at the frequency representation of 83.3 kHz. [Modified and combined from Bruns V (1976) J Comp Physiol 106: 77–97]

extremely narrowband filter at 83.3 kHz, accurately tuned to the Doppler-shifted echo frequency. However, this sharp tuning is not due to intrinsic functions of the inferior colliculus. A specialization of the *peripheral* auditory system determines the filter properties. Sharp tuning for the Doppler-shifted echo frequency is already seen in the audiogram of the cochlear microphonic response, i.e., the receptor potential. Obviously this sharp tuning is based on peripheral frequency analysis at the level of the cochlea. From a morphological point of view the basilar membrane stands out by its development in the range of the Doppler-shifted echo frequency: 83 kHz in the horseshoe bat (Fig. 102). The exact mechanical properties of the basilar membrane that are finally responsible for this extreme filter function have not yet been clarified in detail. Of course, the general frequency analysis in mammals is based upon the properties of a traveling wave moving along the basilar membrane. This is also obtained in the horseshoe bat, however, except

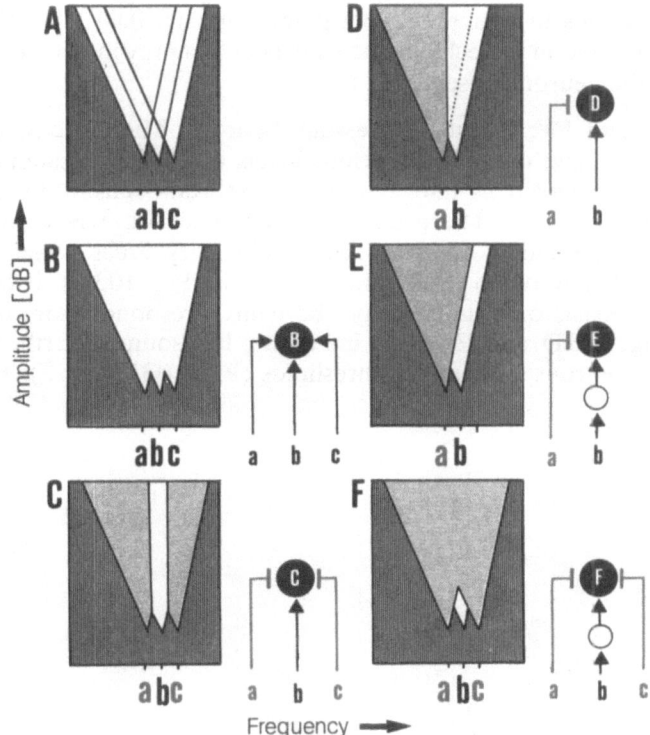

Fig. 103 A–F. Illustration of response characteristics of particular neuron types **(B–F)** from the central auditory system of the bat. The neuron *(black)* may receive excitatory *(arrows)* or inhibitory *(red lines with cross bars)* inputs from three neurons a, b and c, whose tuning curves are shown in **A. B** Neuron with a relatively wide excitatory band *(white)* responding both to *CF* and *FM* sounds. **C** *CF*-specialized neuron with a narrow excitatory band *(white)* flanked bilaterally by inhibitory areas *(red)*, resulting from symmetrical lateral inhibition. This neuron is only activated by pure tones of specific frequency (*CF* specialized neuron). **D** Asymmetrical neuron with a narrow excitatory band flanked unilaterally by an inhibitory area, producing an asymmetrical inhibition. **E** Asymmetrical neuron with upper-thresholds for pure tones. Here the inhibitory input arrives earlier than an excitatory one does, because the excitatory channel has a delay mediated by an excitatory interneuron. This neuron is sensitive to *FM* sounds, and in the present example to those whose frequencies are modulated downward. Because of the inhibitory area this neuron does not respond to *FM* sounds with frequencies modulated upward. **F** *CF* specialized neuron with an upper threshold. The inhibitory inputs arrive at neuron (F) earlier than the excitatory input does, because of a delay circuit. The properties of neuron types (B)–(F) described above are determined by first-order interactions. Second-order interactions may determine the complex properties of *CF* deaf and *FM* specialized neurons. [Slightly modified from Suga N (1973) In: Møller AR (ed) Basic mechanisms in hearing. Academic Press, London New York]

for that part of the membrane tuned at the Doppler-shifted echo-CF. Here local resonators appear to influence the frequency analysis. Auditory nerve fibers tuned to 83.3 kHz sound (horseshoe bat) or to 61 kHz (moustache bat) are extremely sensitive to slightest frequency variations in that range. Variations of only 0.012% – that could be generated in the echo by the wing beat of a prey insect – cause a change of the neuronal discharge rate.

CF and FM Signals Processing Neurons. Microelectrode recordings from single neurons at various levels of the central auditory pathway have provided us information on important sensory processing steps (Fig. 103 B–F). There are narrow-band neurons whose response areas are surrounded by symmetrical inhibitory areas which make them specifically tuned to certain CF sounds (Fig. 103 C). If the inhibitory area exists only unilaterally, the neuron responds more to decreasing (Fig. 103 D and E) or to increasing FM sounds. Furthermore, there are neurons with upper thresholds (Fig. 103 E and F). Higher-order

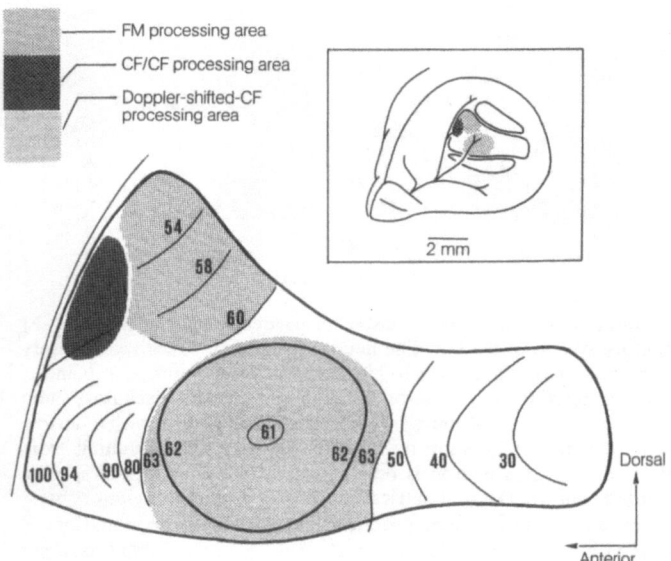

Fig. 104. Tonotopic organization of the primary auditory cortex in echolocating CF-FM bats. The information-carrying frequency bands of the CF part and the FM part of the echoes occupy different areas. *Above,* Dorso-lateral view of the left cerebral hemisphere of the Panamanian moustache bat *Pteronotus parnellii rubiginosus.* Below *red,* Iso-best-frequency contours in the Doppler-shifted-CF processing area; note "over-representation" around 61 kHz. *Grey,* FM processing area, tuned at 50–60 kHz. *Black,* CF/CF processing area. [Combined and modified from Suga N, Jen PH-S (1976) Science 194: 542–544; Suga N, O'Neill WE (1978) Science 200: 778–781; Suga N, O'Neill WE (1979) Science 203: 270–274]

interactions may determine the complex properties of CF-deaf and FM-specialized neurons. These neurons are not necessarily detectors for orientation sounds: CF as well as FM sounds and noise are important carriers of auditory information.

Neurons processing such components may on their own participate in the selective analysis of defined acoustic characteristics. A particular substrate for processing of constant-frequency signals is located in the primary auditory cortex. Here the tonotopic and ampliotopic representations show peculiarities in the band of the Doppler-shifted echo: in the moustache bat around 61 kHz (Fig. 104) and in the horseshoe bat around 83 kHz. This *Doppler-shifted-CF processing area* [101] obviously serves to analyze velocity and subtended angle of a target. Furthermore, there is a discrete *FM processing area* with complex tonotopic representation (Fig. 104). Neurons of this region are processing the main frequency-modulated component for ranging, localization, and characterization of a target. Most recently, in the moustache bat a discrete *CF/CF processing area* has been identified (Fig. 104) [102]. The responses of neurons from this area were found to be facilitated by combinations of two or more harmonically related tones, mainly by those closely associated with the CF component of the orientation sounds (Fig. 99B). Presumably these neurons serve for the identification of signals produced by echolocating conspecifics.

Let us summarize some essential points:

1. The echolocation system of the moustache bat and horseshoe bat represents an optimization of performance to a specific environmental situation. In a number of functions its performance surpasses that of present man-made sonar systems.

2. The extremely sharp tuning of the biosonar system to specific Doppler shifted pure tones is based on the particular properties of the peripheral auditory system.

3. After ablation of auditory cortex some general localization functions remain still intact. Bats obviously reserve this structure for specific identification and localization tasks, thus obtaining information for adaptive behavioral strategies. The principle of echolocation is also found in vertebrates (birds) that do not possess neocortex (neopallium).

4. Certain comparisons may be made between the visual and auditory systems with regard to the first steps in the abstraction of features from sensory input:

a) The frequency band of an auditory neuron corresponds to a specific section of the basilar membrane; in the case of a visual neuron its counterpart is a defined two-dimensional area in the retina of the eye.

b) If the sensory response area of a neuron is large, we speak of a wideband neuron in the auditory system (Fig. 103 B) and of a wide field neuron in the visual system (Fig. 10 B, *E*).

c) If the response area is surrounded by symmetrical inhibitory fields (Fig. 103 C) it could be a CF-sensitive neuron in the auditory system or an *on* -or *off* -center neuron in the visual system (cf. also Fig. 61 a, b).

d) If the response area is asymmetrically surrounded by inhibitory fields (Fig. 103 D and E) a FM-sensitive neuron might be present in the auditory system, comparable to a directional-sensitive neuron in the visual system.

6.4.2 Electrolocation

There are fish that generate around themselves a weak electric field (Fig. 105) and recognize objects in the environment from their perception of the field distortions (Fig. 106). Just as in the case of echolocation, electrolocation has developed phylogenetically as an adaptation to the biotope (turbid muddy water) and to the life style (e.g., nocturnal activity). Here again, we have an active localization system where the required carrier energy for the gathering of information is produced by the animal itself. With the aid of modified muscles (or modified axons) –

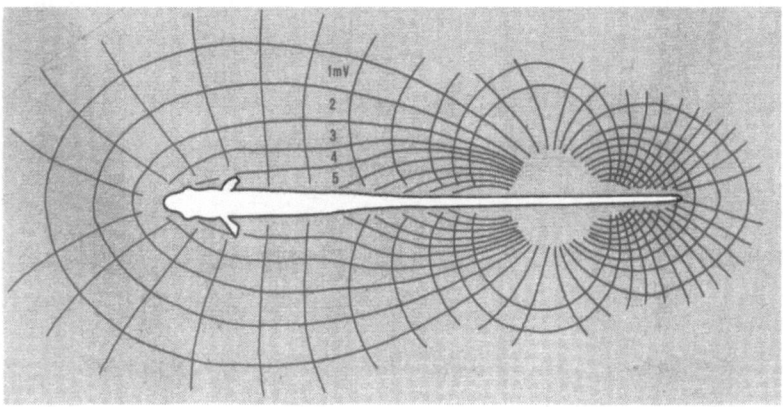

Fig. 105. Isopotential lines and the flow of current in the weakly electric fish *Eigenmannia virescens*. [Modified from Scheich H, Bullock TH (1974) In: Autrum H et al (eds) Handbook of sensory physiology vol III/3 Fessard A (ed) Springer, Berlin Heidelberg New York]

Fig. 106. Typical distortions of current pattern of a weakly electric fish in the neighborhood of a conducting (metal) and a nonconducting (plastic) object

called *electric organs* – the fish generates a weak electric field around itself at brief time intervals (Fig. 107). The voltage is only a few millivolts to volts, head to tail. (Apart from these low-voltage fish, there are high-voltage fish – e.g., the electric eel, *Electrophorus* – which produce up to 900 volts and thereby paralyze or kill prey animals). In this mode of behavior excitation of the membrane does not induce muscular

Fig. 107. Functional principle of the electric organ of an electric fish. The electrocytes are serially stacked. *Above right,* cross section through medulla with the pacemaker center, and cross section through spinal cord with a motor neuron from which the electrocyte is unilaterally innervated. The potentials of the electrocyte membranes are shown for the resting and the excited state. *Below,* current pattern of a simple physical dipole. Further details in text

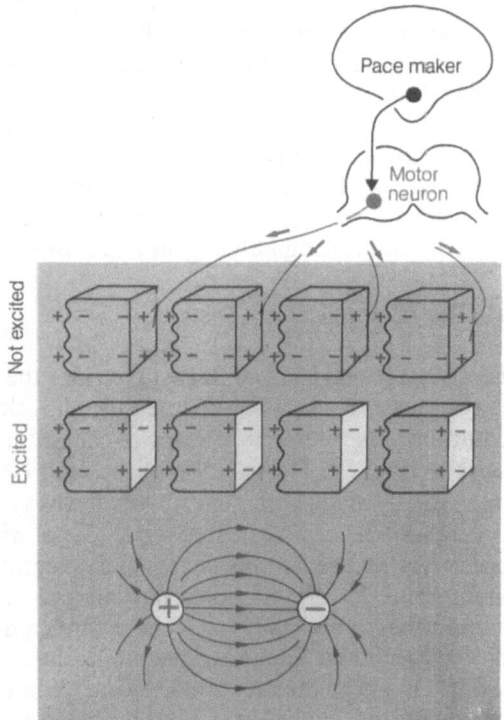

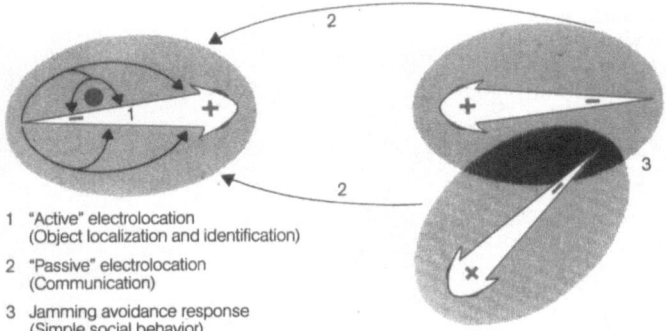

1 "Active" electrolocation
   (Object localization and identification)

2 "Passive" electrolocation
   (Communication)

3 Jamming avoidance response
   (Simple social behavior)

Fig. 108. Three different behavior modes in weakly electric fish. [Modified from Szabo T (1973) Fortschr Zool 21: 190–210]

contraction, but results only in electric discharges. Along the length of its body the weakly electric fish possesses electroreceptors able to detect the electric field. In comparison with the echolocation system of bats, this is an example of direct self-stimulation, and that at the velocity of light. Nearby objects are detected because their conductivity differs from that of water and therefore changes the isopotential lines (Fig. 108; *1*). With the aid of their "electrosystem" these fish can also communicate with each other (Fig. 108; *2*), but neighboring conspecifics may interfere with object localization by superposition of their electric fields (Fig. 108; *3*).

From this constellation of phenomena let us select three questions: (1) What is the neurophysiological basis of "electrical behavior"? (2) How does the fish localize and identify objects in the environment with its "electrosystem"? (3) What are the neuronal processes enabling the fish to avoid interfering fields from a nearby conspecific fish?

a) The Electrosensory System

Electrical Behavior. Let us first consider the electric organ (Fig. 107). It may consist of modified striated muscles that have lost their ability to contract. The elementary unit is called an electrocyte, and these are serially stacked. The membrane of each electrocyte is excited only on one side by signals of a motor fiber whose cell body is localized in the ventral horn of the spinal cord. The whole membrane surface is polarized in the resting state. On excitation of the motor fiber the membrane is first depolarized on the innervated side; all lateral membranes now have, for a short time, small voltage gradients in the direction of their linear arrangement; the electrocytes become batteries [103]. Owing to the fact that they are "connected" in series, the voltage contributions add linearly. During the discharge of the electric organ an electric field develops around the fish, corresponding to its dipole-like character (Fig. 107). For this brief period the fish can scan its environment with its electrosensory system.

In the high-voltage electric fish *Electrophorus* the innervated membrane produces an action potential, whereas the opposite membrane only responds with passive depolarization. The net voltage measured along the electrocyte corresponds approximately to the time course of the action potential, and is only slightly reduced in amplitude by subtraction of the amount of the passive depolarization. In the low-voltage electric fish *Staetogenys* the facing electrocyte membranes (isolated against the others) both generate action potentials. The AP at both sides shows a small delay in temporal occurrence and is slightly different in amplitude. The peak-to-peak amplitude of the net potential across the electrocyte in this case is significantly lower than in high-voltage fish.

The motor neurons of the ventral horn receive excitatory input from a pacemaker center situated in the medulla (Fig. 107). It controls the discharge frequency of the electric organ. In some species it may reach 100–2000 Hz (Type 1), in others 1–65 Hz (Type 2). Whereas the discharge rate is very stable in high-frequency fish, it is subject to considerable fluctuations – depending on new stimulus situations – in low-frequency fish: sound, food, electrically conductive objects may alter the pacemaker frequency. What is the biological significance? Rapid and repeated scanning of the environment increases the flow of information, a phenomenon also known from other sensory systems (cf. p. 178).

Among the South American Gymnotoidei, both Type 1 (wave type) and Type 2 (pulse type) have been developed; *Eigenmannia* is a wave type. The African order Mormyriformes evolved independently and has almost exclusively Type 2 (exception: *Gymnarchus*).

What can the electrosystem do? We know from numerous behavioral training experiments that electric fish can distinguish between food dishes of equal appearance but different material on the basis of the different conductive properties. They are able to avoid obstacles and to navigate according to the form and pattern of their electric field. Recently a behavior designated as an "electromotor response" has been

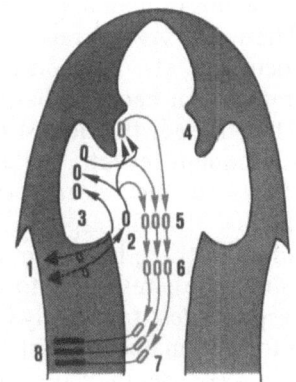

Fig. 109. Important afferent and efferent tracts in the electrosensory system of a weakly electric fish. *1* Electroreceptors; *2* and *3* lateral lobe, *4* torus semicircularis of midbrain, *5* hypothetical relay neurons in the reticular formation, *6* pacemaker neurons in the medulla, *7* motor neurons of spinal cord, *8*, electric organ. [Slightly modified from Bullock TH, Hamstra RH Jr, Scheich H (1972) J Comp Physiol 77: 1–48]

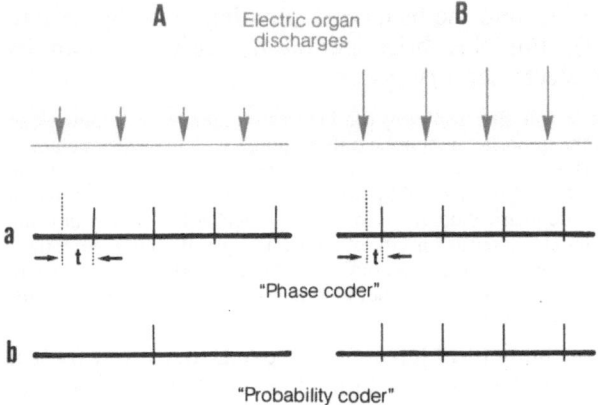

Fig. 110 A and B. Response properties of two electroreceptor types, phase coder *(a)* and probability coder *(b)*. **A** weak, **B** strong electric field. Discharges of the electric organ *(red)* and impulses of receptors *(black)* are schematized. Further details in text

studied in greater detail [104]. If a fish, for instance, is located between two nonconducting plastic rods which are swinging back and forth, it carries out corresponding correcting movements, so that it avoids collision with the obstacles.

What are the relevant receptors and how can they contribute to the processing of electrical images?

Electroreceptors. The electric field that spreads three-dimensionally around the fish forms the basis for object localization. Here, the spatial parameters of the environment are projected to a certain extent onto the body surface. They can be perceived through the electroreceptors of the lateral line system. The receptors are secondary sensory cells. Their excitations are conducted in the lateral line system to the lobus lateralis and from there to the tori semicirculari (homolog of the inferior colliculus of mammals) for further processing (Fig. 109).

There are several receptor types. In the high-frequency fish we shall focus our attention on two of them. The first type responds with an impulse for each discharge of the electric organ (EOD) at a 1:1 ratio. The stronger the field stimulus (μV/cm), the shorter the latency of the subsequent receptor spike (Fig. 110a). In the case of these receptors, therefore, the information on stimulus intensity is contained in the latency of their response; they are called *phase coders* (T-neurons). In addition there is another type which, at a high-stimulus intensity (field amplitude), responds to each EOD, but, at lower stimulus intensity, responds less frequently (Fig. 110b). In this type of receptor, therefore, information about the stimulus intensity is contained in the probability of a response, they are called *probability coders* (P-neurons).

Fig. 111 A–C. Coding of various stimulus parameters by different types of electrosensitive neurons of weakly electric fish. **A** Change of discharge rate of an electroreceptor (recording point is indicated by a *red arrow* on the abscissa) when moving a metal *(red)* or plastic *(black)* plate along side the fish *Sternarchus.* **B** Change of discharge rate of a neuron from the lateral lobe (cf. Fig. 109) during the back- and forth-movement of a metal plate along *Sternarchus.* **C** Phasic-tonic response of an electroreceptor of *Steatogenys elegans* when an object approaches the receptor site *(arrow on left),* maintains its position and finally moves away again *(arrow on right).* [Combined and modified from Hagiwara S, Szabo T, Enger PS (1965) J Neurophysiol 28: 775–799 (A); Enger PS, Szabo T (1965) J Neurophysiol 28: 800–818 (B); Szabo T (1973) Fortschr Zool 21: 190–210 (C)]

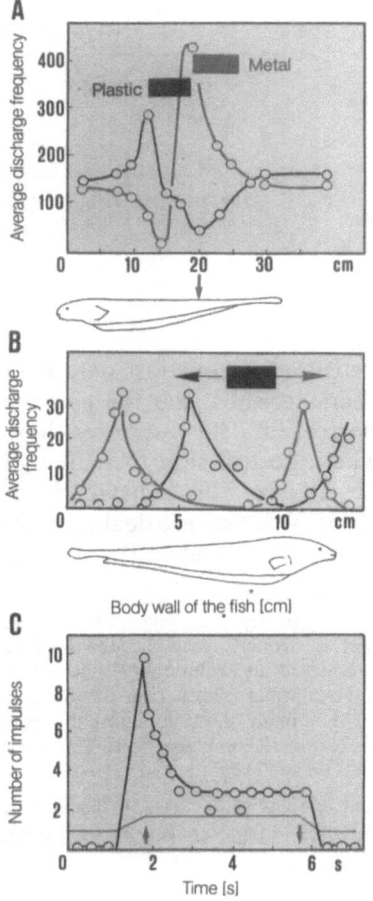

Such receptors acting in concert with adjacent receptors may transfer information to the central nervous system about the nature of electrically conducting or nonconducting objects (Fig. 111 A), or about the position of an object in the vicinity. In addition, at higher levels of the CNS, different neuron types are found which are capable of indicating (1) whether an adjacent object is stationary or moving (Fig. 111 C), (2) in which direction it moves (Fig. 111 B), and (3) where (independently of conductivity) the edge of an object is located.

b) The Jamming Avoidance Response, JAR

What happens when two high-frequency electric fish whose field frequencies deviate slightly from each other meet? In this case the

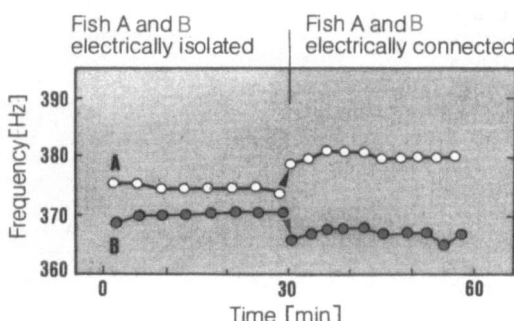

Fig. 112. *Left,* The discharges of the electric organs of two electric fish *Eigenmannia* show similar frequencies when both animals, *A* and *B,* are electrically isolated from each other. *Right,* Jamming avoidance response (JAR) when their electric fields interact. [Modified from Bullock et al (1972) cit. in Fig. 109]

detector neurons just described show abnormalities in their response characteristics, and the performance of electrical image processing is decreased. The responses, however, are again approximately normal when the fish shift their frequencies to a "private" range differing by 15–20 Hz. This behavior is called the jamming avoidance response [105]. What we are dealing with here is a simple form of social behavior (Figs. 108; *3* and 112).

Jamming avoidance responses (JAR) are found in gymnotoid pulse species such as *Hypopygus lepturus, Hypopomus artedi,* and *Parupygus savannensis.* The JAR in pulse species strongly resembles the JAR in wave species (e.g., *Eigenmannia*) and can be considered an evolutionary ancestor of the latter. In contrast to all other investigated wave-emitting electric fish, the genus *Sternopygus* does not exhibit JAR. Interestingly these animals show an unusual "immunity" of their performance in electric image processing to jamming signals. This is possibly the reason that this genus has not evolved a JAR system [106].

Let us now study in detail some neurophysiological principles underlying this avoidance response in the high-frequency electric fish *Eigenmannia virescens.*

Key-Stimuli for the Jamming Avoidance Response. "Electrical behavior" provides several neuroethological advantages for the investigator: on the one hand it can be exactly measured; on the other it may be easily simulated. Thus, the key stimulus for the jamming avoidance response may be examined by replacing the interfering field of the conspecific fish by an artificial dipole. Two electrodes partly submerged in water and a sine wave generator serve this purpose. If the frequency F_s of the simulated conspecific is slightly below the sending frequency F_t of our fish by the value of $(-)\,\Delta F$, then its sending frequency increases and we may drive it further upward by maintaining the value of ΔF with the simulator (Fig. 113B, left). If, conversely, F_s is above the fish sending frequency F_t by $(+)\,\Delta F$, then the fish shifts its frequency downward in an analogous manner (Fig. 113B, right). These frequency shifts can be simulated experimentally (Fig. 113A): the discharges of the electric organ are electronically transformed into sine waves. An "operator"

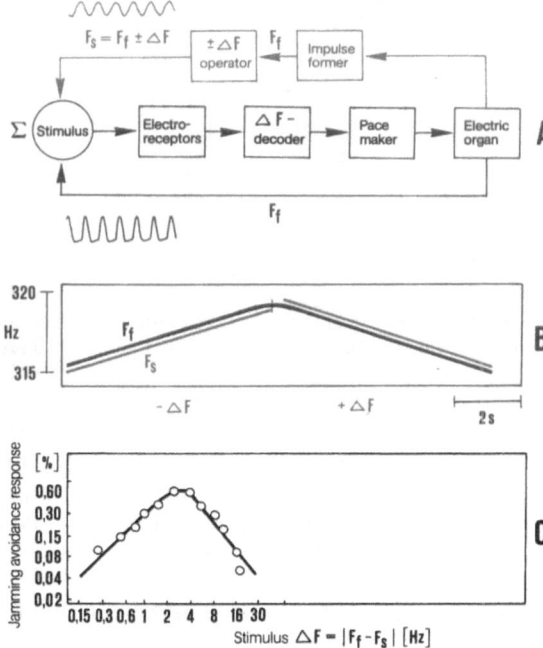

Fig. 113 A–C. Quantitative analysis of the key-stimulus ($\pm \Delta F$) for the jamming avoidance response of the electric fish *Eigenmannia*. **A** Experimental set-up for producing an electric field simulating a conspecific with the frequency $F_s = F_f \pm \Delta F$. *Red,* Simulated conspecific with the frequency F_s; *black,* discharges of the electric organ of the experimental fish with frequency F_f. Further details in text. **B** In the case of $-\Delta F$, the fish's frequency is shifted upward, in the case of $+\Delta F$, it is shifted downward. **C** Dependence of the jamming avoidance response on the magnitude of ΔF. [Combined and slightly modified from Bullock et al (1972) cit. in Fig. 109]

follows the frequency F_f by a constant value ΔF. The frequency of the simulated fish is $F_s = F_f \pm \Delta F$.

The direction in which the fish shifts its sending frequency depends on the sign of ΔF. When the amount of ΔF is changed in successive experiments, it appears that values between 3 and 4 Hz yield the strongest behavioral responses (Fig. 113 C). We thus know the composition of the key-stimulus for the jamming avoidance response.

How does the fish know whether it has to shift its frequency upward or downward? How does it determine the magnitude of the frequency difference $F_s - F_f$ and how does it recognize the sign of ΔF?

The Input Signal. Before searching for possible ΔF detectors we should clarify what the signal reaching the receptors really looks like. An excursion into physics may be helpful in this regard. When two sinusoidal oscillations, F_1 and F_2, whose wavelengths are slightly

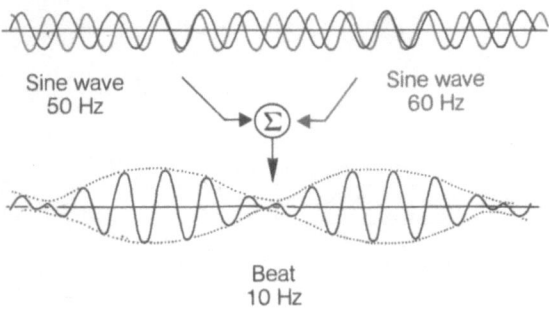

Sine wave
50 Hz

Sine wave
60 Hz

Σ

Beat
10 Hz

Fig. 114. Superposition of two sine waves with the frequencies $F_1 = 60$ Hz and $F_2 = 50$ Hz results in a beat. The beat frequency is $F = F_1 - F_2 = 10$ Hz. See text for further details

different, are superimposed, beats develop with periodical amplitude fluctuations (Fig. 114). The resulting beat frequency is $F = F_1 - F_2$; the smaller the frequency difference between F_1 and F_2 the lower is the beat frequency. The positive and negative beat envelopes are symmetrical.

Applied to our problem, this would mean that the sending frequency F_f of the fish and the jamming frequency F_s form beats by linear superposition, where $\Delta F = F_s - F_f$. The envelopes of these beats (Fig. 115B), however, show deviations from that previously described (Fig. 114). Although here again the positive beat envelopes are symmetrical, the negative envelope maxima are shifted away from the center of the beat cycle: for $(+)$ ΔF they appear later, and for $(-)$ ΔF earlier (see red arrows in Fig. 115B). The reason for this asymmetry is that the electric discharge corresponds to a distorted sine wave (Fig. 115A,a).

Which components of the input stimulus (beats) contain the information about the sign of ΔF? It is possible that the asymmetry of the beat envelope maxima might be responsible. Experimentally this can be tested in an "open loop" experiment.

For this purpose the electric organ is silenced by application of curare. The recorded impulses from neurons of the medullary pacemaker nucleus now trigger a sine wave generator with the pacemaker frequency F_f. The electroreceptors of the fish are then excited by the *symmetrical* beats of two pure sine waves with the frequencies of F_f and $F_f + \Delta F$. Result: although the fish avoids the interference owing to the ΔF difference by changing its pacemaker frequency, the direction of the change cannot be predicted [107]. Apparently the fish is no longer able to recognize the sign of ΔF. The reasons could be the symmetrical beat envelope maxima resulting from the two pure sine waves.

As soon as the stronger sine wave is clipped in a certain polarity (Fig. 115A,a), corresponding to the natural discharges of the electric organ,

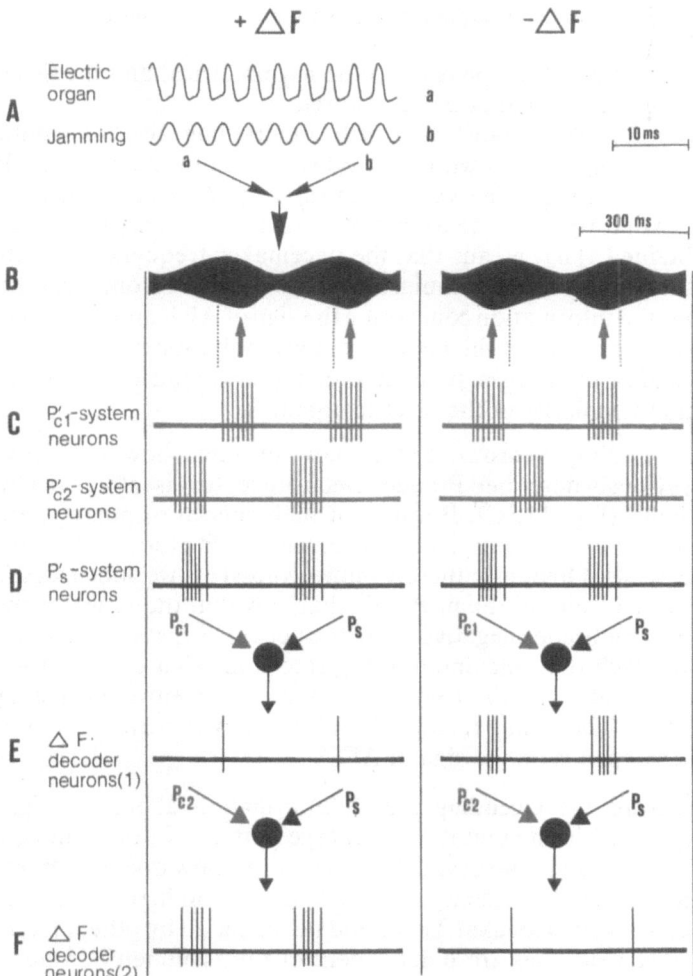

Fig. 115 A–F. A model explaining ΔF recognition in the electric fish *Eigenmannia*. The table lists physical and neurophysiological phenomena related to the sign of ΔF. **A** The discharges of the electric organ of a fish *a* and the waves of a simulated conspecific *b* are superimposed to yield a beat with asymmetrical envelopes. **B** In the negative range, the position of the beat envelope maxima depends on the sign of ΔF (see *arrows*). **C** Two electroreceptive units (P'_{c1} and P'_{c2}) whose responses are related in time to the beat envelope maxima. **D** Electroreceptive unit (P'_s) briefly activated at each ascending phase of the beat envelope and rapidly adapting afterwards. They form a constant time reference system for the P'_{c1}- and P'_{c2}-units. **E** ΔF(1)-decoder neurons receiving subthreshold input each from P'_{c1}- and P'_s-units. Their discharge rate is unaffected at $+\Delta F$ but is enhanced at $-\Delta F$ due to the coincidence of the two inputs from P'_{c1} and P'_s. **F** ΔF(2)-decoder neurons receive subthreshold inputs each from P'_{c2}- and P'_s-units and show opposite responses to $+$ and $-\Delta F$. The ΔF(1,2)-decoder neurons possibly control the frequency of neurons in the pacemaker center and thereby the discharge frequency of the electric organ. A further ΔF recognition system which may be formed both by P- and T-units is also discussed. [Combined and modified from Scheich H, Bullock T H (1974). In: Autrum H et al (eds) Handbook of sensory physiology, vol III/3; Fessard A (ed) Springer, Berlin Heidelberg New York; Scheich H (1977) J Comp Physiol 113:181–255]

the beat envelope peaks become asymmetrical and the fish's ability to recognize the sign of ΔF is restored.

Another experiment is worth mentioning in this context: if by stimulating the fish with electrodes in front of the head and behind the tail, the stronger sine wave is clipped in the opposite polarity – producing a wave form opposite to that of its own electric organ – an *anti-JAR* is obtained. That means that the pacemaker frequency is shifted toward that of the weaker stimulus. We may conclude from these experiments that the information concerning the sign of ΔF is apparently contained in the asymmetry of the beat envelopes of the superimposed sending and jamming discharges. It depends on the asymmetrical course of discharge and the polarity of the electric organ.

ΔF Coding Neurons. Let us now examine how the activity of the previously described P-coder electroreceptors is influenced by such beat stimuli (Fig. 115C). Because of their special response characteristics, the activity of P-units is clearly correlated with the negative beat-envelope peaks. Obviously these P_c-units consist of two subclasses. Representatives of one of them, P_{c1}, discharge within the range of the maximal beat amplitude (negative envelope peaks). Representatives of the other, P_{c2}, reach their maximum firing rate somewhat earlier for $(-)\Delta F$ and somewhat later for $(+)\Delta F$ than the beat envelope maximum (Fig. 115C). The *timing relationship* of the *activity pattern* of both unit types thus depends on the sign of ΔF.

Neurons for Decoding ΔF. If information about the sign of ΔF is contained in the temporal discharge pattern of the P_c-units, how is this pattern read in the brain? In other words, how does the brain of the fish know when the discharges of the P_c-unit occur in relation to the negative beat-envelope peaks? This could be mediated by other electroreceptors whose responses are independent of the asymmetry of the beat peaks. Such P_s-units have indeed been found. In principle they discharge during the ascending phase of the beat envelope and rapidly adapt so that asymmetries in the time course of the beat envelope do not affect their spike frequency. The responses of these units may represent a *constant frame of reference* (Fig. 115D).

If the response characteristics of neurons are now examined at higher levels of the CNS, analogous P_c'- and P_s'-units are recorded from the posterior lobe and the torus semicircularis (Fig. 115 C and D, cf. also Fig. 109). They show a tendency to accentuate the receptor responses described and may together form neuronal systems. However, in the torus semicircularis we find also neurons that are activated predominantly when ΔF is positive, and are almost silent when ΔF is negative (Fig. 115F). We call them ΔF decoding neurons.

How can the response characteristics of such a *decoder* be interpreted? It is assumed that the decoder receives subthreshold excitatory inputs from P'_{c_2}- and P'_s-units (cf. Fig. 115F). If ΔF is positive the activity of both unit types will coincide and their combined inputs will exceed the threshold, thus activating the ΔF-decoder. If ΔF is negative, the inputs of P'_{c_2}- and P'_s-units are out of phase so that their sum is still below the threshold of the ΔF-decoder. Apart from this $\Delta F(2)$-decoder, a $\Delta F(1)$-decoder might be assumed, which receives subthreshold inputs from P'_{c_1}- and P'_s-units and whose characteristics for $\pm \Delta F$ are exactly opposite (Fig. 115E).

Suppose these ΔF decoding neurons are components of a system that controls the pacemaker of the electric organ and that $\Delta F(2)$-decoder inhibits while $\Delta F(1)$-decoder excites the pacemaker. Then these ΔF decoding neurons could be responsible for the modulation of the

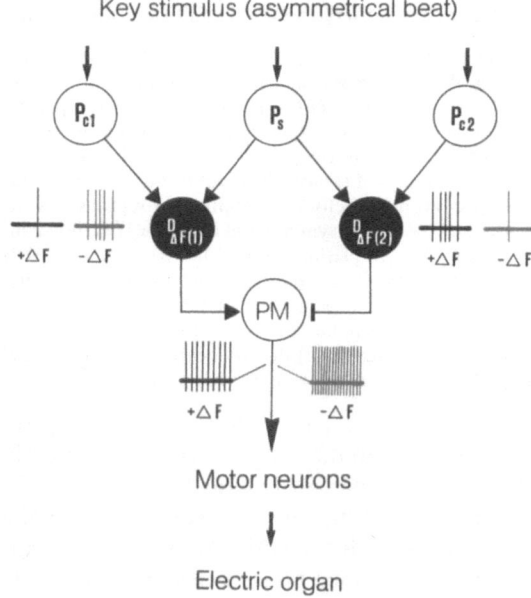

Fig. 116. A possible circuit for the jamming avoidance response of *Eigenmannia*. This scheme summarizes some important information processing steps that might participate in the recognition of ΔF and therefore play a role in the frequency control of the pacemaker. The neurons shown represent neuronal systems. $\Delta F(1)$ as well as $\Delta F(2)$ neurons may be considered as ΔF detectors; they show opposite response behavior to $+$ and $-\Delta F$ on the basis of subthreshold inputs each from P_{c1} and P_s, or from P_{c2}- and P_s-units; see schematized impulse patterns. (Explanation in legend of Fig. 115). The $\Delta F(1)$-decoders may form excitatory, the $\Delta F(2)$-decoders inhibitory synapses with neurons of the pacemaker center (PM). Together they might increase the pacemaker frequency for $-\Delta F$ and decrease it for $+\Delta F$. *Arrows* indicate excitatory, and *lines with cross bar* inhibitory synapses. [Based on the results presented by Scheich and Bullock (1974), Scheich (1977), cit. in Fig. 115]

pacemaker frequency by $\pm \Delta F$ (Fig. 116). The ΔF-decoders may be regarded as detectors for the key-stimulus $\pm \Delta F$ and as vital components of the control system for the jamming avoidance response. Scheich and Bullock (1974) are also discussing a model that considers information both from probability (P) and phase (T) coders.

Are There Still Other Models to Explain Jamming Avoidance Responses? To repeat: if stimuli S_1 (with frequency F_1) and S_2 (with frequency F_2) are added and presented through the same pair of electrodes, no "correct" jamming avoidance responses are elicited, provided both stimuli are pure sine waves. If S_1 and/or S_2 are clipped in one polarity and a sufficiently high intensity is chosen, JAR's or anti-JAR's are obtained, depending upon the polarity of the clipping. However, pure sine waves are sufficient to elicit correct JAR, if S_1 and S_2 are presented with *different field geometries*. This has been demonstrated quite recently by Walter Heiligenberg and his colleagues [108].

The electric organ was silenced by curarization. A train of *strong* stimulus pulses S_1 consisting of pure sine waves was designed to mimic the animal's experience of its own electric organ discharges (EOD); these stimuli were presented via a stomach electrode which sufficiently simulated the animal's EOD field geometry. Pulses S_1 were free-running and *not* phase-locked to the pacemaker. The stimulus S_1 was superimposed by a train of *small* sine waves S_2 (of slightly different frequency). Stimulus S_2 should mimic the EOD's of a conspecific and therefore was applied through a set of lateral electrodes. The resulting sinusoidal beat wave has symmetrical envelopes (as shown in Fig. 114) and its frequency is $F_1 - F_2 = \Delta F$. Surprisingly the fish's pacemaker activity reflects correct jamming avoidance responses for $F_2 \neq F_1$. This means that in this situation, too, the fish is able to detect the magnitude and sign of ΔF. Under these conditions (1) the fish needs no internal reference to the pacemaker and thus extracts all necessary information exclusively from electroreceptor afferents, (2) the fish needs no exact mimic of the EOD wave shape, such as clipped sine waves (pure sine is sufficient).

In these experiments the strong stimulus, S_1, that mimics the animal's own electric organ discharges, is contaminated by a weaker stimulus S_2, which simulates the conspecific: $S_1 + S_2 = S$ (Fig. 117 A). The fact that S_1 and S_2 while being pure sine waves can still elicit JAR's as long as their field geometries differ sufficiently implies that different parts of the animal's body surface must experience different intensities of S_1 and S_2. But how does the fish in this situation solve the dilemma of finding the right sign of ΔF?

Whereas the beat *amplitude modulation*, $|S|$, is independent of the sign of ΔF, *phase modulations*, H, are opposites of one another for positive and negative ΔF. In Figure 117 B amplitude and phase parameters are plotted in a two-dimensional state plane. The direction in which a point moves in this graph is determined by the sign of ΔF.

Evaluation of the phase modulation requires a *zero phase reference*, namely the moments of zero crossings of an uncontaminated stimulus S_1. A direct reference to the pacemaker could fulfill this function. However, since phase locking of S_1 to the pacemaker is not necessary to elicit JAR,

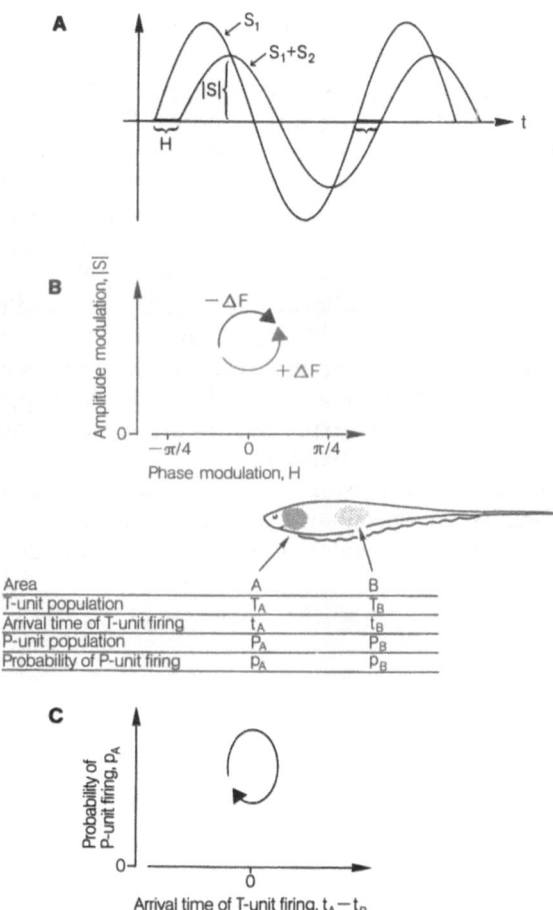

Fig. 117 A–C. A possible model for analyzing ΔF in *Eigenmannia* if both stimuli S_1 and S_2 are presented as pure sine waves with different field geometries on the fish's body surface. (S_1 is a strong stimulus $\triangle$ frequency of fish; S_2 is a weak stimulus $\triangle$ frequency of simulated conspecific). **A** Presentation of S_1 and $S(= S_1 + S_2)$ as functions of time (t). The phase angle (H) is represented by the difference in times of positive zero crossings of S_1 and S (*thick bar* on abscissa). $|S|$ is the amplitude of the mixed signal and known as the beat envelope. **B** Presentation of H and $|S|$ as Lissajous Figure in two-dimensional state-plane. For a sufficiently small value $|S_2|/|S_1|$, which is 1/3 in this figure, the coordinate pair (H, $|S|$), travels in a circular path: counter clockwise rotation is linked to *positive* ΔF. **C** *Above*, Given areas A and B on the animal's body surface are represented in the CNS (torus semicircularis) by higher order T'- and P'-units. S_1 in area A is heavily contaminated by S_2 ($|S_2|/|S_1| \sim 1/3$); S_1 in area B is only lightly contaminated by S_2. *Below*, Arrival time of T-unit firing reflects phase modulation, H, and probability of P-unit firing reflects amplitude modulation $|S|$. [Modified from Heiligenberg W, Baker C, Matsubara J (1978) J Comp Physiol 127:267–286]

the animal could obtain phase reference from an area of the body surface least contaminated by S_2.

Amplitude and *phase relationships* can be coded by P- and T-types of electroreceptors (Fig. 110a and b): P-units coding stimulus amplitude by the probability of firing, and T-units indicating the time of positive zero-crossings of a sinusoidal stimulus. Corresponding T'- and P'-units in the torus semicircularis are *somatotopically* organized. Assume that given areas A, B, ... on the animal's body surface are represented by higher order neurons T'_A, T'_B, ... and P'_A, P'_B, ..., the activity of these neurons reflects the locally averaged phase modulation function, H, and the beat envelope, $|S|$, respectively in the corresponding areas (Fig. 117C, above). A symmetrical interaction between pairs of differently S_2-contaminated areas, A and B, might yield opposite interpretations of the sign of ΔF, as expressed by the sense of rotation of the respective graph (Fig. 117C, below).

Neuronal mechanisms which detect motions in the state plane of P'- and T'-units are speculative at present. Processes comparable to motion detection in the realm of vision might be discucced in this context.

We may recapitulate:

1. Just like echolocation, the electrolocation of electric fish represents an adaptation to particular environments and life styles. By means of their electrosensory system, and corresponding electrical image processing, these animals are able to navigate, recognize objects in the environment, and communicate with conspecifics in the neighborhood.

2. Fish whose electric fields interfere with each other during the evaluation of their environment shift their sending frequencies into private ranges. This jamming avoidance response is a simple form of social behavior. The key-stimuli are the magnitude and the arithmetical sign of the difference (ΔF) between the jamming frequency and the fish's own sending frequency.

3. There are receptor and neuron systems capable of coding the magnitude and sign of ΔF. Recognition of ΔF may be based on the time patterns in the responses of two different types of unit. Thus, in this instance analysis of waveform is probably not achieved by means of frequency filters *(frequency analysis)*; rather the information is processed in the time domain *(time domain analysis)*.

4. New investigations reveal that *stimulus field geometry* – a parameter not included in the initial theory – is an important cue by which an animal distinguishes its own electric organ discharges, S_1, from foreign signals, S_2. Evidence is given that the jamming

avoidance response is controlled in a cumulative manner, by local interactions of neighboring electroreceptive fields on the animal's body surface which, as a consequence of different S_1 and S_2 field geometries, experience different degrees of contamination of S_1 by S_2.

6.5 Attenuation of Self-Stimulation

6.5.1 The Reafference Principle

There are numerous types of behavior that result in self-stimulations of the individual's own sensory systems. On the one hand, this may contribute to the processing of information, but, on the other, this may also interfere with the recognition and localization of environmental signals. For instance, when we look at an object before us, different regions of the retina are stimulated, because of the tracking movements of the eyes. In our perception, however, the position of the object relative to the background remains constant. Obviously, the displacements of retinal images due to the turning of our gaze are screened out or, rather, corrected.

Proper evaluation of the environment is based on central processes governed by the so-called "reafference principle" (Fig. 118 A). We may illustrate this by the example mentioned above. An eye movement command w_E corresponds to a message w_M about a moving retinal image, which is conducted to a center in the brain. A copy w_E of the "eye movement command" also arrives in this center; we call it the *efference copy*. It is subtracted there from the corresponding sensory message and the retinal image is perceived as $w_P = w_M - w_E$; in our example $w_P = 0$.

6.5.2 Eye Lid Shadow Inhibition in Toads

Eye ball retraction in toads may result in a closing of the eye lids. The function is to either protect the cornea against mechanical irritation, or reduce moving images on the retina during certain movements of the animal (e.g., walking, jumping). When a toad darkens its retina by lid closure, those ganglion cells of its retina (classes R3 and R4, see p. 102) that normally respond to sudden darkening of the total visual field, remain silent (recording procedure is shown in Fig. 161). However, R3 and R4 neurons do respond, when an artificial lid shadow – simulating the lid-closure dynamics and the reduction of light – is applied to the open eye. Presumably, the command "close the eye" is coupled with a further command to the retina [109] "inhibit the *off*-reaction of the

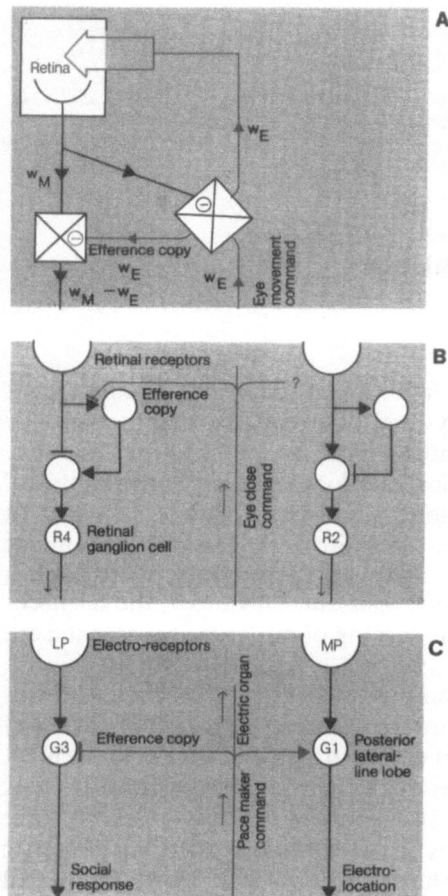

Fig. 118 A–C. Speculations and concepts on the attenuation of self -excitation. **A** The reafference principle, explained for voluntary eye-movement and the perception of the stationary environment; w_E eye movement command, w_M retinal message on moving image. **B** The "eye lid shadow inhibition" in the common toad; $R2$, $R4$ classes of retinal ganglion cells ($R4$ being *off*-units), *interneurons* bipolars and amacrines. **C** Screening of self-induced electric fields in mormyrid electric fish; *LP* large pore electroreceptors, *MP* medium pore electroreceptors; $G1$ neuron functioning as an AND-gate transmitting information in the simultaneous presence of excitatory pacemaker command-related input; $G3$ neuron functioning as an AND-gate subtracting pacemaker command related input from corresponding electric image. *Arrows* denote excitatory, and *lines with cross bars* inhibitory synapses. For explanations see text. [Combined and modified from Holst E v, Mittelstaedt H Naturwissenschaften 37:464–476; in: Hassenstein B (1966) in: Gessner F (ed) Handbuch der Biologie Vol I/2. Akad Verlagsges, Frankfurt/M (**A**); Borchers HW, Ewert JP (1978) J Comp Physiol 125:301–303 (**B**); Zipser B, Bennett M V L (1976) J Neurophysiol 39:693–721 (**C**)]

corresponding ganglion cells" (Fig. 118 B). If, however, the neuronal *off*-activation has been induced by darkening of the visual field while eye was open, it can no longer be inhibited by a subsequent lid closure. This might be an indication that the presumed efferent inhibition due to lid closure reaches a point within the retinal network before the *off*-activity is generated (Fig. 118 B, left). By the "lid shadow inhibition" the toad brain should be able to focus on changes in brightness occurring only in the visual environment. Such a system would, however, guarantee that moving retinal images of a stationary stimulus, caused by slight eye ball retractions due to *distinct* motor commands, can be evaluated.

6.5.3 Screening of Self-Induced Electric Fields

In mormyrid electric fish so-called AND-gates have been identified in their CNS (posterior lateral line lobe) that transmit only information when electrical organ discharges originate from conspecifics. The corresponding neurons belong to the G3 type (Fig. 118 C). They are inhibited by the "pacemaker command" precisely at that time when excitation by the animal's own electric organ discharge is expected to arrive. The G3-system, which receives inputs from particular receptors (large pores), is used for the control of social responses. In addition there are distinct AND-gates, G1, receiving inputs from other receptor groups (medium pores). In contrast to those described previously, these gates will transmit information only when, at the same time as the receptor input, a pacemaker command-related signal is present as required for electrical image processing. The G1-system forms an important basis for electrolocation of environmental objects.

6.5.4 Attenuation of Self-Stimulation During Vocalization

The auditory system is also excited by the individual's own vocalizations. Self-stimulation may be important, e.g., for the control of vocalization; however, only weak excitatory inputs are required for this purpose. In humans, cats, and bats it is known that the middle ear muscles contract synchronously during vocalization, and thereby reduce the self-stimulation. There is also attenuation by neural mechanisms.

As we have seen previously, bats of the genus *Myotis* emit frequency-modulated orientation sounds for echolocation. If these sounds were to stimulate their own ears directly, the perception of the echo would be strongly diminished at short distances. To avoid this, the response to self-produced sounds is decreased neuronally [110]: synchronously with vocalization the activity of neurons in the nucleus of the lateral lemniscus – an auditory pathway – is reduced by about 15 dB. Together with the middle ear muscle contraction, a stimulus decrement of altogether 35–40 dB results.

In primates certain areas of the limbic cortex (precallosal cingulate cortex) and the midbrain central grey play an important role in the production of vocalization. By means of combined stimulation and recording techniques it can be shown in squirrel monkeys that these limbic structures have inhibitory influences on neurons of the auditory cortex (gyrus temporalis superior) which are involved in signal processing. Many of these neurons are strongly activated by playbacks of tape-recorded self-produced vocalizations. However, during the animal's own vocal activity the responses of the same auditory neurons are strongly damped [111].

To recapitulate:

Sensory localization and recognition systems must have input channels that are free from interference by self-induced stimulation in particular situations. Attenuation of excitation due to self-stimulation is found in various sensory systems. They may be based on the reafference principle. We already have some examples that suggest neuronal substrates of this phenomenon.

7 Neuronal Circuits for Fixed Motor Behavior Programs

The electrical brain stimulation experiments in crickets and toads suggest that programs exist in the CNS for the triggering of species-specific behavior patterns. It may be assumed that such motor programs are based on fixed neuronal circuits which, when activated, initiate a spatially and temporally coordinated motor pattern. Such programs could be initiated spontaneously via commands originating in the brain itself, or they could be released by signals from the environment, following appropriate stimulus identification. It is conceivable that at the end of the relevant information processing chain there is a particular neuronal system that, when appropriately stimulated, activates the corresponding motor program.

Do We Have Any Experimental Proof For Neuronal "Command Systems" and "Program Circuits"? Such systems have been characterized in invertebrates such as lobsters and certain marine slugs. Gastropod animals are particularly suitable for such studies because their whole nervous system consists of relatively few neurons. The brain of man is made up of about 100,000,000,000 cells, the CNS of higher invertebrates contains only around 100,000 cells [112]. Moreover, the cell bodies and fibers of invertebrates are sufficiently large – up to several hundred microns – to be detected sometimes without a microscope. Some of the neurons can be identified by their distinctive pigmentation (black, white, red, yellow). Furthermore the same cells have the same locations in every preparation (concept of "identified neurons") [113]. The experimental advantages of this system are thus considerable.

7.1 Example of a Neuronal Program Circuitry

As an example of a fixed action pattern we shall study the escape reaction of *Tritonia diomedia*, which is a marine nudibranch gastropod.

The Escape Motor Pattern. *Tritonia* is about 12 cm long. Its enemies are members of a certain starfish species. When *Tritonia* comes into contact with them (a drop of concentrated salt solution may be used in the laboratory as key-stimulus), it responds with a sequence of stereotyped movements lasting about 10–90s: retraction of all body appendages (rhinophores, gills); paddle-like swimming movements due to alternating contractions of the flattened dorsal and ventral body sides. There are altogether 1–8 contraction cycles. This escape swimming simply removes the slug from the site of stimulation. The time from stimulation to the onset of swimming varies between 2 and 6 s.

Brain Organization. The CNS of *Tritonia* consists essentially of three closely grouped ganglion pairs. The ganglia receive messages from sensory organs, e.g., the chemoreceptors of the skin, via sensory fiber tracts; outputs to various muscle groups are mediated by motor fibers.

Intracellular Stimulation Experiments. Electrical stimulation experiments using microelectrodes provide insight into the behavioral relevance of single neurons. During the experiment the animal is freely moving, only the area surrounding the brain is fixed by holders for electrical stimulation (Fig. 119). When certain neurons are stimulated intracellularly, specific effects may occur: retraction of body appendages, contraction of the dorsal or ventral muscles. However, there are also neurons which initiate, upon brief stimulation (500 ms) the complete sequence of behavior lasting on the average 30 s. These trigger neurons appear to interact with each other via electrically conducting junctions. This allows appropriate sensory input to be magnified into a discrete burst or into a short train of self-limiting bursts.

These observations show some parallels to the brain stimulation experiments in toads: Various components of the prey-catching behavior could be triggered by stimulation of certain points in the brain.

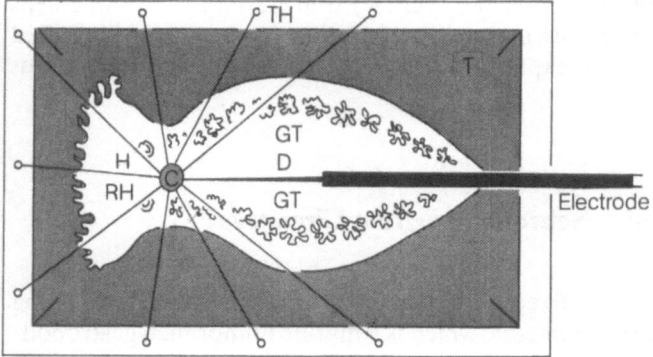

Fig. 119. Experimental set-up for electrical brain stimulation and cell recording in the marine snail *Tritonia*. The experimental animal is in a tank filled with sea water. The area surrounding the brain is immobilized by fixing it with threads. The other body parts are freely movable. *C* cerebral ganglion, *D* Dorsum, *GT* gill tufts, *H* head, *RH* rhinophores, *T* tank, *TH* thread hook. [Modified from Willows A O D (1973) Fed Proc 32:2215–2223]

There were also regions which in response to brief stimulation brought about the whole behavioral sequence in the correct order. Whereas, however, in the toad we were able to correlate only relatively gross brain areas to single behavioral components, it is possible with invertebrates to study analogous relationships at the level of single neurons.

Intracellular Recordings During Various Behavior Phases. Recordings from individual cells in the freely moving animal permitted the identification of different neuron types whose activation patterns are correlated with specific behavior phases. There are five main types:

1. Neurons activated between stimulation and movement.
2. Neurons activated during retraction of body appendages, but inhibited during the swimming movements.
3. Neurons strongly discharging generally during the movement phases.
4. Motor neurons excited during dorsal flexion.
5. Motor neurons excited during ventral flexion.

A Neuronal Circuit for Escape Swimming. On the basis of these results how might the neuronal circuit be organized for escape behavior in

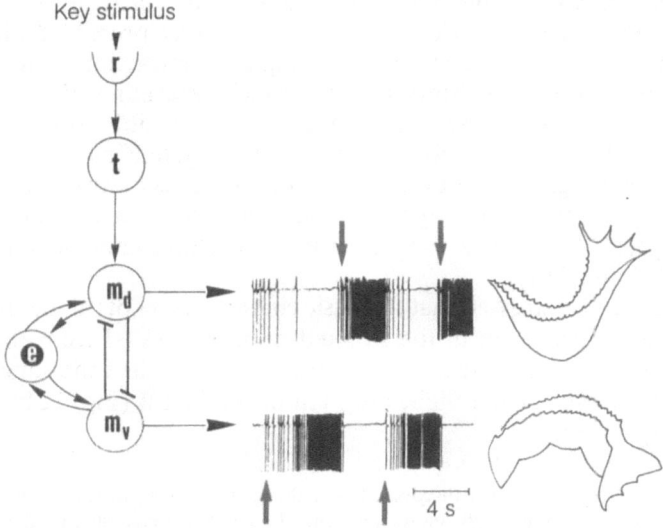

Fig. 120. Simplified neuronal circuit controlling ventro-dorsal flexion cycles in escape-swimming of *Tritonia*. The scheme accounts for identified neuron types; they stand for neuronal networks. *r* Chemoreceptors, *t* trigger system neurons, m_d motor neurons driving dorsal flexor muscles, m_v motor neurons driving ventral flexor muscles, *e* system of interneurons in which excitation may be stored by positive feedback. *Arrows* denote excitatory, and *lines with cross bars* inhibitory synapses. The two impulse patterns shown are original recordings of the two motor neuron types during ventral-dorsal flexion cycles; both types (see *red arrows*) are alternately excited and inhibited. [Modified from Willows (1973) cit. in Fig. 119]

Tritonia? We can illustrate the principle in a rough scheme. The neuronal types shown in Figure 120 are representatives of populations that may themselves be interconnected in a network.

Let us begin with the question of the *key-stimulus* and its identification. Certain surface-active chemicals are found to be key-stimuli, e.g., a drop of concentrated soap or sodium chloride solution may simulate an enemy. The sensitivity for this is distributed over the entire body; the minimum stimulus area comprises only a few mm^2.

Where is the *stimulus filter* located? Neurophysiological studies show that basic processes of stimulus identification – similar to that in odor specialists of insects – occurs at the level of the receptor membrane due to its functional specialization.

Neurons that are activated by the *recognition system* constitute a trigger system that – when excited – brings about the discharge of motor neurons (Fig. 120 m_d, m_v) which cause alternating contractions of the dorsal and ventral body sides, according to a definite time program. The output of the *trigger system* obviously acts as a signal for the subsequent self-sustaining process, which is itself independent of the key-stimulus. Essentially three different populations of neuron types probably participate in the neuronal escape *motor program* (Fig. 120). The program characteristics are based on: (1) reciprocal inhibition between the motor antagonists guaranteeing the contraction cycles; (2) positive feedback via excitatory neurons which determines the duration of the motor pattern. The trigger system serves mainly to initiate the motor program. Since neurons of the trigger system are also active during swimming, the question arises as to how far activity in trigger neurons is responsible for re-triggering the swim oscillator network cycle – by – cycle [114]. Hence, these neurons may fulfill command functions (cf. pp. 213 and 214).

It must be stressed that the basic characteristics of this neuronal circuit is maintained even in the isolated brain separated from the body. This proves that the escape program of *Tritonia* – like the song programs of the cricket or the flight program of the locust (Fig. 121) – is centrally organized.

Recently, it has been discovered that the mating dance of the moth *Bombyx mori* is controlled by the flight motor pattern. The dance is performed by the male. The most characteristic component of the dance – wing vibration – plays an essential role in the olfactory location of the female. Males sniff the air to test the pheromone by performing the wing vibration during mating dance. What is the neural basis of the dance pattern? By means of recordings from the indirect flight muscles it could be shown that the elevators and depressors of the wing fire orderly during flight, whereas disorderly during mating dance. The flight motor pattern becomes disordered as a result of inhibition of the depressors. Thus, the mating dance is a disorderly inhibited flight pattern. Inhibition is produced by varying mechano-sensory inputs from the leg regions which fluctuate in intensity during the mating dance. The mating dance is *initiated* by the sex pheromone. There are descending interneurons in the cervical connectives which respond specifically

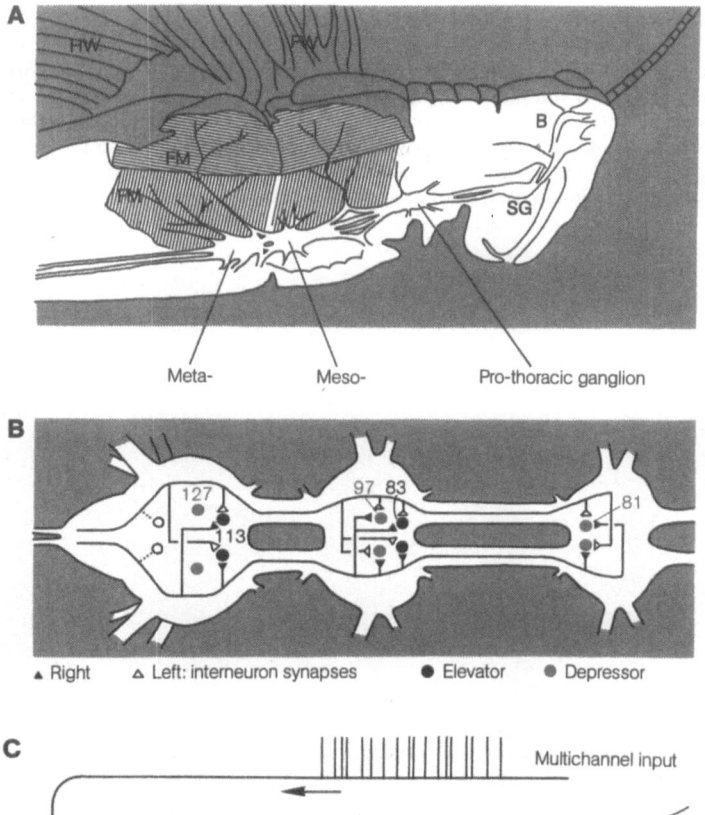

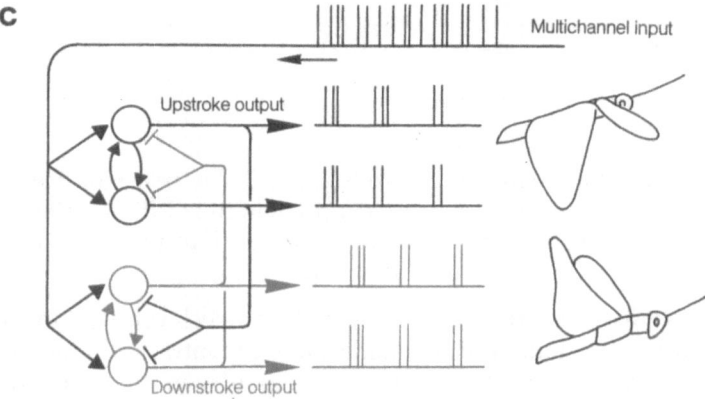

Fig. 121 A–C. Hypothetical circuit for the control of wing upstroke and downstroke in the flying locust. **A** Schematic view of the locust showing the brain *(B)*, subesophageal ganglion *(SG)*, thoracic ganglia (Meta-, Meso-, Pro-), flight muscles *(FM)*, and wings *(HW, FW)*. Upstroke muscles are *vertically* and downstroke muscles *horizontally striped*. **B** Simplified representation showing synaptic connections made upon some flight motor neurons, elevators and depressors, by interneurons. About 30 flight motor neurons, located in the three thoracic ganglia, have been identified individually in the locust *Schistocerca gregaria*. **C** Simplified possible circuit for the control of alternating wing *upstroke* and *downstroke* cycles, based on (1) tonic multi channel input (command), (2) cross excitation (time program), (3) cross inhibition (motor antagonist coordination). The principle may be similar to that in Figure 120. *Arrows* indicate excitatory and *lines with cross bars* inhibitory synapses. [Modified from Wilson D M (1968) Sci Am 218:83–90 (A, C); Burrows M (1975) J Exp Biol 63:713–733 (B)]

to the sex pheromone with a long-lasting characteristic firing pattern. The corresponding fibers project to the thoracic ganglion and they are supposed to be associated with a command system for the mating dance. With the onset of copulation the mating dance is immediately *inhibited* as a result of ascending sensory inputs from the genital organs. Experimentally the dance can be inhibited by electrical stimulation of this pathway (Y. Obara and Y. Uondo pers. comm., Vancouver, B.C. 1979).

To recapitulate:

There are in the animal kingdom examples in which fixed motor programs and corresponding trigger or command neurons can be analyzed at a cellular level. Essential components of these programs are positive feedback systems *(time program)* and systems of reciprocal inhibition *(antagonist coordination)*. Such neuronal circuits produce alternating bursts of activity. Central pattern generators form the basis for a variety of autorhythmic behavior patterns, e.g., for the coordination of locomotion (cf. Figs. 120 and 121).

There are motor programs which appear to be organized and controlled exclusively by the CNS. Depending on the tasks and the environmental conditions, motor programs may also exhibit varying degrees of interaction between central and peripheral components. Finally, certain motor programs may be generated only with the involvement of peripheral feedback loops.

7.2 Command Neurons

Fixed neuronal circuits controlling particular behavior patterns with rapid time courses have been identified in the CNS of different animal groups. These motor programs are called into play by preconnected command cells or cell systems [115].

Giant Fibers in Crayfish. If the tail of a crayfish is briefly touched by a disturbing stimulus, the abdominal muscles contract very rapidly and the animal moves away. Four principal groups of neurons are involved in the circuitry for the tail-flip escape response (Fig. 122): (1) mechanoreceptors (pit-hair receptors of the carapace), (2) different types of interneurons, (3) command neurons, and (4) motor neurons.

The rapid escape response is mediated by electrical synapses and thick, fast-conducting command fibers. They are called giant fibers. The lateral giant cells are segmented and tightly coupled – via contralateral axons – by *electrical synapses* which make them behave functionally as a fast system. A single action potential conducted in this system elicits the tail-flip response. In the re-extension, delayed sensory feed-forward and movement-induced sensory feedback systems are involved. Special

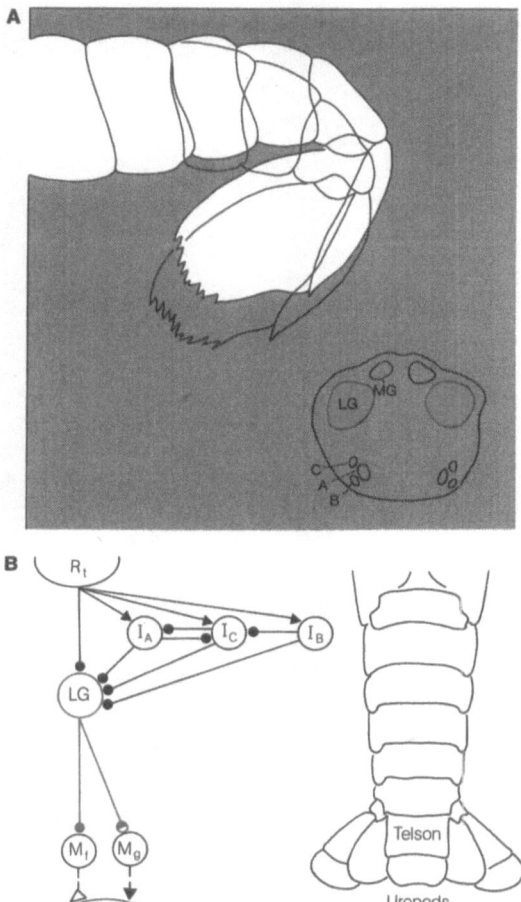

Fig. 122 A and B. Simplified circuit for identified neurons mediating escape response in crayfish *Procambarus clarkii*. **A** *Upper left,* Abdominal flexion response; *lower right,* section through the abdominal nerve cord showing lateral giant axon *(LG)*, median giant axon *(MG)*, and tactile interneurons (*A* unisegmental, *B, C* multisegmental). **B** Circuit of neurons involved in the escape response. R_t tactile receptors, $I_{A,B,C}$ interneurons, *LG* lateral giant neuron, M_f fast flexor motor neurons, M_g giant motor neuron. Only excitatory connections are shown: ▲ Plastic, antifacilitating chemical synapses that undergo depression with repeated stimulation, ● electrical junctions, ◖ unidirectional (rectifying) electrical synapse, Δ facilitating chemical synapse. [Modified from Zucker R S, Kennedy D, Selverston A I (1971) Science 173:645–649]

chemical synapses *(plastic synapses)* between mechanoreceptive input and interneurons are responsible for habituation of the behavior in the case of repetitive stimulation, a phenomenon already described in the example of the marine gastropod *Aplysia* (p. 49).

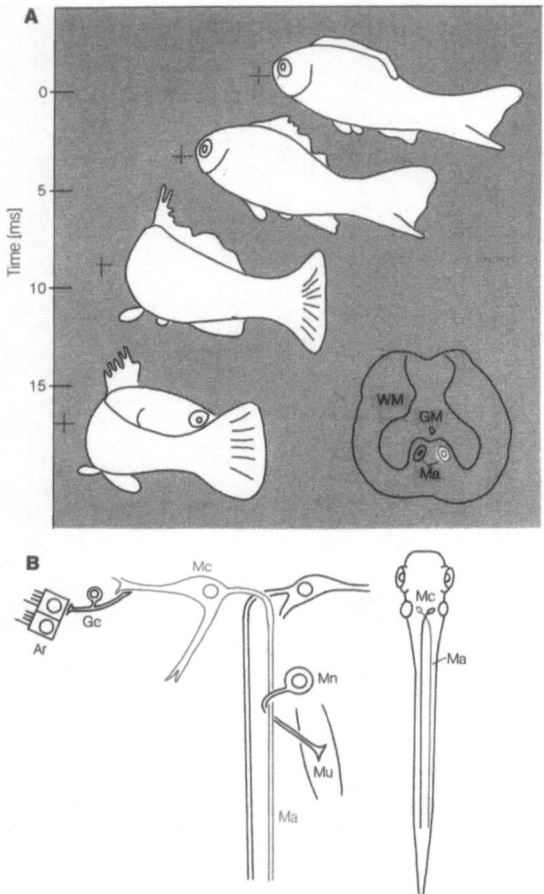

Fig. 123 A and B. Mauthner cell initiated startle response in teleost fish *Paralabrax clathratus* (kelp bass). **A** *Top,* A vibratory stimulus occurs at 0, the fish responds with fast-body-bend. Representative behavior patterns of a frame by frame analysis; crosses indicate a fixed reference point. The fish is about 6 cm long (head to tail). *Bottom right,* Section through spinal cord showing grey matter *(GM),* white matter *(WM)* and Mauthner axons *(Ma).* **B** Simplified neuronal circuit mediating startle response. *Ar* auditory receptors, *Gc* 8th ganglion cell, *Ma* Mauthner axon, *Mc* Mauthner cell, *Mn* spinal motor neuron, *Mu* muscle. [Modified from Eaton R C, Bombardieri R A, Meyer D L (1977) J Exp Biol 66:65–81; Eaton R C The Mauthner neuron system in embryos and larvae of the zebra fish *Brachydanio rerio.* Ph D Dissertation Univ Calif Riverside]

Mauthner Cell in Teleost Fish. Are there also comparable command neurons in vertebrates? If one taps on the side of an aquarium, the fish turn in a flash and swim in the opposite direction. An adequate stimulus situation in the biotope might be caused by diving or striking predatory birds. The stimulus reaction time of the fish is very brief; around 10 ms from the onset of the vibrational stimulus to the first body bend (Fig. 123 A). This action pattern is controlled by four sequentially connected neuron types: (1) hair cells responding to the vibrational stimulus, (2) 8th cranial nerve ganglion cells, (3) Mauthner cell, and (4) motor neurons. One of the two morphologically well-developed Mauthner cells functions as a command neuron. Their somata are arranged as a pair in the hindbrain and their axons lead after decussation to the spinal cord. In the zebra fish the Mauthner cell responds with a single action potential 6–17.5 ms after onset of the vibrational stimulus. The latency between the Mauthner cell spike and the first muscular contraction is about 2 ms. In this case a pool of contralateral spinal motor neurons is activated via chemical synapses by an action potential conducted in the Mauthner cell axon. Only the first phase of behavior ("fast-body-bend") is commanded by the Mauthner cell spike. This initial contraction ("startle response") turns the head away from the side of the vibrational stimulus. The subsequent motor patterns (return-flip and swimming) are controlled by other neuronal systems. Within 100 ms after the beginning of the startle response, fish are displaced 0.5–1.5 body lengths from their initial position. Habituation of the response to repetitive sensory stimuli occurs in larvae after hatching.

Does the Mauthner neuron mediate unique behavior? This question was recently investigated in zebra fish larvae with lesions of the Mauthner cells induced by irradiation at the gastrula stage [116]. The answer is that animals continue to show startle responses even when both Mauthner cells are missing. But the presence of the Mauthner cells increases the probability of eliciting the short-latency fast-body-bend. In animals with only one Mauthner cell a higher proportion of short-latency responses is observed to the side possessing the Mauthner axon than to the opposite side. Thus, these results indicate that there must be other neuronal circuits capable of producing startle responselike behavior in the absence of Mauthner cells. We do not know at present whether such wiring "arises" as a result of loss of Mauthner cells.

Mauthner cells also exist in larvae of anuran amphibians. This system, however, disappears after metamorphosis and transition to a terrestrial life.

7.3 Command Functions in the Primate Brain

There are experimental results that indicate that in the parietal lobe of the association cortex of primates, neural command circuits for directed visual attention have been developed. They fulfill integrating tasks and exert command functions regarding hand and eye movements. In contrast to the systems described above, these do not mediate relatively simple input-output relationships, rather, these command systems are far more complex. They depend on motivation and the particular situation, or, expressed in general terms, they are state-dependent.

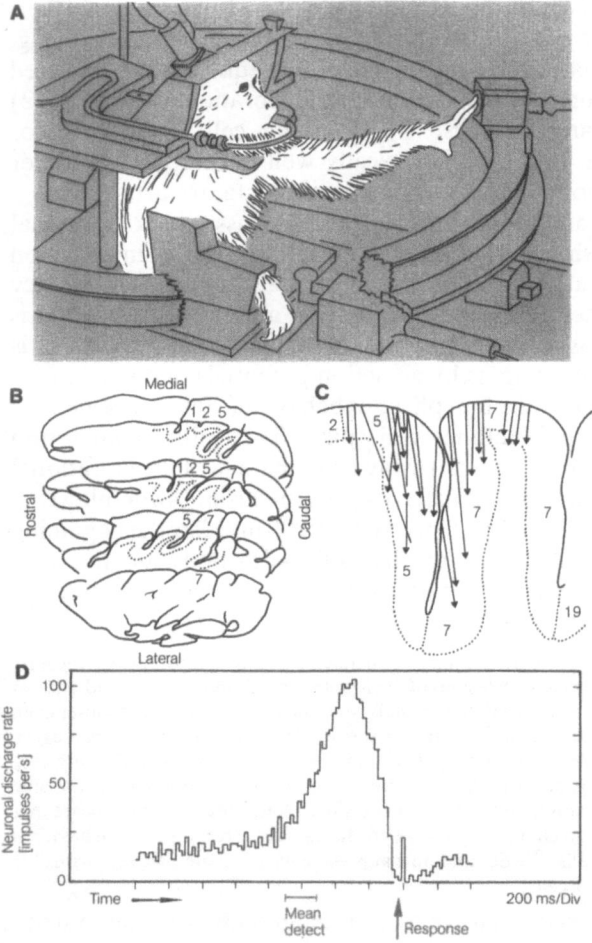

Fig. 124 A–D. Command functions in the parietal cortex of speciosa monkey *M. arctoides*. **A** Procedure for recording single neurons of the parietal cortex in the freely moving monkey, showing head fixation apparatus with electronics and reward tube. The animal has just released a key and projects that arm and hand forward to contact a lighted switch. The latter is mounted on a moving carriage that can be moved from any preset position in either direction. **B** Sections through the left cerebral hemisphere showing cytoarchitecture of the parietal lobe (around areas 5 and 7). **C** Reconstruction of recording electrode positions in areas 5 and 7, identified in serial sections. **D** Histogram of the discharge rate of an "arm projection neuron" from area 5. The activity of the neuron accelerates before release of the detect key, reaches a peak as the arm moves through the air toward the target switch and decreases to zero before the hand touches the target switch. [Combined and slightly modified from Mountcastle V B, Lynch J C, Georgopoulos A, Sakata H, Acuna C (1975) J Neurophysiol 38:871–908]

This can be illustrated with some examples. Single cell recordings from area 5 of the parietal lobe in the awake monkey have identified some neurons that strongly discharge only when the monkey reaches forward with his arm, so as to grasp a rewarding object (*arm projection neuron,* cf. Fig. 124 D). The neuronal activity begins to increase before the arm is moved, it reaches its maximum during the stretching of the arm and declines before the hand has reached its target. The temporal coincidence of the activity of these arm projection neurons is specific for the behavior pattern, since, during other movements (e.g., aggressive actions) these neurons are not activated, not even when the same muscle groups are utilized. There are also neuron types that discharge only when the animal executes certain manipulations with the fingers *(hand-manipulation neuron).* Particular neurons from area 7 have been recorded which selectively respond when the monkey fixates a rewarding object with the eyes *(visual fixation neurons).* Others were only activated when the monkey followed such an object with its eyes *(visual tracking neurons).* Representatives of the above described neuron types are neither sensory nor motor in the usual sense.

Motivation, too, plays an integral role in their activation. If, for instance, an animal is satiated and a banana is placed in the fixation field, there is no corresponding increase in the activity of the visual fixation neurons. Evidently in this case a command for fixation has not been issued.

Mountcastle [117] develops the hypothesis which suggests that the mammalian brain contains many sources of commands for action. The commands themselves, motivation-dependent and therefore state-dependent, are formed by integration of outputs from neuron populations which are not yet known in detail.

The command neuron concept:

1. *A command neuron* is a neuron (interneuron) whose excitation is both necessary and sufficient to activate a central program (motor pattern generator), and thus elicit a given behavior. These criteria can be tested by [118]:

a) *establishing the response pattern of the putative command neuron during presentation of a given stimulus and execution of a well-specified behavior;*

b) *removing the neuron and showing that the response is no longer elicited by the stimulus (necessary condition;* but functional recovery by activation of other neuronal elements must be considered for this criterion; cf. recent results on Mauthner neurons, p. 210);

c) *firing the neuron in its normal pattern and showing that the complete behavioral response occurs (sufficient condition).*

Command neurons participate functionally in the decision process or command of behavior and they are active during the behavioral act in question. Of course the definition of the command neuron mentioned above is very restrictive. Even the neurons considered to be the best candidates (crayfish giant fibers and fish Mauthner cells) have not been tested against all three criteria. At present we might call many neurons "command" neurons that may not meet these criteria. But we should use the term "command" with the prefix "putative" in all those cases where the neuron's behavioral significance has not been adequately demonstrated [118].

2. *Command systems* are formed by several neurons, each of them connected with the motor pattern generator. There are many possible circuits of which two examples will be mentioned: (1) Activation of each of the identical neurons can elicit the entire behavioral pattern. In this case a neuron will meet criterion (c) but fail to fulfill criterion (b). We might call each neuron a *command element,* and their interaction would resemble an "OR gate". (2) Simultaneous activity of all command elements is necessary to activate the motor pattern. Here a neuron will meet criterion (b) but fail criterion (c) In this case the command elements interact like an "AND gate". When treated as a unit in both cases, (1) and (2), the neurons will meet the criteria of a command neuron. The notion "command element" has a much broader applicability than the notion "command neuron". In those cases where a behavioral sequence is associated with activity of a number of different command neurons (or systems) – each of which eliciting a component of the entire sequence – these neurons may be collectively referred to as a *multiple-action system* [118].

3. *Trigger neurons (systems).* In contrast to the command neuron, the responses of trigger neurons can be very brief and still release (initiate) a movement sequence of much greater duration. In both the command and trigger systems, the neural control of a motor pattern is based on the properties of a *neural center* [119].

4. A *neural center* is an assembly of neurons whose coordinated action produces a stereotyped movement or series of movements. It is characterized by three essential features [119]: (1) a sensory analyzer at the input side which – via command neurons or elements – will assure the activation of the center if, and only if, a specific pattern of input occurs; (2) an intrinsic pattern of neuronal connectivity that upon activation is capable of producing a consistent distribution of spatio-temporal excitation and inhibition; (3) the output of the center has privileged access to the required motoneuronal pools.

5. The command neuron concept was originally developed in the context of hierarchical organization of neuronal information flow.

But it is becoming increasingly apparent that there is feedback from the motor neuronal network to the putative command neurons. Hence, it is imperative that the evolving concept of command neurons/systems incorporates the notion *feedback interaction* between a command neuron/system and the motor pattern generator driven by it [115].

6. In those cases where command systems are comprised of more than a few electrically uncoupled cells it is as yet technically impossible either to stimulate or hyperpolarize all the cells simultaneously to check conformity to the sufficiency and necessity criteria. Therefore, it might be unwise to expect rigorous applicability of these criteria to *neural centers* [119] that have more than a countable number of neurons, such as in the vertebrate CNS.

8 Central Representation of Behavior Motivation

Suppose a chicken stands next to a stuffed polecat (Fig. 125), and pays hardly any attention to this potential predator. However, as soon as an area in the diencephalon is stimulated electrically by implanted electrodes the chicken assumes a fighting posture, threatens the stuffed polecat, and lunges at him. When the electrical stimulus is stopped, the chicken remains standing and threatens only mildly (Fig. 125 A, below). However, if the stimulus continues, after temporary adaptation a change may finally occur; the chicken hesitates and flies away screaming (Fig.

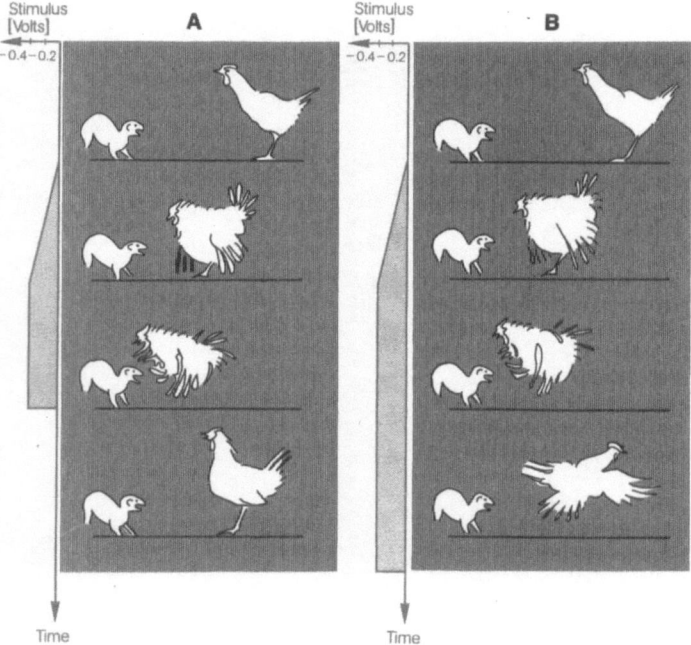

Fig. 125. A During electrical stimulation of particular sites of the brain stem a chicken attacks a stuffed polecat. When the electrical stimulus is stopped the chicken remains standing. **B** If in a similar experiment the electrical stimulus is maintained, the chicken may fly away screaming. [Modified from Holst E v, St Paul U v (1960) Naturwissenschaften 47:409–422]

125 B, below). This experiment shows, on the one hand, that the level of motivation for a particular mode of behavior is not constant, and on the other, that it presumably cannot be correlated with stimulation of one definitely demarcated area in the brain. Thus, in electrical brain stimulation experiments it can be shown that the stimulated area may at times be "silent" and at other times "active" in terms of triggering a behavior pattern. Furthermore, from one excited area different types of behavior can be activated over a period of time. When excitation is induced by two electrodes in two distinct areas, from each of which a different behavior pattern is normally triggered – we shall call them A and B – the following effects might occur: superposition of both patterns

(A + B), averaging $\left(\dfrac{A + B}{2}\right)$, alternating (B, A, B) disappearance

(A + B = O), change to another pattern (A + B = C), or suppression of one of the patterns (A or B, respectively).

On the basis of experiments involving the *electrical control of behavior*, Erich von Holst (1960) [121] at first discussed "drive" as a concept to explain specific actions, the mechanisms of which are unknown (Fig.

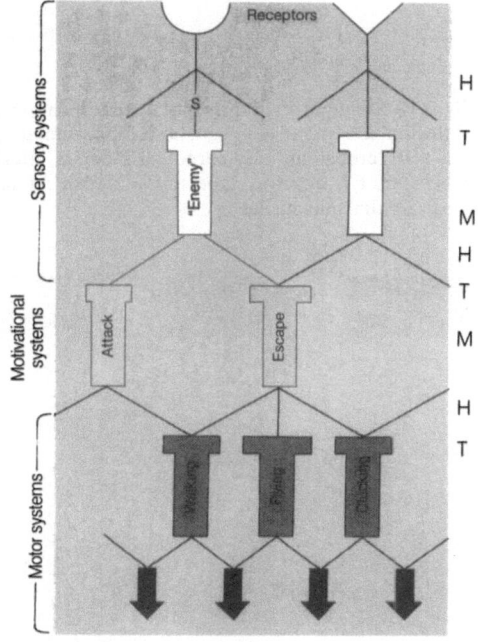

Fig. 126. Piece of a functional framework for control of drives, based on electrically elicited behaviors in the chicken. Receptors and effectors may represent common pathways for different behaviors. The sensory part of the system is plastic and depends on motivation; motor systems appear to be more rigid. The part in this scheme denoted by *red* may refer to the electrical brain stimulation experiment mentioned in Figure 125, *S* indicating the stimulating site. Important functional properties of the network are: habituation, change of motivation, thresholds (the latter being denoted by the length of the horizontal bar). Further details in text. [Modified from Holst v, St Paul v (1960) cit. in Fig. 125]

H   Habituation
T   Threshold
M   Change of motivation

126). This model is helpful in elucidating important concepts of descriptive ethology, such as displacement activity, vacuum activity, change of mood and spontaneity.

What are the neuronal substrates of behavior – such as feeding, drinking, escape, defense, aggression, and sex – that serve to preserve the individual and the species?
This question was first pursued in *mammals* by systematic brain stimulation and lesion techniques in conjunction with histological techniques. It was found that behaviorally effective structures are by no means sharply delineated and restricted to a single area, but may be identified in diverse areas of the brain. Insights into their functions were thus obtained in combination with neuroanatomical micromethods. In

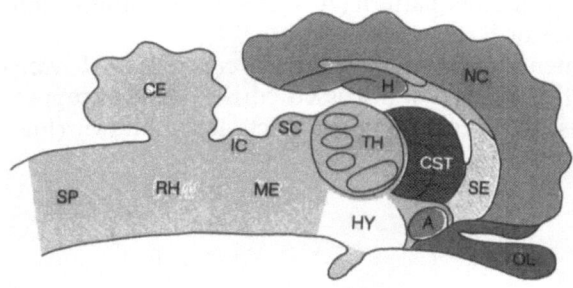

Fig. 127. Mammalian brain, sagittal view, highly schematized. *Red areas:* some structures of the limbic system; *A* amygdala, *H* hippocampus, *SE* septum; *white: HY* hypothalamus; *grey: CE* cerebellum, *CST* corpus striatum, *IC* inferior colliculus, *ME* mesencephalon, *NC* neocortex, *OL* olfactory lobe, *RH* rhombencephalon, *SC* superior colliculus, *SP* spinal cord, *TH* thalamic nuclei

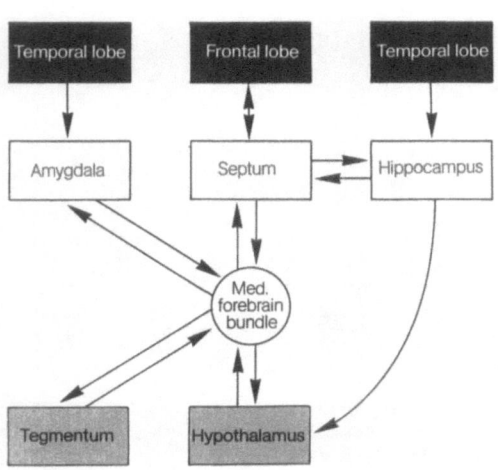

Fig. 128. Important connections between structures of the limbic system *(red)*, hypothalamus and tegmentum *(grey)* via medial forebrain bundle. *Black,* Connections to the cerebral cortex. [Modified from Raisman G (1969). In: Martini L (ed) Integration of endocrine and non-endocrine mechanisms in the hypothalamus. Academic Press, London]

mammals the brain structures relevant for behavioral motivation are arranged in a kind of "border", called the *limbic system* (*limbus,* Latin, = border). Phylogenetically older parts of the telencephalon – such as the *amygdala, septum,* and *hippocampus* – belong to the limbic system. It is functionally connected via specific fiber tracts – especially the *medial forebrain bundle* – with the *hypothalamus* of the diencephalon and the *tegmentum* of the mesencephalon (Figs. 127 and 128). Together with the hypothalamus they form circuits (Fig. 128). At different levels of neuronal integration these systems control motivated behavior and fixed action patterns [120]. We shall study this more closely in a few selected examples (For experimental techniques, see Chap. 9, Methodological Appendix).

Further components of the limbic cortex are: bulbus olfactorius, tuberculum olfactorium, regio praepiriformis, regio periamygdalaris, subcallosal and fronto-temporal cortex, gyrus dentatus, subiculum, fimbria hippocampi, gyrus parahippocampalis, gyrus cinguli; further subcortical structures are: epithalamus, nucleus anterior thalami, corpus mamillare.

8.1 Food Intake: Hunger and Thirst

Important functions involving food intake appear to be controlled by certain hypothalamic structures, among them the ventromedial and lateral nuclei – abbreviated as VMH and LH (Fig. 129, above). Their behavioral relevance may be analyzed by a combination of various experimental techniques [122].

8.1.1 Brain Lesion Experiments

If in cats or rats the VMH is destroyed by coagulation their hunger is increased to such a degree that they may eat themselves to death (Fig. 129, below right). Presumably the normal function of the destroyed area consists in the inhibition of food intake. This area of the brain may be called a "satiation center" (Fig. 130); the syndrome observed after the lesion is called *hyperphagia.* Interestingly, lesions in the LH have the opposite effect (Fig. 129, below left): the animals refuse food and die of starvation. We may call this brain area "feeding center"; its elimination induces *aphagia.*

However, there are problems with the concept of dual centers of feeding control: (1) Lesions in VMH also destroy fibers – the ventral *noradrenergic bundle* – originating in the midbrain locus coeruleus and traveling rostrally to VMH and preoptic structures. Cutting this tract outside the VMH – for instance at midbrain level – also produces a *hyperphagic* syndrome. The noradrenergic system is supposed to be an activation system that modulates the level of "central excitation".

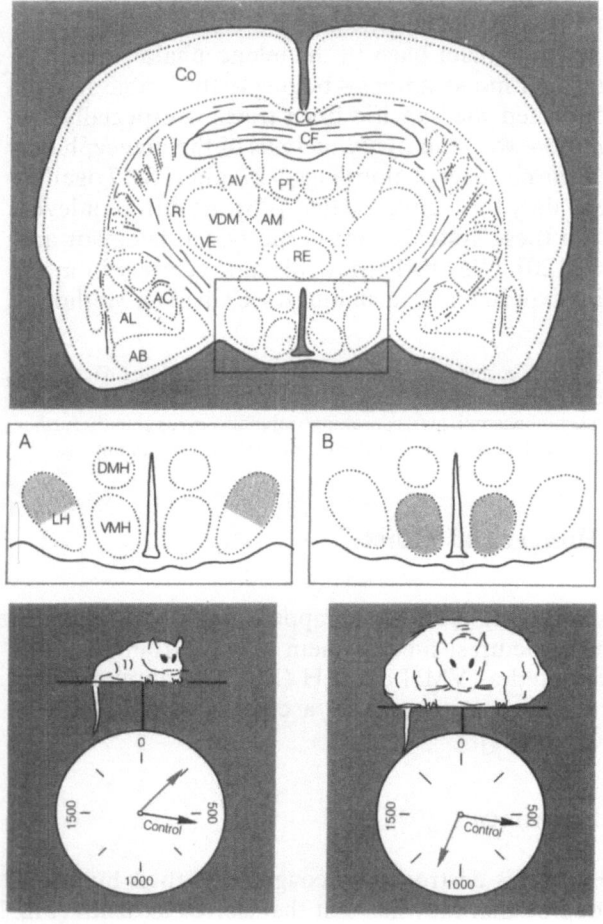

Fig. 129 A and B. Influence of hypothalamic lesions on feeding in rats. *Above,* Cross section through a rat brain at N° AP 1 · 5 (stereotaxic atlas). **A** Lesions *(red)* in *LH* lead to aphagia. **B** Lesions *(red)* in *VMH* lead to hyperphagia. *AB* nucleus basalis amygdalae, *AC* n. centralis amygdalae, *AL* n. lateralis amygdalae, *AM* n. anteromedialis thalami, *AV* n. antero-ventralis, *CC* corpus callosum, *CF* commissura fornicis; *CO* cerebral cortex, *DMH* n. dorsomedialis hypothalami, *LH* n. lateralis hypothalami, *PT* n. parataenialis, *VE* n. ventralis thalami, *VDM* n. ventralis pars dorsomedialis, *VMH* n. ventromedialis hypothalami, *R* n. reticularis, *RE* n. reuniens. [Modified from Stevenson J A F (1969). In: Haymarker W, Anderson E, Nauta W J H (eds) The hypothalamus. Ch C Thomas, Springfield Illinois]

(2) Lesions in LH usually damage the nigrostriatal bundle containing *dopaminergic fibers* and running from substantia nigra to striatum and prefrontal cortex. These fibers are involved in mediating sensory and

Fig. 130. "Dual center" hypothesis for the control of food intake in the mammal by ventromedial *(VMH)* and lateral hypothalamus *(LH). Arrows* denote excitatory, and *lines with cross bars* inhibitory influences. Further details in text. [Combined from Anand B K, Dua S, Shoenberg K (1955) J Physiol (London) 127:143–152]

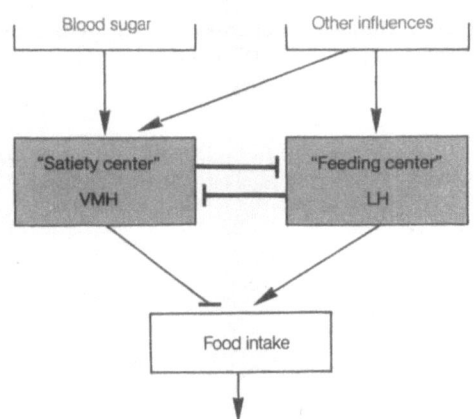

motor functions. Damage to these pathways outside the LH leads also to an *aphagic* syndrome [123].

Furthermore, the concept of a "center" seems to be of limited usefulness when the results of studies in adjacent areas of the limbic system are examined. Lesions of specific portions of the amygdala and the temporal lobe produce changes in feeding patterns. The effects are stronger in the monkey than in the cat and stronger in the cat than in the rat. Generally, however, the deficiencies are not as lasting as after hypothalamic lesions. Therefore, if we want to continue using the terms feeding and satiation centers, we must be aware of the fact that we are dealing with a physiological correlate of a complex functional structure that extends over large areas of the brain.

8.1.2 Single Cell Recordings with Microelectrodes

We now go a step further in our study of functional structures that may control feeding and ask whether neuronal activity in certain hypothalamic regions is correlated with the *nutritional state*. The answer is positive; in hungry animals the discharge rate of particular neurons recorded from LH is increased compared to others recorded from VMH, and vice versa. Presumably the activity in both areas is reciprocally inhibited (Fig. 130).

Recently single cell recordings in the awake monkey were made during visual discrimination learning of food. By measuring the latencies of neuronal activity (in response to seeing food) from *sensory, limbic,* and *motor* structures we may analyze how information is processed (Table 8).

Table 8. The latencies of activation of neurons of different areas of the brain in response to seeing food [Rolls ET (1978) TINS 1: 1–3]

System	Neuronal substrate	Latency (ms) of activation	Response related with
Sensory	Inferotemporal cortex	100–140	Visual stimulus (independent of its significance)
Limbic (subcortical)	LH and substantia innominata	150–200	Food (and other behaviorally relevant stimuli)
Motor	Globus pallidus	300	Movement made by the monkey

Indeed a population of neurons identified in the LH and the substantia innominata is activated while the monkey is looking at the food. These neurons only respond to food, if the monkey is hungry, and their activity becomes associated with the sight of food during learning. Thus, neurons cease to respond to food as the monkey becomes satiated. In other words, in the hungry monkey the response of these neurons *precedes* and *predicts* the food-related motor pattern. The response characteristic of these neurons is different from those recorded in the parietal cortex (p. 211) according to the direction (orientation) of visual attention and the performance of particular motor patterns.

8.1.3 Selective Cell Lesions

What governs neuronal activity of the "appetite centers"? Apparently it is the *blood glucose level,* or rather the degree to which glucose can be utilized (which in turn depends on insulin). Indeed, manipulation of glucose level changes the neuronal activity in a way similar to what we find in animals fed at various levels: a high blood sugar level increases activity of neurons in the VMH [124].

In order to learn something about the *glucose sensitivity* of the neurons, we augment the blood sugar with goldthioglucose, GTG. The organism is unable to distinguish this substance from glucose. However, as soon as GTG is taken up by cells with glucose receptors it destroys them. The gold component of GTG is toxic and kills cells. In this way neurons with glucose receptors in the brain may be selectively eliminated and at the same time identified. GTG is taken up only by neurons of the VMH. They are destroyed and the animal exhibits hyperphagic behavior. Therefore, only cells of the VMH have a special affinity for glucose [125]. However, at present there appears to be no convincing evidence for insulin sensitivity in this brain area.

Of course, it should not be assumed that hunger is controlled exclusively by the blood glucose level and the two antagonistic pancreatic hormones, insulin and glucagon. Additional effects might be due to the level of fatty acids and amino acids. We must also consider the possibility that other detectors which measure and monitor the amount of stored nutrients may be located both inside and outside the brain [126]. Furthermore, there are *gastric, duodenal,* and *hepatic factors* to be taken into account. The liver appears to be an especially important source of signals of both hunger and satiety. Liver receptors monitor metabolic rate (not just glucose level) and relay this information to the brain.

8.1.4 Electrical Point Stimulation of the Brain

What happens when the neuronal activity in VMH or LH is increased by electrical stimulation via chronically implanted electrodes? During stimulation of LH, feeding activity starts immediately, even in a satiated animal. Upon stimulation of the VMH, hungry animals immediately stop feeding; they may even vomit [127].
Stimulation of different parts of the limbic system, such as septum and hippocampus, may also have appetite-stimulating effects. Electrical point stimulation of specific regions of the amygdala, moreover, indicate *complex relationships* between hunger and thirst. Stimulation of the rostral amygdala inhibits the appetite and produces thirst in the animal. Upon ablation of this area, although thirst is "quenched", hunger persists. Stimulation of the caudal amygdala "quenches" both hunger and thirst; ablation of the stimulated region has the opposite effect. At present these results cannot be satisfactorily interpreted within an unified scheme.

8.1.5 Chemical Brain Stimulation: Relation Between Hunger and Thirst

The hypothalamus is rich in acetylcholine (neurotransmitter of the parasympathetic nervous system) and in epinephrine and norepinephrine (neurotransmitters of the sympathetic nervous system). We might ask therefore, what happens when $1-4$ μg crystals of these transmitters, or their blocking agents, are implanted in the LH or just above the hypothalamus.
The result is surprising: norepinephrine restores the appetite in satiated rats so that they feed again. In the same brain area, acetylcholine (or synthetic carbachol) has the effect of inducing rats – which previously had fed and drunk to satiation – to drink immoderately.
Possible interconnections between a hunger and thirst system which were only vaguely demonstrated upon electrical stimulation of the

amygdala emerge much more clearly in experiments where the hypothalamus is chemically stimulated. Micro-injections of *adrenergic* substances enhance food intake, whereas *cholinergic* substances appear to enhance thirst (Fig. 131). Thirst and hunger systems seem to exert a mutual inhibition: rats treated with acetylcholine eat less; animals treated with norepinephrine drink less (Fig. 131) [128].

The hypothesis of a cholinergic system for *water intake* and of an adrenergic system for *food intake* is supported by experiments with appropriate blocking agents (atropine or ethomoxane, respectively) (Fig. 131); their application produces corresponding but inverse effects. However, it must be emphasized that the results briefly reported here are still controversial. There are contradictory findings and other interpretations of the pharmacological effects must be also considered [129].

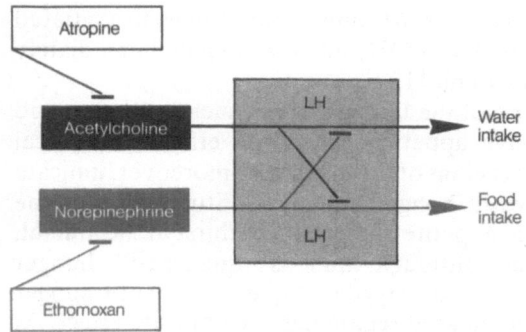

Fig. 131. Influence of drugs on water and food intake in rats. Strongly schematized survey of the action of transmitters (acetylcholine, norepinephrine) and blocking agents (atropine, ethomoxane) injected just above the hypothalamus. *Arrows* denote activating, and *lines with cross bars* inhibiting effects. The results are based on chemical brain stimulation experiments. Further details in text. [Results by Grossman S P (1964). In: Wayner M J (ed) Thirst. Pergamon Press, Oxford]

8.1.6 Influence of Angiotensin on Thirst

How is the thirst system normally aroused? Loss of fluid from the extracellular compartments – called hypovolemia – leads to an increased blood level of the octapeptide angiotensin II:

$$H - asp - arg - val - tyr - ile - his - pro - phe - NH_2$$

Renin which is secreted by the kidneys converts blood angiotensinogen to angiotensin. There is also a cerebral renin-angiotensin system. It is known that angiotensin II controls the blood volume [130]. With regard to central actions this hormone appears to be detected by neurons located in the periventricular region of the anterior hypothalamus. Experimentally it has been shown that injection of angiotensin II into this brain region causes (1) increased thirst, (2) increased sodium appetite, (3) secretion of aldosterone and antidiuretic hormone, and (4) rise in blood pressure. Briefly, this hormone may act as an input to the complex neural circuits controlling thirst.

The anterior hypothalamus appears to be a substrate for the "convergence" of neural circuits mediating *hypovolemic* and *osmometric thirst.* Osmoreceptors are located in the hypothalamic nucleus circularis and in the lateral preoptic area. Injection of hypertonic saline solution into these regions produces drinking. Lesions of these areas abolish drinking, even after systemic injection of hypertonic saline.

Finally, another part of the limbic system has to be considered in the neural circuitry of thirst: the septum. The function of particular septal structures appears to be inhibitory with respect to thirst. Thus, septal lesions enhance water intake when thirst is activated by an angiotensin injection. In this context we should keep in mind that the septum is an important area for mediating the effects of learning.

8.1.7 Self-Stimulation Experiments

Experimental results indicate that rage and fright may be elicited from certain areas of the limbic system. The limbic system (hippocampus) also plays an important role in learning behavior and memory. It guarantees the animal's adaptability to constantly changing environmental influences. We shall return to this point later.

In training tests, food acts as a reward for the performance and maintenance of certain learning tasks. The reward may be replaced by electrical stimulation of certain points of the "feeding center" (cf. also p. 244). Rats or cats for instance then learn to stimulate themselves by pushing a lever (Fig. 146). Such a brain stimulation has a stronger reward effect than the actual food itself. Similar results may also be achieved by appropriate chemical stimulation.

Some remarks on the question of "drive centers":

1. In the light of the current state of knowledge, the impression is gained that the so-called primary drive circuits for *hunger, thirst, sexual behavior,* etc. are arranged in a "parallel pattern" in the limbic system and the hypothalamus (Fig. 132).

2. Apparently the specificity of such a circuitry has a chemical basis. The term "center" must be interpreted in this context. Thus, in the

male rat, from the same area in a certain part of the hypothalamus, thirst may be elicited by stimulation with *acetylcholine*, hunger with *norepinephrine* and maternal behavior, such as nest building, with *testosterone*.

3. It appears therefore that the evidence from electrical brain stimulation experiments alone is rather limited. Chemical stimulation is selective and provides more detailed results (Fig. 133). There are "chemical messengers" in the brain [147].

4. The situation is even more complicated, since what has been found for the *rat* may not necessarily be applicable to analogous functional structures of the *cat:* here acetylcholine injection in a corresponding brain area does not induce thirst but sleep. It is claimed that the *sleeping* circuit of the cat in the limbic system is similar to the thirst circuit of the rat. Moreover, the cat is supposed to possess an *arousal* system wired in parallel to the sleeping circuit; it is activated by injection of norepinephrine.

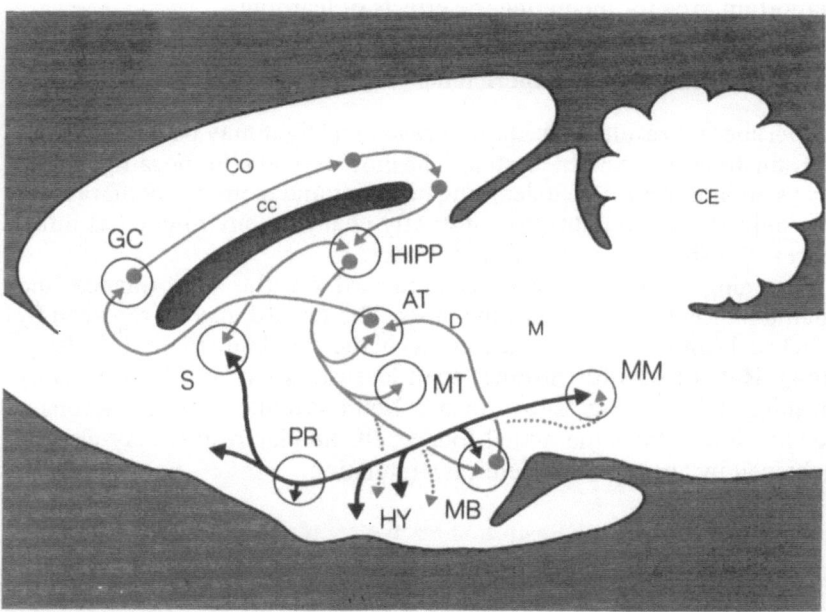

Fig. 132. Simplified representation of a neuronal network of the "thirst system" in rat brain (sagittal view, schematized). *Red:* the so-called Papez-circuit. *Black:* medial forebrain bundle, *AT* anterior thalamus, *CC* corpus callosum, *CE* cerebellum, *CO* cerebral cortex, *D* diencephalon, *GC* cingulate cortex, *HIPP* hippocampus, *HY* hypothalamus, *M* mesencephalon, *MB* mamillary body, *MM* medial midbrain, *MT* medial thalamus, *PR* preoptic region, *S* septum. [Modified from Fisher A E (1964) Sci Am 485]

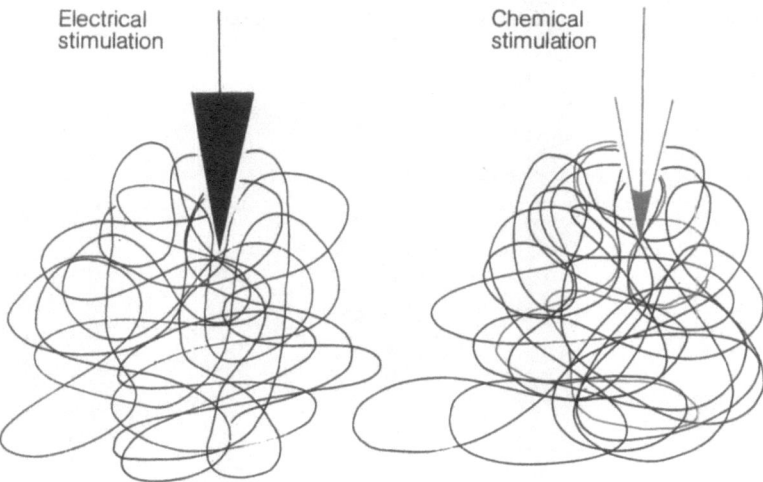

Electrical
stimulation

Chemical
stimulation

Fig. 133. Schematic representation showing selectivity of chemical brain stimulation *(right)* compared with electrical brain stimulation *(left)*. Electrical stimulation frequently activates several neuronal circuits *(black lines)*, simultaneously releasing various behavioral responses and syndromes, resulting in overlapping effects. Chemical stimulation, on the other hand, activates mainly the circuit *(red line)*, for which the chemical stimulus is specific. [Modified from Fisher (1964) cit. in Fig. 132]

8.2 Sexual Behavior

Sexual behavior is represented in wide areas of the brain, especially in the limbic system and its connections. Here, too, the use of different experimental methods provides important insights into the central framework.

8.2.1 Brain Lesion and Electrical Stimulation Experiments

Sexual behavior in response to peripheral signals may be affected by stimulation or ablation of specific areas of the limbic system. After bilateral lesions in the amygdala, sexual behavior of a male cat can be enhanced to such a degree that it attempts to copulate with inappropriate mates such as a dog or a monkey. This *hypersexuality* may again be dampened by further lesions in the septum and the anterior-medial area of the hypothalamus.

Table 9 summarizes effects of brain lesions on sexual behavior in rats:

Table 9. Influence of brain lesions on sexual behavior in rats

Brain area	Effect of lesion	Supposed function
Medial preoptic area	Inhibition of male sexual behavior. Facilitation of lordosis in females	Expression of male or female behavior. Control of lordosis
Ventromedial hypothalamus	Disruption of lordosis in females	
Caudal ventromedial hypothalamus	Hypersexuality in males	Control of sexual behavior
Anterior medial hypothalamus, medial amygdala	Inhibition of male (and female) sexual behavior	Control of sexual behavior
Temporal lobe	Inability to choose appropriate sex object	Analysis of goal objects
Frontal lobe	Functional decrease mainly in males	Integrative functions pertaining to motor patterns

At present we are far away from a complete picture of the central framework of sexual behavior in rats [131], but we may summarize some essential points. (1) The female's display of sexual receptivity, when she is mounted by the male, is marked by a "lordosis response": arching the back and elevating the pelvis. The basic circuitries for *lordosis* in females and *genital reflexes* in males are located in the spinal cord. They are controlled by certain areas of the hypothalamus. (2) Both the medial preoptic area and the ventromedial hypothalamus appear to be involved in the *expression* of male-type or female-type behavior. (3) Caudal hypothalamic and amygdaloid structures may *modulate* sexual behavior. (4) The temporal lobe is involved in the analysis of *sexual goal objects*. (5) The frontal lobe is important for the *performance* of sexual behavior patterns, which are in rodents rather more complex in the male than in the female.

Electrical point stimulation of certain areas of the limbic system and its connections in the male squirrel monkey triggers penile erection (Fig. 134). Corresponding stimulations of the female lead to an enlargement of the clitoris. The behavioral reactions represent more than simply a general expression of sexual arousal. The *genital display* of primates is a gesture directed at conspecifics (Fig. 134). It is a component of highly developed social behavior. The phallic display occurs in the squirrel monkey within a few days after birth while the proper sexual behavior develops much later.

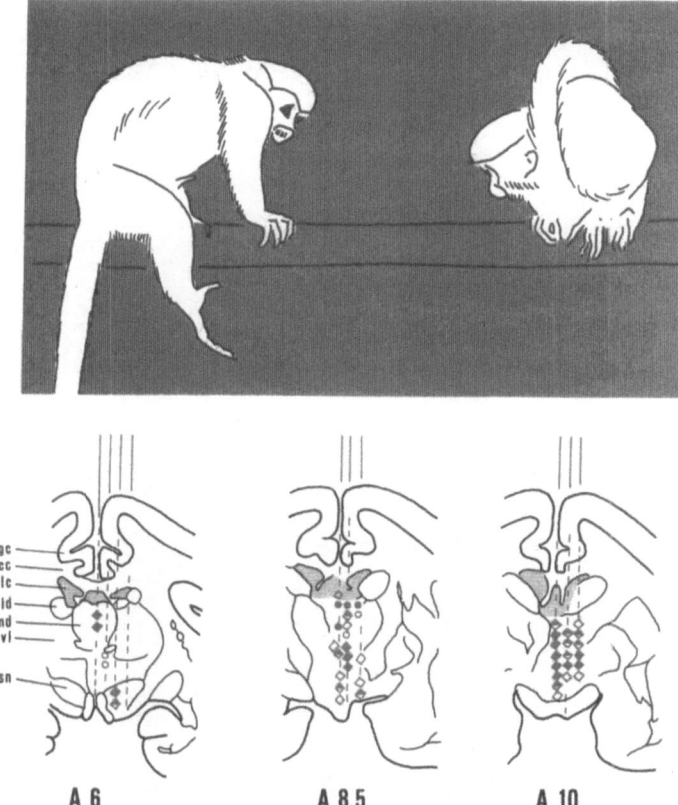

Fig. 134. Genital display in the squirrel monkey *Saimiri sciureus. Above,* displaying partner *(left)* facing addressed partner *(right).* [Slightly modified from Ploog D (1972). In: Gadamer H G, Vogler P (eds) Neue Anthropologie. DTV & Thieme, Stuttgart]. *Below,* Cerebral representation of genital function in the squirrel monkey. The *red* symbols in the brain cross sections – corresponding to the stereotaxic coordinates A6, A8.5, A10 – designate sites where electrical point stimulation elicited penile erections in the freely moving animal: slight (◇), medium (◈), strong (◆). From some points (●) erection occurred only after cessation of the stimulus. Vertical dashes (l) designate sites which did not elicit an erection. *cc* Corpus callosum, *gc* cingulate gyrus, *ld* nucleus lateralis dorsalis thalami, *md* nucleus medialis dorsalis thalami, *sn* substantia nigra, *vl* nucl. ventr. lateralis thalami, *vlc* lateral ventricle. [Modified from McLean P D, Ploog D (1962) J Neurophysiol 25:29–55]

8.2.2 Brain Stimulation with Sex Hormones

Testosterone is a male sex hormone. If in a male or female rat it is injected into a central area of the preoptic region (located in front of the hypothalamus; Fig. 135b) using a fine injecting instrument and a thin canula, rats of both sexes respond with maternal behavior within a

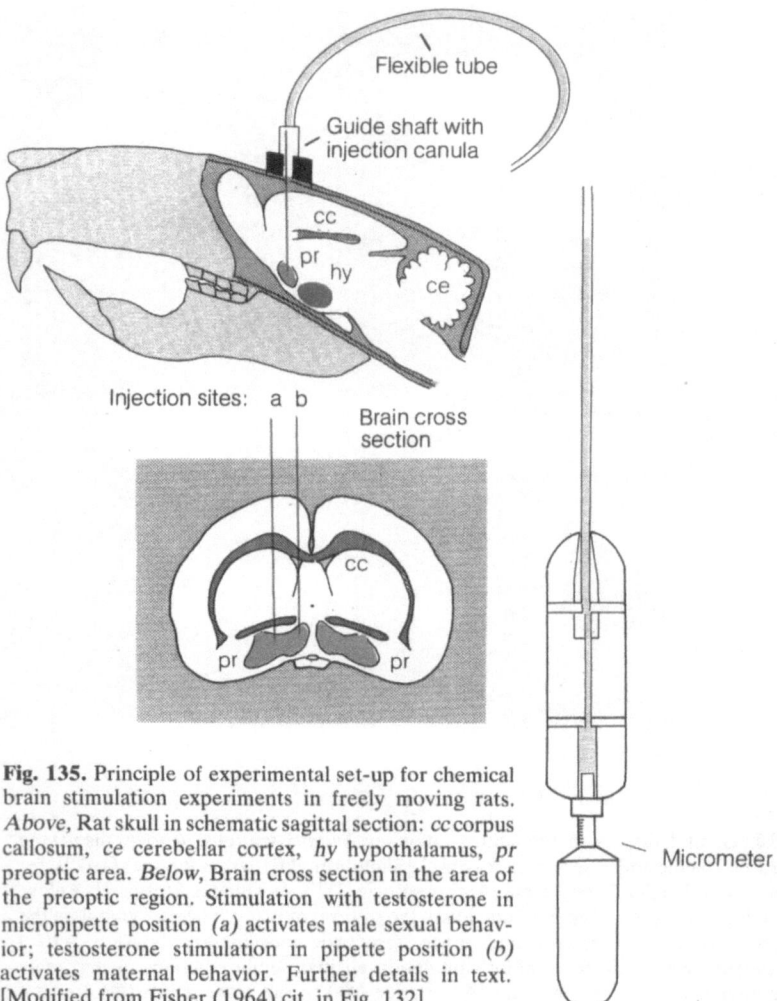

Fig. 135. Principle of experimental set-up for chemical brain stimulation experiments in freely moving rats. *Above,* Rat skull in schematic sagittal section: *cc* corpus callosum, *ce* cerebellar cortex, *hy* hypothalamus, *pr* preoptic area. *Below,* Brain cross section in the area of the preoptic region. Stimulation with testosterone in micropipette position *(a)* activates male sexual behavior; testosterone stimulation in pipette position *(b)* activates maternal behavior. Further details in text. [Modified from Fisher (1964) cit. in Fig. 132]

surprisingly short time. They build a nest and carry scattered rat pups into the nest. If there is no pup, an adult conspecific of either sex may serve as substitute. If testosterone is injected slightly lateral to the central area of the same general brain region (Fig. 135a), males as well as females exhibit male sexual behavior; they try to copulate with any partner.

Testosterone is also effective in certain other regions of the hypothalamus. For instance, testosterone injected between the central and the lateral hypothalamus of the rat induces both maternal and male behavior. Males as well as females take care of their young, carry a rat pup around in their mouth and at the same time they mount adult conspecifics of either sex!

How can it be explained that neurons apparently taking part in the control of male or female behavior respond to the same sex hormone testosterone?

8.2.3 Determination of Sexual Behavior

Whereas testosterone is a male sex hormone, progesterone is involved in the female with pregnancy and therefore with maternal behavior. Both are steroid hormones (Fig. 136). Many observations suggest that, depending on the dose, steroid hormones are able to mimic one another's actions. The observations described above might therefore be interpreted as follows. Normally the male has little or no progesterone. Its sexual hormone is secreted in smaller concentrations that guarantee male behavior. If, now, as in the brain stimulation experiment, testosterone is provided in relatively high concentrations directly to the cells, it might reveal progesterone potency and stimulate neurons relevant for the female behavior. How then is the relevant behavior pattern determined?

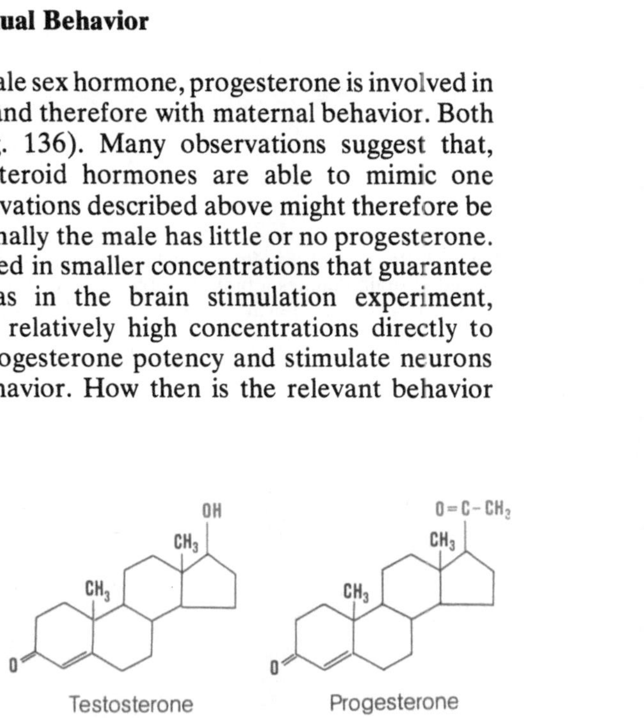

Fig. 136. Chemical structure of testosterone (male sex hormone) and progesterone (female sex hormone)

Testosterone   Progesterone

There is experimental evidence to show that during early ontogeny the mammalian brain, unless exposed to testosterone at a critical phase, develops female characteristics. However, if there is sufficient testosterone in the blood stream at that critical period, the brain will develop male characteristics [132]. During early development the *sex chromosomes* cause indifferent gonads to develop into ovaries or testes. This phase occurs in Guinea pigs and monkeys before birth, and in rats shortly after. Consequently rats have become important subjects for hormone experiments. (1) When in young male and female rats ovaries and testes are mutually transplanted during the critical phase (Fig. 137) the males develop into females and the females become masculinized. (2) If a female is injected with testosterone at birth ("neonatally

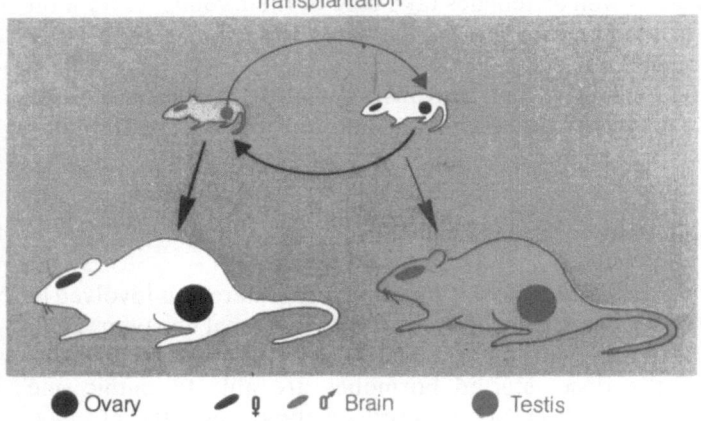

Fig. 137. Reversal of sex after mutual transplantation of testes and ovaries in neonate rats. ♂ and ♀ brain denotes a brain capable of generating either a male or female behavior pattern respectively. Further details in text. [Modified from Levine S (1966) Sci Am 214:84–90]

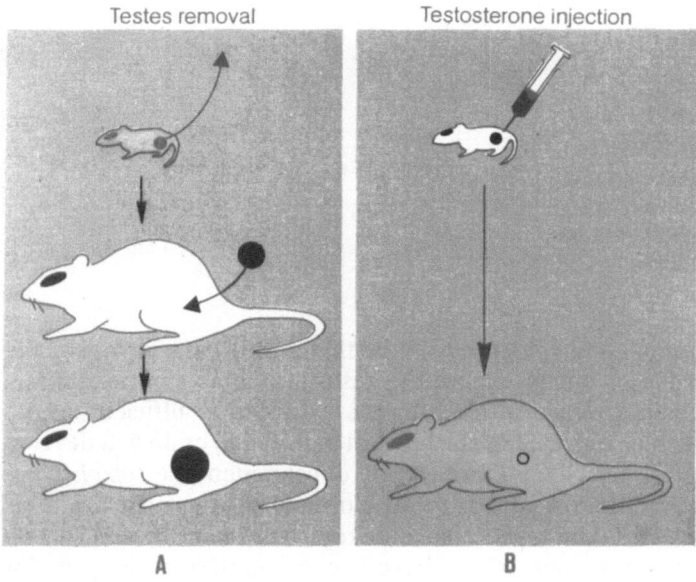

Fig. 138. A When a neonate male rat is castrated, a female type of brain develops. Ovaries implanted into this animal become functional. **B** After injection of testosterone in the neonate female rat, a masculinized brain develops: ovaries undergo atrophy. Further details in text. [Modified from Levine (1966) cit. in Fig. 137]

androgenized", Fig. 138B) and reinjected at maturity, it exhibits male behavior and the ovaries degenerate. (3) Males castrated at birth develop female physiology and implanted ovaries become functional (Fig. 138A).

In the newborn male rat testosterone is converted into estradiol by enzymes located in neurons that are targets of sexual differentiation. Thus, estradiol (one of the female hormones!) gives rise to the male pattern of brain circuitry. The neonatal female rat is "protected" against its maternal estrogen by an estradiol-binding alpha-fetoprotein. In females this substance (identified only in rats and mice) is present in blood and cerebrospinal fluid during the first three weeks of postnatal life [132].

8.2.4 Sexual Dimorphism in the Brain

The brain of a male is not precisely the same as that of a female. This concerns specific morphological structures as well as the way in which these may respond to sex hormones. Thus, the preoptic area of a male rat does not respond to estrogen by stimulating the pituitary gland to produce luteinizing hormone, LH, which causes ovulation in females. Similarly, electrical stimulation of the preoptic area (or the arcuate nucleus) elicits ovulation in the female rat, but no comparable effect can be seen in neonatally androgenized females.

In the normal female rat ovulation occurs every four days. The follicle grows in response to the follicle-stimulating hormone, FSH, that is released by the pituitary gland, and the follicle starts to secrete estrogen. Estrogen inhibits the production of FSH and excites neurons of the preoptic area which in turn stimulates neurosecretory cells of the arcuate nucleus in the basal hypothalamus to produce LH-releasing hormone. Thereby the anterior pituitary gland is stimulated to secrete LH. Ovulation is triggered at a certain ratio of LH to FSH, and the follicle (corpus luteum) is made to produce progesterone – a pregnancy-promoting hormone – which inhibits further pituitary release of LH.

Are there also indications for *anatomical* differences in male and female brain structures? In the female rat there are at least twice as many dendritic spines in the mid-dorsal part of the preoptic area as there are in the male. These structural differences depend on the presence or absence of androgenous steroid hormones during a critical neonatal period.

Especially impressive examples of sexual dimorphism have been found in birds. In male *songbirds* certain songs, which are learned by reference to auditory information, represent important male behavior patterns. In zebra finches and canaries these songs are essentially controlled by different areas of the forebrain (RA, HVC, X in Fig. 139) as well as by the hypoglossal nucleus of the medulla (N.XII in Fig. 139). Those *vocal control areas* are larger in males than in females with respect to cell size, density of the cellular layers, and cell mass (Fig. 139). The X-region is particularly well developed in males of both species, but less so in female canaries and it is almost not recognizable in female zebra finches. Females of both species do not normally sing! However, when

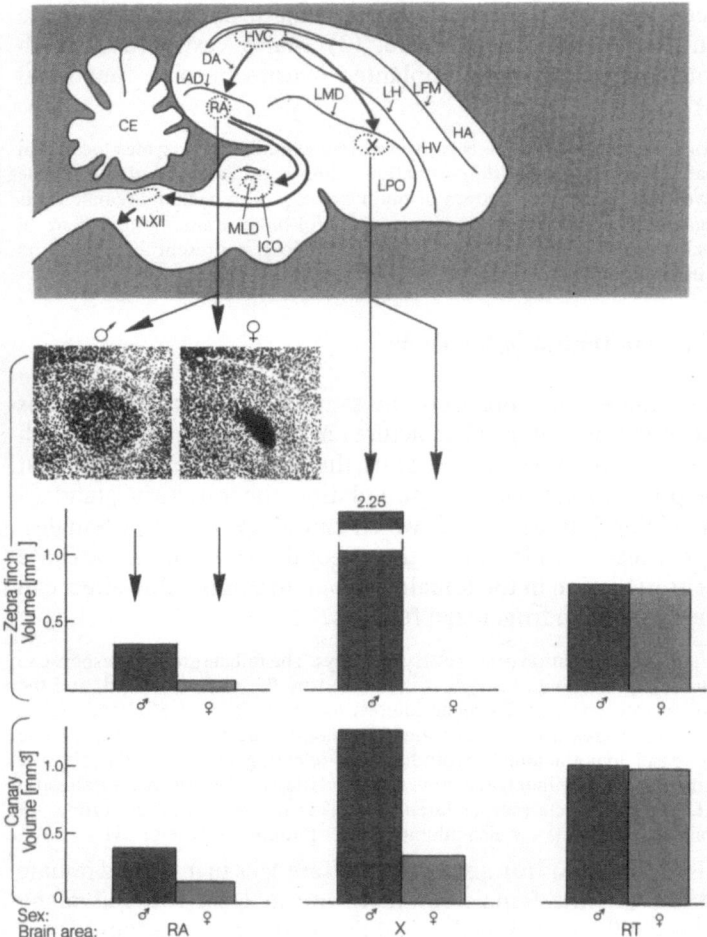

Fig. 139. Sexual dimorphism in vocal control areas of zebra finch *Poephila guttata* and canary *Serinus canarius. Above,* Schematic sagittal section through the songbird brain. Unilateral lesions of *HVC* or *RA* in canary result in major disruptions of song. The arrows indicate degenerating fibers traced from such lesions (cf. Chap. 9, Methodological Appendix). *Below,* Volume of vocal areas *RA* and *X* of males *(grey)* and females *(red)* in zebra finches and in canaries. For comparison cf. also relationships in nucleus rotundus *(RT)* which is not associated with vocalization. The photographs show distributions of cells in vocal control area *(RA)* of the right hemisphere of a male and a female zebra finch. *CE* cerebellum, *DA* tractus archistriatalis dorsalis, *HA* hyperstriatum accessorium, *HV* hyperstriatum ventrale, *HVC* hyperstriatum ventrale pars caudale, *ICO* nucleus intercollicularis, *LAD* lamina archistriatalis dorsalis, *LFM* lamina frontalis suprema, *LH* lamina hyperstriatica, *LMD* lamina medullaris dorsalis, *LPO* lobus parolfactorius, *MLD* nucleus mesencephalicus lateralis pars dorsalis, *N. XII* nucleus nervi hypoglossi, *RA* nucleus robustus archistrialis, *X* area X. [Modified from Nottebohm F, Stokes T M, Leonard C M (1976) J Comp Neurol 165:457–486; Nottebohm F, Arnold A P (1976) Science 194:211–213]

pellets of testosterone propionate are implanted in adult female canaries they are able to sing, possibly because of the presence of X-region, feebly developed though it may be. In adult female zebra finches, which do not possess a well-developed X-region, testosterone implants have no corresponding effect [133].

We summarize some essential points:

1. Primarily, mammalian sexual behavior is programmed to develop the female type.

2. The male pattern of brain circuitry is determined by testosterone activity during a defined critical phase which is prenatal in most mammals (neonatal in rats).

3. There is evidence of anatomical and functional sexual dimorphism in specific brain areas.

4. There are neuronal circuits for male and female behavior patterns in the brain of both sexes.

5. In the adult male testosterone in high concentrations may imitate the action of progesterone and thereby activate neuronal circuits relevant for female behavior patterns.

8.3 Aggression: Attack and Defense

Aggression as a mechanism for the preservation of the individual is represented at various levels of integration in the brain. As regards differentiation and adaptability, we usually distinguish three levels: (1) midbrain level (periaqueductal grey), (2) the hypothalamic level, and (3) at the highest level, limbic structures of the cerebral cortex.

8.3.1 Brain Lesion Experiments

The first indications of brain structures that influence aggressive behavior were obtained by lesion experiments. If, by means of knife cuts or small lesions, the cerebral cortex is partly or completely separated from the brain stem in cats, monkeys, or dogs (Fig. 140), aggressive behavior may be disinhibited; the animals are then in a nearly constant state of irritation and react to the slightest peripheral stimulation. In contrast to the natural behavior, the fits of "sham-rage" are unspecific and practically show no fatigue.

Presumably the ablated brain areas contain structures that normally damp aggression. This effect was impressively demonstrated in an electrical brain stimulation experiment (Fig. 141) by the Spanish neuroscientist José Delgado. He was able, by telestimulation, to interrupt the behavior of a fighting bull that was charging a red cloth and have it peacefully retreat.

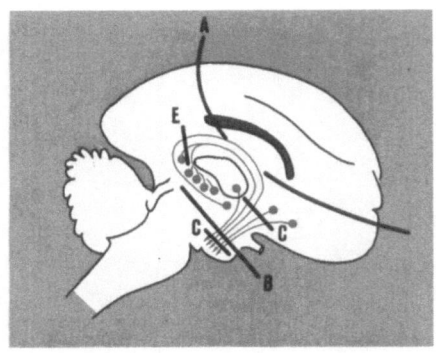

Fig. 140. Schematic sagittal section of cat brain. Planes for surgical sections *A–E* leading to sham-rage are shown. [Modified from Poeck K (1970) Umschau 70:33–36]

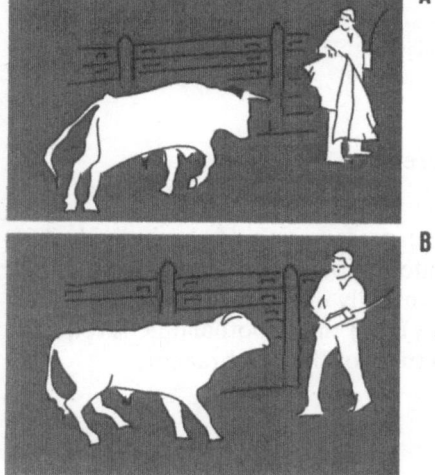

Fig. 141 A and B. A Spanish bull charging a red cloth (**A**) changes its behavior after telestimulation of aggression-damping brain regions **B** [Slightly modified from photographs in Delgado J (1971) Gehirnschrittmacher. Ullstein, Frankfurt]

8.3.2 Combined Brain Stimulation and Ablation Techniques

When behavioral changes released by electrical point stimulation of certain brain areas and results of lesions in these areas are compared, a preliminary picture of the central representation of aggressive behavior is obtained. During these experiments the analysis is focused mainly on a certain type of aggressive behavior such as predatory attack, fear-induced aggression, inter-male aggression, or territorial defense.

Mechanisms are investigated in terms of the probability of the occurrence of the behavior. When judging the results it must be kept in mind that postsurgical effects may depend strongly upon the general pre- and postoperative motivational state of the animal.

Electrical stimulations of various points in the brain of the squirrel monkey show, for instance, that sites for *directed vocal aggression* are located in certain areas of the limbic complex (Fig. 142, red).

Brain stimulation and brain ablation experiments in cats suggest that certain structures of the amygdala, the medial hypothalamus and the periaqueductal grey of the midbrain are responsible for the control of *aggression* and *defense*. Both modes of behavior may be triggered, for example, by electrical point stimulation of certain amygdaloid or hypothalamic areas.

A cat, which was calm before, becomes irritated with the start of hypothalamic stimulation, arches its back with hair bristling, hisses, shows its claws and reacts even to harmless objects with violent aggression which may suddenly switch to panic flight (Fig. 148).

The following *ergotropic syndromes* appear: Increase of heartbeat rate, respiratory rate, blood pressure and muscle blood flow, as well as a decrease of blood circulation in skin and

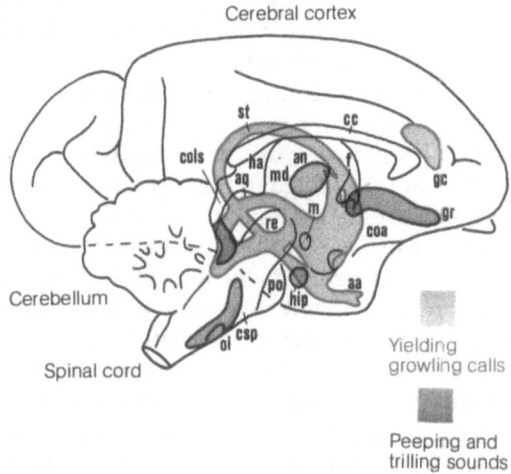

Fig. 142. Structures active in vocal aggression *(red)* in the brain of squirrel monkey *Saimiri sciureus*. In the schematic sagittal section, areas are shown from which specific types of vocalization such as directed aggression (yielding and growling calls) can be elicited by electrical point stimulation in the freely moving animal. *aa* Area anterior amygdalae, *an* nucleus anterior, *aq* substantia grisea centralis, *cc* corpus callosum, *coa* commissura anterior, *cols* colliculus superior, *csp* tractus corticospinalis, *f* fornix, *gc* cingulate gyrus, *ha* nucleus habenularis, *hip* hippocampus, *m* corpus mamillare, *md* nucleus medialis dorsalis thalami, *oi* nucleus olivaris inferior, *po* griseum pontis, *re* formatio reticularis tegmenti, *st* stria terminalis. [Slightly modified from Jürgens U, Ploog D (1970) Exp Brain Res 10:532–554]

intestine. This syndrome is produced by stimulation of neuron populations of the sympathetic (autonomic) nervous system responding to norepinephrine that is discharged into the blood stream by activation of the adrenal medulla. Excitation of the hypothalamo-pituitary system leads to the mobilization of humoral factors: ACTH is secreted by the anterior pituitary (cf. Fig. 147) leading to the secretion of corticosteroids from the adrenal cortex. The activation of aggressive behavior by electrical point stimulation of the hypothalamus results in a mobilization of factors that preserve the physical integrity and increase the level of physical performance – similar to what is observed in a corresponding natural situation. Electrical stimulation of adjacent areas of the hypothalamus may cause sleepiness accompanied by *trophotropic syndromes*. They subserve the recovery of the organism and contribute to the conservation of its energy: food intake, decrease of blood sugar level, decrease of heartbeat and respiratory rate, increase of blood circulation in skin and intestine and decrease of muscular blood circulation. These syndromes result from the activation of the parasympathetic nervous system which uses acetylcholine as its neurotransmitter.

Rat-killing by cats shows two types of behavioral patterns; type (1) *affective attack* which is accompanied by strong emotion, and type (2) *quiet-biting attack* which is similar to that seen during normal hunting, but not necessarily related to hunger. Both types of behavior can be elicited by electrical stimulation of the medial hypothalamus (type 1), the lateral hypothalamus (type 2) and the midbrain central grey (types 1 and 2) [134]. After elimination of the corresponding hypothalamic structures types 1 and 2 can still be elicited. Obviously both motor patterns are organized at midbrain or lower levels, and the circuits are controlled by hypothalamic fiber systems. The hypothalamus itself receives two projections from the amygdala. A cortico-medial division of the amygdala projects via stria terminalis to hypothalamic and forebrain structures; influences from basolateral amygdala are conveyed by the ventral amygdalofugal pathway to the hypothalamus, preoptic area, midbrain tegmentum and central grey. Thus, attack as well as defense can also be elicited in cats by electrical stimulation of the basolateral division of the amygdala. The effects of these stimulations, however, are absent if particular hypothalamic and midbrain regions have previously been destroyed. Lesions of the amygdala do not have marked effects on the defense reactions induced by hypothalamic stimulation. Whereas hypothalamic lesions do not seem to have a lasting effect on aggressive behavior, bilateral ablation of corresponding structures in the midbrain may result in a lasting extinction of any aggressivity. Septal lesions facilitate aggressive behavior, depending on the animals' previous experience.

The experiments performed on cats suggest a multiple brain representation of *affective defense* behavior [135]: (1) the midbrain is, so to speak, involved in the control of basic motor patterns, (2) the hypothalamus contributes an object orientation to the behavioral complex and presents a connection between somatic, autonomic and endocrine reactions, (3) the amygdala subserves the adaptation of behavior to the changing environmental situations.

8.3.3 Pathological Observations

Aggression is disinhibited in the course of certain disease processes in deep areas of both temporal lobes of the cerebral cortex, as for example, in advanced stages of rabies. During epileptic seizures excessively strong discharges may be observed in certain regions of the amygdala, leading to unrestrained violent reactions. Pathological violent behavior may be prevented by surgical stereotaxic lesions in the area of the amygdala *(amygdalotomy)*. But the final results of such a lesion are not always unequivocally predictable: apathetic inertia, but also a tendency toward hypersexuality may supervene [136].

When a monkey is bilaterally deprived of the poles of the temporal lobe and the amygdala, he becomes extraordinarily tame and peaceful, hardly excitable and almost apathetic. On the other hand he may exhibit hypersexuality. What effect do such behavioral changes have on the *social group?* When these brain lesions are carried out on the leader, he may still be accepted as long as the syndrome is slight, but the social cohesion of the group deteriorates. In the case of strong behavioral changes he is increasingly avoided by members of the group because of his sexual dissipations and in time he is relegated to the lowest rank.

To recapitulate:

1. Aggressive behavior, too, contributes to the self-preservation of the individual. Aggression is subject to adaptation to environmental influences. Aggressive behaviors are controlled within a central frame work of drives by multiple representations at different central levels of integration and interaction.

2. Spatial "overlapping" of various functional structures in the limbic system considerably reduce the expectations of success in brain surgery *(psychosurgery)* in man. A decision always has to be made as to what is of greater consequence for the patient; the pathological behavior or the behavioral changes resulting as side-effects of the surgery – in as much as the results are at all completely predictable [136].

8.4 Learning: Storing, Recalling, Reinforcing

In connection with electrical self-stimulation it has already been mentioned that the limbic system (hippocampus) plays an important role in the learning and memory functions of the brain. From this complex field – where our knowledge is still quite incomplete – we shall select a few aspects.

8.4.1 Principles of Information Storage

Memory has at least two basic forms: *short-term* and *long-term* storage. Short-term storage of excitation lasting several seconds may be due to reverberation loops of interconnected neurons with positive feedback, in a more redundant version as has been sketched previously (Fig. 22 D). Such circuits might be triggered by neurons with excitatory synapses and again be inactivated by others with inhibitory synapses. This inactivation would be definitive unless the information had been previously transmitted to long-term storage and has been fixed as a lasting *engram* with a "stable" biochemical basis.

The transition from short-term to long-term memory is called *consolidation*. Obviously such a process takes time (consolidation time lasts from a few seconds to minutes). It can be disrupted by particular events, such as by trauma after head injury. Experimentally, consolidation can be influenced by electroconvulsive shock or hypothermia, to name two examples. Experimental results indicate that the shorter the time interval between learning and subsequent shock, the lower the probability that information can be transferred to long-term memory. However, when memory has reached the "long-term store" it does not appear to be influenced by shocks.

As has been shown in brain lesion experiments, consolidation is not restricted to a certain brain area, but is widely distributed. It can be assumed in such a case that, on the one hand, the same neuronal assembly may be involved in the storage of different information and, on the other, that the same information may be "fixed" in different neural systems. Presumably wide parts of the cortex are involved in the storage of many different engrams. Indeed a rat's performance at a learning task decreases with the amount of cortex ablated, a phenomenon which Lashley (1950) called "mass action". Hence, the cortical representation of information has been compared to a holographic process [137].

8.4.2 Neural Correlates of Storing and Recalling Information

Reverberatory Activity in Isolated Cortical Patches. Suppose a slice of cortex is separated from its neuronal connections by undercutting it from the underlying white matter [138]. Sufficient blood supply must be preserved in this isolated patch for the following experiments: a stimulating electrode penetrates the tissue and a recording microelectrode is positioned close to it. Normally neurons are silent, but after sufficient stimulation with a train of electrical impulses, long-lasting spike activity (up to a few minutes) can be recorded from the isolated slab. This experiment indicates that information fed into nervous tissue generally can be stored for a while, probably due to circulating activity in networks with positive feedback loops.

The Conditioned EEG Reaction. Brain potentials may be recorded from the scalp in the form of the electroencephalogram (EEG) with relatively large surface electrodes (see Chap. 9, Methodological Appendix; Fig. 156 C). It was found that the α-rhythm is always blocked and replaced by a higher frequency β-rhythm when impulse patterns are being processed in the brain. This may occur, for example, when solving a mathematical problem, but it is also seen in response to photic stimulation (Fig. 143 A). If a light stimulus is applied repeatedly in combination with an indifferent signal – e.g., a sound – then, after repeated combinations, the sound alone may finally block the α-rhythm and induce the β-rhythm (Fig. 143 B and D). Thus, a conditioned reaction has appeared, linking an external stimulus with brain activity (measured by the EEG). The brain has revealed evidence of learning.

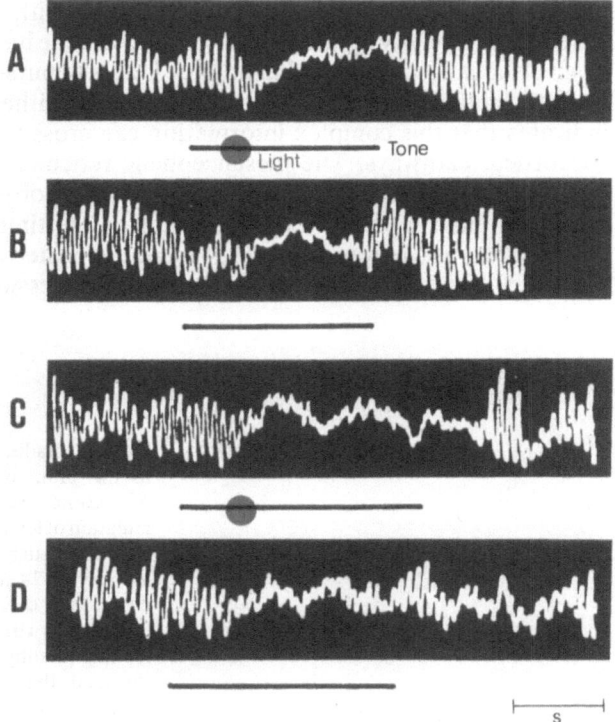

Fig. 143 A–D. Development of a conditioned EEG reaction in man. During stimulation by a point of light *(red dot)* the α-rhythm is substituted by a β-rhythm. After repeated combination of light point *(red dot)* and an indifferent sound *(black line)* (**A** and **C**), the sound alone may induce the β-rhythm (**B** and **D**). [Modified from Popov N A (1953) C R Acad Sci Paris 236:125]

Neurons Participating in Pattern Recognition and Object Reward Association. The EEG is an important indicator, but it has not revealed anything much about neurophysiological processes. In the infero-temporal cortex responses have been recorded from single neurons or neuron populations showing obvious correlations with *known* (associated) and *new* stimuli. Neurons may be silent in the case of new stimuli but can be readily activated by known stimuli. The inferior temporal lobe in monkey is an area associated with modality specificity and learning of new stimuli to which the animal has been exposed. We have evidence that the inferior temporal lobe makes and has contact both with striate cortex and limbic system [139].

Recalling of Information. Our knowledge about storage and recalling of information is still scant. Neurophysiological observations suggest that information fed into wide areas of the brain may be recalled from certain sites in the cerebral cortex. For example electrical point stimulation of area 18 in a macaque monkey can recall the image of a butterfly: the monkey pursues it with his eyes, captures it with his hand and then carefully and slowly opens his fist to check what he has caught (Fig. 144). Even after callosal section the monkey is able to pursue with either hand objects hallucinated through stimulation of either brain site. This indicates that this complex information can cross the midline by deep subcortical pathways. The mesencephalic reticular formation seems a likely system for the transfer of "brain-created" objects.

During brain surgery in man electrical point stimulations of certain cortical regions (between area 17, 42 and the posterocentral gyrus) may "mobilize" visual or auditory memories. Impressions gained several

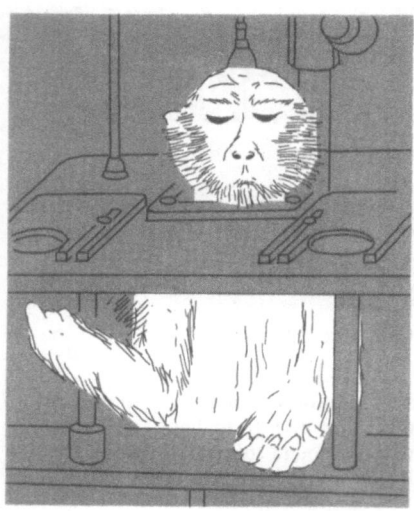

Fig. 144. Hallucination of a "butterfly" in the brain of a macaque monkey *Macaca nemestrina* upon electrical stimulation of left area 18 just posterior to the lunate sulcus (at 0.2 mA, 50 cps): monkey M-24 moves the eyes steadily down; then there is a sudden catching movement with the left hand, and, looking intently as if to see what it has captured, the monkey carefully opens its fist – just after having caught a "butterfly". From photograph. [Doty R W (1967). In: Rusinov V S (ed) Electrophysiology of the central nervous system. Science Press, Moscow 1967; translation (1970) Plenum Press, New York]

days before may thus be activated and be, as it were, relived. Of course these examples do not explain how information is normally recalled.

Thalamo-Cortical Circuits. What is the relation between long-term and short-term storage? We may postulate that there is a free exchange of information between both, since it is the only way to establish a connection between the present and past. In mammals the two-way connections between cerebral cortex (frontal lobe) and the diencephalon (dorsomedial thalamus) appear to be important sources for short-term memory. Among the various connections in the brain the importance of such thalamo-cortical circuits for short-term memory is revealed in the behavior of patients with certain mental diseases. Obsessive neuroses such as agoraphobia, compulsive washing, etc. might be traced back to the fact that important components of the short-term storing sites have been damaged and the normal neuronal activity preempted by a pathological pattern thus disrupting the patient's normal interactions with the environment.

It is possible to liberate the patient from his compulsions by the operative section of some of these tracts – at a high price however: the information exchange between long-term and short-term memory is strongly reduced. The connection between *present* and *past* is abolished for the sake of the *future*. The patient lives almost completely in the present. The surgical frontal lobe incision is called prefrontal leucotomy [140] and was introduced by E. Moniz (1936).

The "Hippocampal Selection-Loop". Not all information reaching short-term memory is transmitted to long-term memory. Thus, a selective process controlled by the level of motivation at that time decides as to which engrams are consolidated in long-term memory.

Those brain areas connected to structures controlling motivation are especially suitable for selection of information. In the mammal these are particular components of the limbic system, known as *Papez-circuit* or "cingulate gyrus-hippocampus" loop (Fig. 145). It connects long-term and short-term memory and receives inputs from the motivational systems (hypothalamus, amygdala, orbital cortex).

After disruption of this loop or damage to any components of the Papez-circuit, information may only be retained for a few seconds or minutes, although both storage sites are intact. In man this amnesiac syndrome is called Korsakoff's syndrome. In these patients recall from long-term memory acquired prior to the disease is quite normal. However, new events will be completely forgotten.

An important function in the interaction among short-term memory, long-term memory and motivation may be ascribed to the frontal cortex. Fixed behavior programs or previously learned ones may be adapted there to suit individual experiences. Patients with lesions in this brain region are no longer capable of adapting to changing situations,

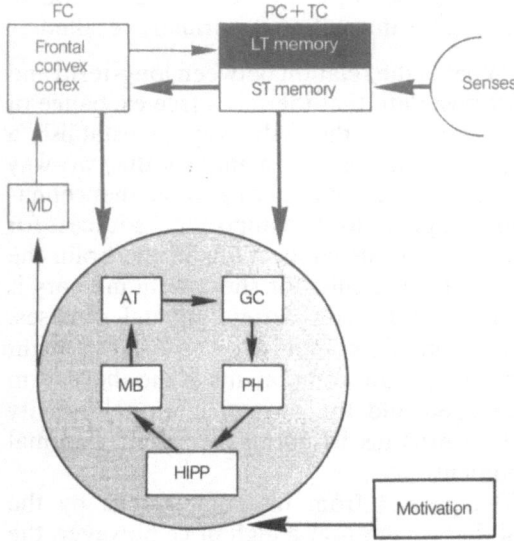

Limbic system "selection unit"

Fig. 145. The "hippocampal loop" as a unit for selecting information from short-term *(ST)* memory to long-term *(LT)* memory. The flow of information is shown by *arrows*. Sensory information is apparently fed into the hippocampal selection-loop *(red)* directly from short-term memory or via the frontal cortex, and in turn is conducted to long-term memory (sensory association areas) via mediodorsal thalamus and frontal convex cortex. Explanations in text. *AT* Anterior thalamic nucleus, *FC* frontal convex cortex, *GC* cingulate gyrus, *HIPP* hippocampus, *MB* mamillary body, *MD* mediodorsal thalamic nucleus, *PC* parietal cortex, *PH* parahippocampus, *TC* temporal cortex. [Modified from Kornhuber H H (1973). In: Zippel H P (ed) Memory and transfer of information. Plenum Press, New York London]

developing behavioral *strategies* and applying new *tactics* to overcome varying adverse conditions (i.e., to exhibit behavioral plasticity).

8.4.3 Reinforcement Systems

The discovery of brain structures whose stimulation is felt as either *reward* or *punishment* by the animal has greatly contributed to the formation of our neurobiological concepts of motivation and reinforcement.

Brain areas relevant in this context may be localized experimentally by means of electrical *self-stimulation,* a technique introduced by Olds and Milner (1954) [141]. Rats are the usual experimental animals: fine wire electrodes are implanted in specific brain regions. A lever contact present in the floor of the experimental cage is activated when the animal accidentally runs over it and thus causes an electrical self-stimulation through the electrodes (Fig. 146); the switch is connected to a counter. If the electrode is located in an indifferent brain region, such accidental contacts occur about 25 times per hour. Depending upon the electrode position the rate of self-stimulation per hour may be increased up to 200–7000 or decreased to 2–3. Obviously in these cases the stimuli are strongly associated with either *reward* or *punishment*. If animals are allowed to administer their own brain stimulation in such positive areas, the rewarding effect appears to be much stronger than any natural reward. Reinforcing by brain stimulation is also more rapid and it persists longer. Natural drive appears to be "added" to brain stimulation effects. However, drive and reward systems are not identical. There are a variety of electrode positions where stimulation elicits (1) eating but no self-stimulation, (2) only self-stimulation, or (3) eating and self-stimulation. Corresponding results were found in higher mammals and to some extent also in man.

Where are these reinforcement zones located? Stimulation of the limbic system in areas of the medial prefrontal cortex, the septum, the lateral hypothalamus and certain midbrain regions are perceived as reward; they are interconnected by the medial forebrain bundle (Fig. 128). On the other hand, stimulation of areas rostral to the midbrain tegmentum and leading to periaqueductal areas of the midbrain acts as punishment.

It is believed that the positive reinforcement system is predominantly associated with *adrenergic* transmission and the negative reinforcement system with *cholinergic* transmission. Numerous studies have shown that learning processes may be accelerated and the recalling capacity

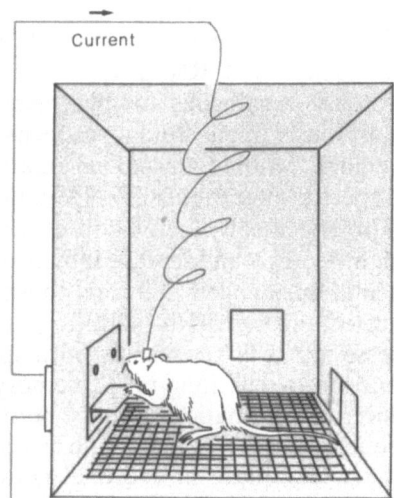

Fig. 146. Experimental set-up for electrical self-stimulation. By pressing a lever the experimental animal (a rat in this case) may electrically stimulate a specific point in its brain via a chronically implanted electrode. Further details in text. [Modified from Olds J (1956) Sci Am 30]

prolonged if the animals are pretreated with certain drugs (e.g., *amphetamines*) which release norepinephrine. If, however, the pretreatment is with substances that block the release of norepinephrine (e.g., chlorpromazine) or inhibit norepinephrine synthesis (e.g., reserpine) the opposite effect may be seen. The significance of *dopamine* as precursor of norepinephrine for the reward aspects of electrical self-stimulation is still partly controversial. Lesion experiments support the hypothesis that the ipsilateral dopamine system must be intact for electrical self-stimulation to be rewarding, even when the neuronal substrate at the electrode tip is itself not dopaminergic.

In this connection it is noteworthy that after administration of so-called *minor-tranquillizers*, the negative reinforcing effects are reduced or even abolished.

By combined brain stimulation and brain lesion experiments it was shown that positive reinforcement structures of the limbic system have an inhibitory effect on negatively reinforcing structures and that the latter again have an inhibitory effect on the positively reinforcing structures in the lateral hypothalamus and midbrain.

Ploog and Gottwald (1974) raise the question of whether such control mechanisms for positive *(reward)* and negative *(punishment)* reinforcement have their origin in analogous structures of the amphibian brain for *approach* (orienting) and *avoidance*. Similar interactions of these behavior systems were found in the toad's brain structures (cf. Fig. 69 and p. 120).

8.4.4 The Problem of "Memory Molecules"

The *genetic information* of the cell is transmitted by a chemical code of deoxyribonucleic acid, DNA. It is read chemically and copied by ribonucleic acid, RNA, which controls the synthesis of species-specific proteins. Analogous to this "genetic memory", characterized by an enormously high storage capacity, there might be an "ontogenetic memory" where *external information* reaching the brain possibly leads to specific alterations of RNA in neurons and glial cells.

More specifically, activation of a neural circuit (short-term memory) might lead to an increase in neurotransmitter that, via activation of a second messenger, induces phosphorylation of nonhistone proteins located on the DNA strands (p. 39). This enables the synthesis of messenger RNA which in turn mediates protein synthesis (possible contribution to long-term memory [142]). In such a case particular pieces of information might be permanently fixed in corresponding base sequences of RNA and amino acid sequences of proteins. Such a process would take place in neuronal networks where neuronal elements are working in relation to adjacent units. Essentially three approaches are possible for the experimental exploration of these problems [143]: (1) investigation of protein synthesis activity in the brain before and during defined stages of learning, (2) study of possible correlations between learning and administration of substances that inhibit the synthesis of RNA or protein, (3) injections of brain extracts from trained animals to naive animals.

Approach 1: If the protein synthesis activity in various brain regions is examined in rats during a learning process, a significant shift of synthesis rates from limbic (hippocampus) to cortical structures is observed. Synthesis activity is increased in comparison to control animals. *Problems:* It is difficult to decide whether this change is specific to learning (memory) or caused by "side-effects" produced by variations in physiological states.

Approach 2: If inhibitors of RNA synthesis (e.g., actinomycin-D) or of protein synthesis (e.g., puromycin, cyclohexamide) are applied at different periods before and during the learning process, different effects may be found: if the drugs are injected into the brain before the appearance of successful learning or immediately after training, a long-term memory may no longer be established. If drugs are injected 1–3 h after consolidation of learning, they appear to have no comparable effects on memory. *Problems:* The inhibitors do not block completely and their actions are not specific. Consolidation of memory can be influenced secondarily by toxic side-effects.

Approach 3: The question arises as to whether transfer of memory "from one animal to another" is possible. Such experiments were initially performed with the flatworm *Planaria,* later also with fish, mice, and rats. For example, rats are trained by electroshocks to avoid dark hiding places, contrary to their natural inclination. When their brain extracts are injected in naive rats, it is possible to show that these animals learn the "dark avoidance" more quickly. An active compound could be isolated from brain extracts of trained rats and was subsequently synthesized in the laboratory. This substance denoted as "scotophobin" (fear of the dark) is a polypeptide containing 15 amino acids and having its maximum effect in rats on the sixth day of training [144]. At this phase it shows the following distribution: 37.0% in fronto-parietal cortex, 29.6% in temporal-occipital cortex, 30.2% in brain stem and cerebellum, and 4.5% in subcortical regions. The action of scotophobin

H-ser-asp-asn-asn-gln-gln-gly-lys-ser-ala-gln-gln-gly-gly-tyr-NH$_2$

is reported to be diminished in a synthetic product where amino acids in position 2,5 or 11 have been exchanged by other amino acids, such as in 5,11-glu-scotophobin(desacetyl scotophobin) [145].

Problems: The specificity of scotophobin is still doubtful. It is known, for instance, that pituitary peptides with definite amino acid sequences positively affect learning behavior of rats in *active* and *passive avoidance tests.* After administration of such substances the extinction of the learned task is delayed. Among these substances are ACTH and analog peptides, such as the decapeptide $ACTH_{1-10}$:

H-ser-tyr-ser-met-glu-his-phe-arg-try-gly-OH

and vasopressin and related peptides, such as lysine vasopressin:

H-cys-tyr-phe-gln-asn-cys-pro-lys-gly-NH_2

Whereas the effect of lysine vasopressin after a single injection may continue for several weeks, it declines 6 h after the administration of ACTH analogs. The active core of ACTH-related peptides in both "fear"-motivated tests appears to reside in the amino end of the molecule

$$ACTH_{1-13} > ACTH_{11-24} < ACTH_{4-10}.$$

The steric configuration (D,L form) of amino acids also has an effect on extinction of the avoidance response. The effect of all-trans-form $ACTH_{1-10}$ may reverse when L-phenylalanine is substituted by D-phenylalanine in position 7.

The effect of ACTH is not restricted to avoidance learning (i.e., extinction) in mammals. ACTH administration in common toads may also delay the extinction (recovery) after habituation of prey-catching in response to repeated stimulation with the same prey dummy [146] (p. 83).

It is interesting to note that synthetic desacetyl scotophobin in the rat also increases the resistance to extinction of active and passive avoidance responses, thus resembling the effects of ACTH and vasopressin related peptides. Desacetyl scotophobin appears to be active for 5–7 days in these tests [147].

At present we can hardly judge whether polypeptides like scotophobin are "memory code words", and if they are, what exactly are they doing. One of the main doubts in the discussion of "memory transfer" and related substances concerns the question of their specificity. However, most recently in honey bees the specific properties of a learned *time signal* could be transferred from one individual to the other [148]. Presumably bees will become new subjects for investigation of the question of "memory molecules".

To summarize:

1. Memory is based on short-term and long-term storage. A free exchange of information appears to take place between them.

2. Selection of information for long-term storage in the mammal might be accomplished by specific functional structures of the limbic system. These are linked to motivation systems, which themselves contain structures that mediate positive and negative reinforcements.

3. Information is fixed in wide areas of the brain, but may be recalled from particular sites.

4. There are indications that rates of biosynthesis (RNA, proteins) are enhanced during learning. However, so far it has not been possible to demonstrate that specific memories (engrams) are "fixed" in specific molecules.

8.5 Social Stress

For each individual the animal next to it is part of its environment. Social interactions may more or less determine its behavior and thereby also its physiological state. This is especially important for animals living in a social community; changes of social structure may have an impact on the physiology of all members. Among the external factors that can affect social structure are population density, loud noise, struggle, or fear. Stress can also be released by other stimuli, such as temperature extremes, oxygen deficiency, side-effects of drugs. These stimuli are called stressors. The physiological state is called stress. Stress is not a disease, but a stress situation may develop during a disease.

8.5.1 How is Stress Perceived?

Stress leads to an activation of the sympathetic (autonomic) nervous system and to the discharge of pituitary hormones (CRF and ACTH), and adrenal hormones (catecholamines, corticoids) (cf. Fig. 147). Within a very short time there is a rise in heartbeat and respiration rates, blood pressure, and blood circulation in skeletal muscles (at the expense of blood circulation in gastrointestinal and renal systems); the level of blood sugar also increases to provide the organism with additional readily-available metabolic energy.

Stress is an important survival mechanism; it defends the body against "physiological insult". Furthermore, we know that the pituitary-adrenal system – which responds to stressors – may also play an important role in

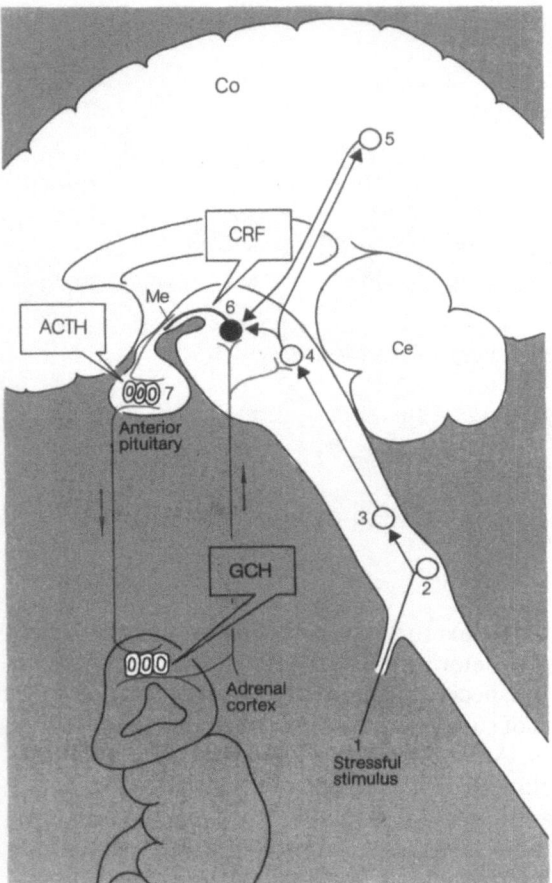

Fig. 147. Stress and its connections to the pituitary-adrenal system. Neural messages on stressful stimuli *(1)*, e.g., cold, are conducted via subcortical *(2–4)* and cortical areas *(5)* to the hypothalamus *(6)* and stimulate the neurosecretory cells to release corticotropin-releasing factors, CRF. These are carried through hypothalamic-hypophyseal portal vessels *(red)* to the anterior pituitary gland *(7)*, which in turn releases adrenocorticotrophic hormone, ACTH, into the general circulation. When ACTH acts on the adrenal cortex, glucocorticoid hormones, GCH, are released into the general circulation and they reach all of the tissues (e.g., neurosecretory cells in the hypothalamus and higher order neurons that modulate production of CRF). Secretion of ACTH is suppressed when GCH reaches a certain level. *Ce* cerebellum, *Co* cerebral cortex, *Me* median eminence. [Modified from Scharrer E (1966) Endocrines and the central nervous system. Williams and Wilkins, Baltimore]

learning (cf. p. 248) and development of behavior. It is even suggested that effective behavior depends on some optimal level of stress. Rats "deprived" of any stress during early life develop a pituitary system which responds slowly and ineffectively.

The response of the organism to stress stimuli is characterized by three phases according to the situation: (1) *"Alarm reaction"* in response to sudden but not frequently occurring stressors. It is characterized by a shocklike syndrome (e.g., rise in blood pressure, etc.), and an increased secretion of ACTH. (2) *"Resisting"* if stressors are long lasting. The activation of the adrenal cortex reaches its maximal value due to continuous stimulation by ACTH. The organism develops a nonspecific resistance, while the symptoms of the shock syndrome are disappearing. (3) *"Exhaustion"* after extremely frequent and lasting repetition of stress situations. The organism may lose its adaptive capacity. Despite a normal food intake the body weight declines, and the sex glands can be inhibited. Finally it may be fatal. These phases are not separated clearly from each other and the conditions for (3) might depend also on the species as well as the disposition of the individual.

8.5.2 How Can Stress be Measured?

There is a mystifying quality about stress in that the response to it cannot be easily detected in each individual. However, in some mammals there are certain stress indicators. A familiar example is "goose-pimples" that occur in certain anxious situations. In mammals fear is frequently connected with a bristling of the body hair. Raising of the hair is caused by contraction of the hair follicle smooth musculature that is innervated by the autonomic nervous system. The hair bristling can be released in dogs, cats, and monkeys by electrical point stimulation of certain areas of the central autonomic system in the limbic complex, such as the hypothalamus (Fig. 148). They are often accompanied by other features of autonomic syndromes, such as pupil dilatation, perspiration, rise in blood pressure, etc. (cf. p. 237). Therefore stress, too, has a representation in the CNS.

The bristling of hair is particularly marked in tree shrews *Tupaia belangeri*. These are diurnal mammals about the size of squirrels. They live alone or in pairs in the forests of South East Asia. Their tail is remarkable and in this context especially important for us. Normally the tail hairs are laid flat (Fig. 149 A). Upon stimulation of the autonomic system the hairs immediately rise (Fig. 149 B). This is an indication of stress. For instance the duration of tail bristling (TB) can be measured as a percentage of a 12-h observation period: the tail bristling value ("%TB") is a measure of the duration of activation of the autonomic system. This value is constant as long as the environmental situation of the animal is constant. By means of TB the ecological basis of stress situations may be explored quantitatively [149]. Here we shall focus on two stressors: population density and fear.

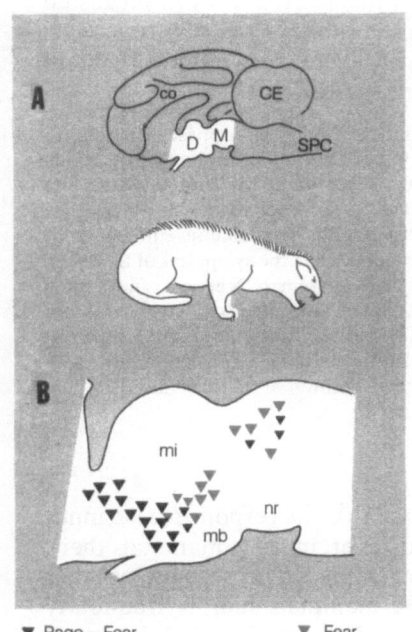

Fig. 148 A and B. Schematic representation of points in cat hypothalamus whose stimulation by means of a chronically implanted electrode elicits rage-fright and fright. The bright area in the schematic brain sagittal section above **(A)** is shown below **(B)** as an enlarged section. *CO* cerebral cortex, *D* diencephalon; *M* mesencephalon, *CE* cerebellum, *SPC* spinal cord; *mb* mamillary body, *mi* massa intermedia thalami; *nr* red nucleus. [Modified from Hess W R (1954) Das Zwischenhirn, 2. Aufl. Schwabe, Basel]

▼ Rage – Fear   ▼ Fear

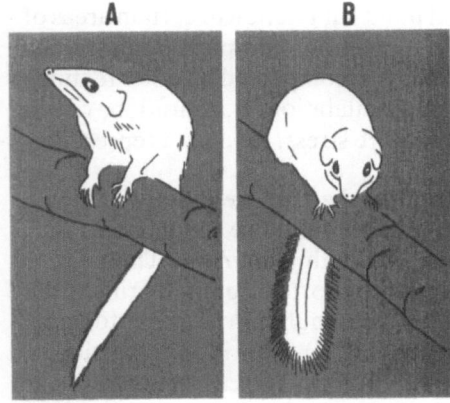

Fig. 149 A and B. Adult female tree shrew *Tupaia belangeri.* **A** Normally the tail hair is laid flat. **B** "Tail bristling": following any disturbance the sympathetically innervated hair follicle musculature in the tail contracts, thus raising the hair. The tail then takes a bushy appearance. [From photographs by Holst D v (1969) Z Vergl Physiol 63:1–58]

8.5.3 Population Density

Young tree shrews live with their parents in a family. The TB of the parents is constant for rather long periods of time. It increases, however, as soon as a young tree shrew (at an age of 60 days) becomes sexually mature and starts marking the territory with specific secretions (urine and sternal gland secretions). Sexually mature males stress only the father, sexually mature females only the mother.

The increase in TB values of the parents appears to be linked to the sex hormone level of the young. The influence of testosterone can be studied in castration experiments. If the young males are castrated before they are sexually mature, the TB value of the father does not rise. After injection of testosterone in the castrated males, the TB of the father increases. Obviously the TB values of the father reflects the scent marking (chinning) activity of the young males. Chinning activity of the males in turn depends upon their level of testosterone (Fig. 150).

As soon as the TB value of the mother is increased by more than 20%, she devours her offspring. How can this behavior be explained? Normally a sternal gland produces a secretion with which she labels her offspring. This secretion protects the young from other conspecifics, including the mother herself. However, if the mother experiences stress, her gland stops working: now the young are unprotected, even from her.

What is the biological significance of this density effect? It avoids the further increase of population in that territory without recourse to aggression. How is the population density recognized? Tree shrews apply scent marks; the higher the population density, the larger the number of scent marks per unit of territory. An increased frequency of contact with these marks produces social stress that finally may inhibit further increase of the population. TB values also increase if foreign conspecifics have marked the territory.

Fig. 150. Chinning activity of a male tree shrew before and after castration (see *arrow*) and after injection of testosterone propionate *(second arrow)*. The marking activity was measured in an experimental cage scented by the scent marks of a fertile male conspecific. Further details in text. [Slightly modified from Holst D v, Buergel-Goodwin U (1975) J Comp Physiol 103:123–151]

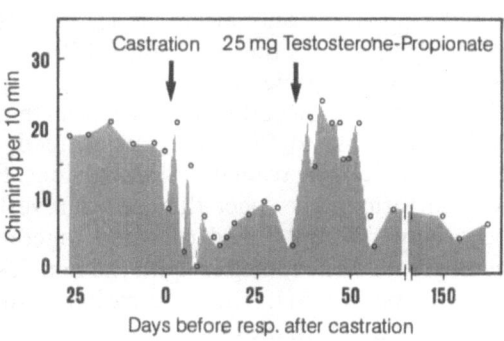

8.5.4 Dominance Relationships

Tree shrews may have very violent rank order fights with foreign conspecifics as well as with family members of the same sex. The combats are short and bloodless. Whereas the winner no longer cares about the loser, the latter is still under the influence of the fight and retreats into a dark hiding place. Its TB values are very high. As long as visual contact with the winner is avoided, the loser is able to recover. However, if the dominant animal is visible, it becomes a very serious hazard for the vanquished. The TB values increase further ($>90\%$ TB), the animal loses weight and dies within 2–16 days by uremia, which results from ischemia or even anemia of the kidneys.

The trigger for this stressful experience is the visual contact with the winner and the concomitant "consciousness" of being the loser. The memory of defeat stored in the CNS by the "hippocampal loop" (Fig. 145) represents the link with the activation of the autonomic nervous system. The image of the victor does not allow the event to fade. More recent studies indicate that "dominance" is represented in certain areas of the limbic system.

8.5.5 Stress in Human Society

Man also reacts to stress produced by family disputes, lack of professional success, fear of the boss – and also to that created by gripping thrillers. However, these reactions do not always develop into a permanent threat to the mental and physical well-being of a person. Unfortunately at present we hardly know about any firmly established causal relations. The reason for this may be that the efficacy of such stressors depends to a large extent on the individual and his disposition. For instance it is difficult to evaluate the relative roles of adaptation and resistance in response to overcrowding, such as that occurs in slums. The actual damage, which could be hidden behind what we call adaptation, may be apparent only after a long time. Here, too, an example from ethology illustrates this possibility [150].

Rats living in colonies show certain feeding habits. They eat their food alone, preserving a certain distance from their neighbors. Experimentally it is possible to force the rats to feed, crowded together, from a communal feeding trough for long periods. But if these rats are returned to their former free space they do not revert to their old feeding habits: now they will only feed crowded together. Months later a serious behavioral deficit comes to light: the females neglect their offspring and cease to build nests. The procreation instinct is irreversibly disturbed and the population dies out.

A reflection:

"The findings of biology suggest possible dangers and pose questions which have to be studied seriously from the point of view of sociobiology, psychology, and physiology with reference to the human condition. It is a prerequisite that the studies be oriented in accordance with biological findings and derive their working methods therefrom. The temptation for us is too great to assume the 'singularity' of man and to refuse to take serious note of such 'animal' physiological mechanisms or to believe that we can solve all of the problems by reason and medical help alone" [151].

9 Methodological Appendix

Answers to the problems concerning neural mechanisms of behavior can be obtained only by the use of diverse experimental methods. The research techniques are mostly derived from *behavioral physiology, electrophysiology, neurochemistry, neuroanatomy* and *engineering.* The results obtained may contribute like pieces of a mosaic to the understanding of the system of the object studied. Many methods often require several years' experience. The collaboration of research groups therefore has proved to be very fruitful, especially in neuroethology.

9.1 Prerequisites for Behavior Analysis

The techniques of ethology – which must be adapted to a particular animal and a special problem – are numerous. Therefore, we shall discuss only some general prerequisites.

9.1.1 The Experimental Animal

It is an experience of physiologists and ethologists that the experimental exploration of a problem stands or falls, as it were, with the choice of the experimental animal. Thus, the fruit fly, *Drosophila,* on the strength of its transparent chromosomal make up and its short maturation cycle has become the "pet" of geneticists; the cat because of its physiology and constant brain/skull relationship belongs, among mammals, to brain physiologists; since Galvani the frog is the favorite of neurophysiologists; the series might be continued with numerous examples. In the domain of neuroethology we are confronted with the problem of finding experimental laboratory animals equally suitable for behavioral and neurophysiological investigations. With respect to the neural basis of behavior it is obviously of little advantage to study, for instance, the behavior of the *bee,* to investigate the neurophysiological systems of the *fly* and to gather histological data on comparable substrates in the *ant* – simply because each of these animals is specially suited for the exploration of the particular problem. Strictly speaking such a procedure would produce only fragments which, a priori, restrict the deduction of causal statements about the functional structure.

In actual fact this is frequently the situation in which the neuroethologist finds himself. An animal accessible for the experimental approach of one area of study may not share these requirements for another area or vice versa. Such differences might have many reasons. Among the most important in the electrophysiological experiments is the physiological condition of the experimental animal after preparation and the particular type of electrode used. Here there is much room for improvement.

Appropriate experimental animals for neuroethological investigations are found especially among the lower vertebrates and the invertebrates. Many types of behavior proceed according to a constant or relatively rigid program in these animals. A neurophysiological advantage in invertebrates may be the relatively small number of neurons comprising the central nervous system and the thickness of some nerve cells and fibers forming in certain animal groups the so-called giant fibers. Flies, crickets, crayfish and some species of marine snails have thus become "pet animals" in many neuroethological laboratories.

9.1.2 Animal Care and Experiment

Not every animal exhibits its natural repertoire of behavior in captivity. In this case comparative experiments in the laboratory and in the natural habitat are indispensable. Keeping of animals usually requires many years of experience. Sufficient care of laboratory animals is the prerequisite for behavioral experiments and comparative neurophysiological studies. Gross differences in animal maintenance between research groups using the same experimental animal may be an unrecognized cause of the criticism that one and the same experiment is not always reproducible in every hand.

9.1.3 Behavior and Activity Periods

A precondition for the investigation of a particular behavior is knowledge about the natural context in which it occurs. This means that we also have to know as much as possible about its entire behavior repertoire. For this purpose all observable modes of behavior shown by the animal spontaneously or in response to external natural stimuli, including to conspecifics, are usually listed in a so-called *ethogram*. It must be borne in mind that all animals are subject to diurnal and seasonal periods of activity. Unless the behavior during the main activity period is accurately known, experiments performed outside this period are relatively useless. They may be considered later on for the sake of comparison.

Hibernating animals are often distinguished, according to the season, as "summer" and "winter" animals in laboratory jargon. A temporal mutual overlap of ethological and neurophysiological experiments in the main activity periods is indispensable in experimentation. In these animals the search for neurons controlling motivated behaviors will have little chance of success if performed in the winter time. Even if the animals are allowed to pass the winter under constant laboratory conditions, they go through a stage of hibernation, owing to the activity of endogenous factors.

9.1.4 How Can Behavior be Measured?

When investigating stimulus–response relationships, it is important to measure quantitatively the key-stimulus for releasing a type of behavior. If the stimulus efficacy is paralleled by the intensity of the behavior response, a suitable unit of behavioral activity must be established. In principle, if a behavior reaction is released by a dummy stimulus at brief intervals over a long period, the number of responses per unit of time may be defined as a measure of stimulus efficacy. If the behavior response occurs more infrequently, it may be sufficient to record the relative frequency of stimulus responses. The capability of an animal to distinguish one stimulus from another may be tested in a number of ways: training, habituation, and two-choice experiments, to name three examples.

In addition to the numerical approach to behavior responses, there are other measuring methods that, however, are always adapted to special problems and relate to a particular experimental animal. For instance, the responsivity of some fish to particular stimuli may be recognized by the type and the intensity of their body color. The spectral sensitivity of the *Guppy* under oblique illumination may be determined from the angular deviation of its body axis from the perpendicular.

The activities recorded from individuals tested one at a time in different stimulus situations mostly yield unequivocal results regarding stimulus–response relationships. False conclusions may be reached if animals are tested in a group. We choose a simple example. Experiment 1: a piece of food is shown to a group of animals; suppose one animal responds. Experiment 2: two pieces of food are shown to the group; suppose two animals respond. The possible conclusion that two food objects represent a more intensive stimulus for a *single* animal than one food object is not cogent; e.g., stimulus- response relationships may be influenced by social interactions.

There are several useful tools for data collection. Among the simplest, yet most important tools, are the stop-watch and checklist (Fig. 151). Electronic counters and event recorders are often used. Tape recordings are especially suited for experiments in the field.

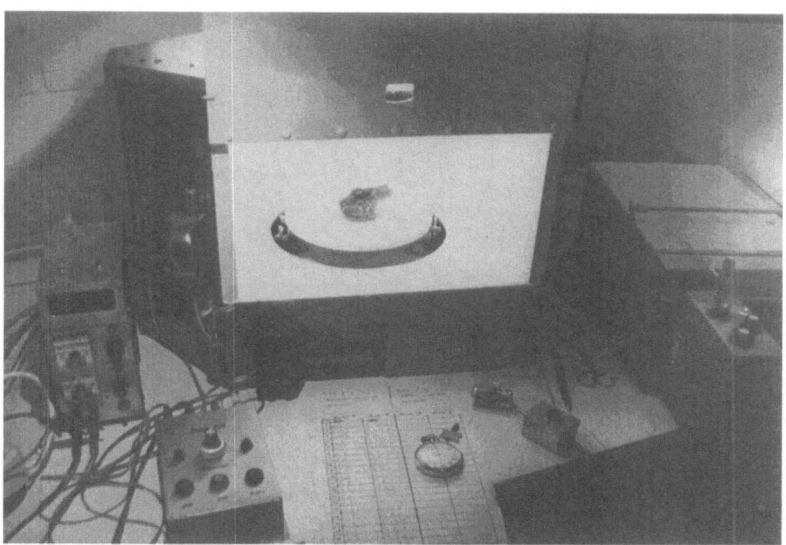

Fig. 151. Example of a simple set-up for quantitative behavior experiments. Using this experimental arrangement the effect of different dummies (pieces of cardboard) on the prey-catching activity of the common toad may be measured. The dummy is moved mechanically at constant angular velocity in the concentric slot invisible to the toad. The number of orientational turning movements by the toad per dummy revolutions within a one or half minute period is defined as a measure of prey-catching activity (cf. p. 73 and Fig. 42). Stop-watch and check list (in the foreground of the picture) are all that is needed to register the data. The time sequence of the stimulus responses may also be recorded by pushing the key of an event-recorder. (Photo H. Burghagen, 1975)

9.2 Investigation of Behaviorally Relevant Brain Areas

9.2.1 The Brain Stimulation Technique

Electrical brain stimulation for the systematic delineation of structures eliciting behavior was first performed by R. W. Hess in freely moving cats. This technique has now become a routine tool of neurophysiologists, neurologists, and neuroethologists.

Preparation Techniques. The procedure is as follows: (1) cutting the scalp in the anesthetized animal, (2) removing portion of the skull by means of a special drill, (3) exposing a part of the brain surface by cutting the dura mater. Two or more electrodes, each insulated except at the tip, are lowered into the brain region to be examined and fixed to the skull with a quick-hardening substance; dental cement is suitable (Fig. 154). In large animals a head holder with adjustable electrodes is used (Fig.

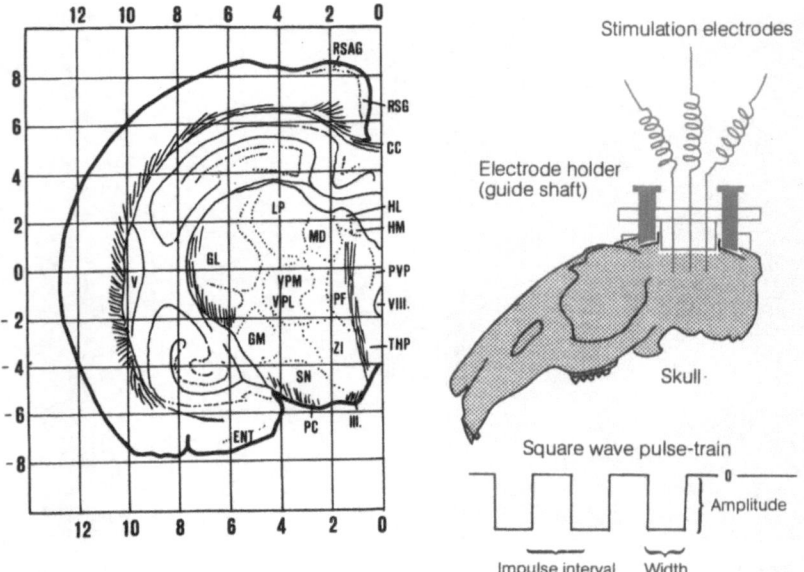

Fig. 152. A page from the stereotaxic atlas of the rabbit brain. *Left,* The map shows cross section AP-5 through the left telencephalon-diencephalon region. *Right,* Illustration of an electrode holder in the skull. *Bottom,* Parameters that can be varied when stimulating with direct current pulses. [Modified from Bureš J, Petráň M, Zachar J (1967) Electrophysiological methods in biological research. Academic Press, New York London]

152, right). A calibrated mechanism allows electrode movements along the three axes.

Stereotaxis. Brain stimulation experiments require a precise anatomical knowledge of the particular brain structures. Stereotaxic brain maps are available for some vertebrate animals and show the important nuclear and fiber regions, in a system of stereotaxic coordinates in relation to fixed points on the skull (Fig. 152, left). It is possible with the aid of these maps (stereotaxic atlas) to drive the electrode "blindly" by means of a special micromanipulator (stereotaxic apparatus) into a particular brain structure. The experimental animal does not feel any pain since the brain itself is insensitive to pain – with the exception of the brain meninges which, however, have previously been removed under anesthesia from that particular brain region.

In *neurosurgery* (Fig. 153) brain lesions – perhaps in order to dampen the activity of a certain nuclear group – are carried out while the patient is fully conscious. During the penetration of the brain tissue by a coagulation electrode the neurosurgeon stimulates the nervous tissue via the same electrode with electrical impulses at definite intervals. He is then able to exactly determine the position of the electrode from the stereotaxic coordinates and the behavior response of the patient. Landmarks are also given by X-ray ventriculograms.

Fig. 153. Stereotaxic guiding device
for psychosurgical operations in man

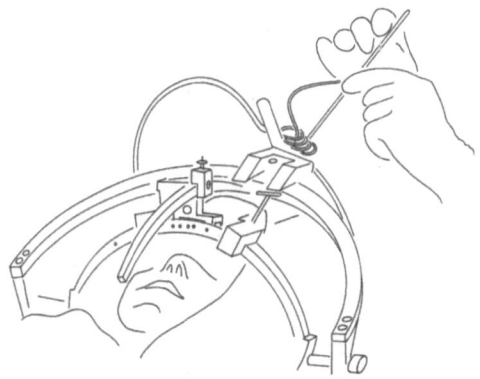

Electrodes and Stimulus Impulses. What kind of electrodes are used in animal experiments? They are electrolytically sharpened fine metal needles (stainless steel, tungsten, platinum-iridium) insulated with a special lacquer except at the tip (length: 10–50 µm, tip diameter: 5–20 µm). Stimulation is carried out either bipolarly between two adjacent (or concentric) electrodes or, more often, by the monopolar method with a stimulating electrode inside the brain and a large surface reference electrode. A small metal plate in contact with the skull of the animal usually serves as reference electrode. The thin connecting wires of the electrodes lead to an impulse generator. In order to avoid mistakes in "drilling" of the wires due to movements of the experimental animal, they may be led through mercury sliding contacts above the experimental site (Fig. 154).

Stimulus impulses usually are sequences of negative square wave impulses whose width, amplitude, and frequency are regulated by the generator and may be checked on an oscilloscope screen. For the elicitation of behavioral modes in the freely moving animal impulse trains of 50–100 Hz, 0.5–5 ms pulse width and a current of 20–200 µA are especially effective (Fig. 155).

The Threshold Current. In electrical brain stimulation experiments the threshold current plays an important role. It is the minimum current flowing between the active and reference electrodes sufficient to trigger a particular type of behavior. It is currently accepted that different behavior patterns are often characterized by varying levels of response thresholds in the CNS. This is shown for example by brain stimulation experiments in the toad. The electrical threshold for release of escape behavior is clearly lower than that for prey-catching (Fig. 155). Furthermore, within a single mode of behavior, such as the prey-catching behavior, the meaningful sequence of the individual compo-

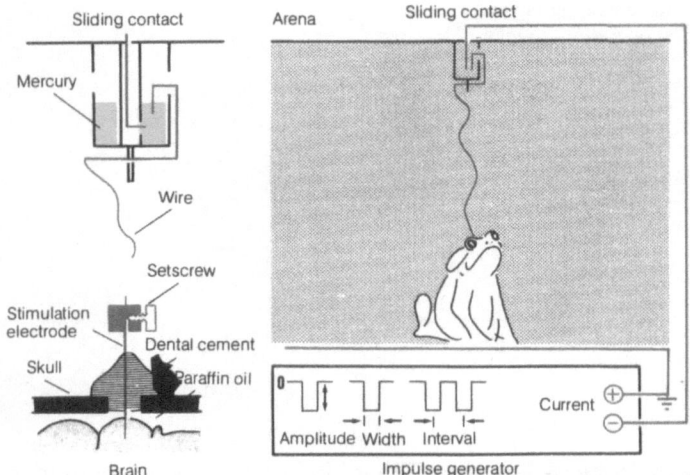

Fig. 154. Simple set-up for electrical brain stimulation experiments in lower vertebrates. [From Ewert JP (1967) Z Vergl Physiol 54:455–481]

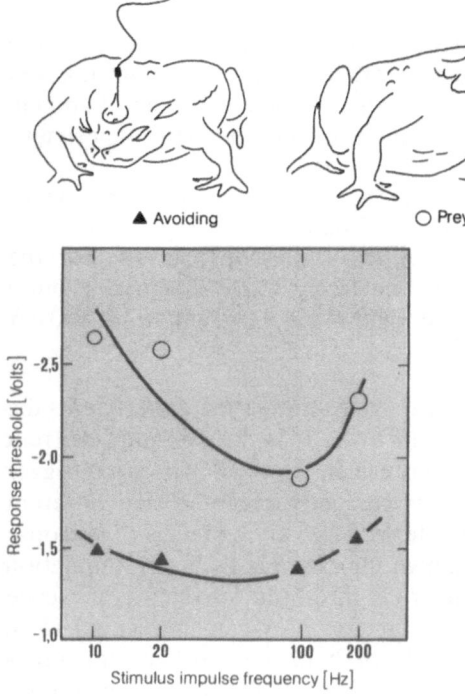

▲ Avoiding   ○ Prey-catching

Fig. 155. Activation of prey-catching and escape behavior in common toads by electrical point stimulation of the optic tectum (○) or the caudal thalamic-pretectal region (▲). Relation between threshold voltage and stimulus impulse frequency [Ewert, 1968]

nents – orienting, snapping, swallowing – is assured by different and graded thresholds.

The Site of Stimulation. With respect to the localization of the electrical stimulus it must be noted that the current flowing between the two electrodes depends both on the resistance of the electrode and the brain (precisely the whole animal). Although the current in the immediate neighborhood of the electrode tip is most dense, it must be considered that, with respect to the whole extent of the electric field, not every site of stimulation, which is able to activate a mode of behavior, is situated near the electrode.

Furthermore, if electrical stimulation of a site in the brain elicits a particular behavioral response, we cannot conclude *solely* from this observation that the corresponding region is activating this behavior. For example, ascending and descending pathways may have been stimulated. Furthermore, ortho- and antidromic stimulation effects must be taken into account. (Orthodromic is the normal direction of impulse propagation in the nerve fiber: from dendritic region of the neuron toward axonal termination). The electrically stimulated area could also be inhibitory with respect to the behavior under investigation, and the elicited response may be activated indirectly (e.g., by disinhibition).

It must be emphasized that the electrical stimulus does not simply duplicate the neuronal events. Since it is known that electrical stimulation does elicit coordinated behavior patterns, we must assume that the artificial stimulus acts as a kind of trigger for the recall of a fixed program. This may be possible at different sites in the brain.

Chronic Electrode Implantation. At the end of an experiment the hole in the skull may be closed. The animals remain alive and – if prepared carefully – do not show any impairment in their behavior. In some experiments, however, it may be important to test the animal over a long period of time. In this case corrosion-free permanently implanted electrodes are used. These chronic implants may stay in the brain indefinitely. At the end of an experiment the electrical contacts at the skull are unplugged. For special problems techniques of telemetric stimulation have been developed (Fig. 141). Among other things this method has the advantage that such stimulation can elicit defined, reproducible behavior patterns while the animal is interacting within its social group. Thus, the effect on conspecifics may be studied.

Chemical Stimulation. The electrical brain stimulation method by itself has only a limited significance for studying the central representation of behavior modes. As we have discussed, electrical stimulation does not unequivocally show whether the stimulated area takes part either directly or indirectly in the activation of a behavior pattern (Fig. 133, left). Therefore, the search for a possible "chemical coding" of behavior has led to the development of chemical brain stimulation techniques (Fig. 133, right). In these methods, neurons are affected more selectively in their activity by local application of drugs, neurotransmitters, blocking agents, or hormones have frequently been used. They are applied to the brain area to be studied either by a fine micropipette (Fig. 135) or in the form of very small (few µg) crystals. The chemical stimulation techniques add a new dimension to the search for the central representation of behavior [147].

Evaluation of Experiments. The behavior patterns activated by brain stimulation are filmed, and simultaneous comments are recorded on magnetic tape. For accurate electrode localization in the brain, small electrolytic lesions may be produced by passing an anodal direct current through the stimulation electrode. If steel electrodes have been used, the resulting iron deposit at the site of the electrode can subsequently be revealed by the Prussian blue reaction in Paraplast- or paraffin-embedded brain sections after fixation in formalin. The same sections can be stained with the Nissl or Klüver-Barrera method, for example, to identify the cellular and fiber architecture surrounding the lesion sites. Stimulation sites are recorded in brain maps and characterized by symbols signifying the particular stimulus effect (e.g., Fig. 134).

9.2.2 Brain Lesion Techniques

Brain Ablations. The oldest method for the experimental investigation of brain function is the lesion technique [152]. For this purpose certain regions of the brain are destroyed and their function is deduced from the abnormal behavior of the lesioned animal. The ablation may be carried out "free-hand" by means of *knife cuts*. In experiments with ablations of varying sizes the localization of a function may be pinpointed from the set of effective sections. Small lesions can be produced by *aspirating* brain tissue with the aid of a pipette and a suction device (vacuum pump). Localized discrete lesions are produced with a lesion maker passing *electrical current* through an electrode of stainless steel or platinum-iridium, electrically insulated except for a part of the tip. The electrode may be guided stereotaxically with the aid of a micromanipulator. Two kinds of current can be applied to the brain tissue: (1) anodal direct current, DC, produces lesions by initiating *chemical reactions* from electrolytically released ions. The products kill cells surrounding the electrode. (2) Alternating current, AC, in the radiofrequency range destroys cells with heat *(thermocoagulation)*. Most recently Laser techniques have been developed for certain problems.

Chemical Lesions. The electrical activity of populations of neurons may be reversibly eliminated by application of filter paper soaked in 3 or 5 M

KCl to the brain surface. This method is called *spreading depression*. It is possible to produce a local selective elimination of nerve cells – while sparing the fiber tracts passing through such areas – with the neurotoxin *kainic acid* [153]. The substance is applied electrophoretically, close to the cells, by a stereotaxically guided micropipette. The neurotoxin is taken up postsynaptically by glutamate receptors at the soma and dendrites and it destroys cells after hyperexcitation of their membranes. The exact mechanism of neurotoxicity is unclear at present. A specific presynaptic neurotoxin is *6-hydroxydopamine* [154]. It is selectively taken up by catecholaminergic neurons because of its structural similarity to dopamine and norepinephrine. This substance is administered by systemic or local injection and is taken up by the neurons presynaptically together with the catecholamines after their synaptic action. At neutral pH 6-hydroxydopamine undergoes spontaneous decomposition, liberating the strong oxidant, hydrogen peroxide.

Brain Lesion and Function. The lesion method has, in principle, the opposite effect to the stimulation method. In the latter case neurons in particular areas are brought to (hyper) activity, in the former their effect is abolished. Destruction of a brain area may indeed lead to the abolition of a type of behavior which previously could be elicited by stimulation of this area. The effect of brain lesions can be evaluated in conjunction with suitable behavioral tests. It must be kept in mind here, that modifications or deficiencies of behavior do not have to be necessarily attributed to the eliminated structure under consideration.

When critically evaluating lesion experiments, it must also be noted that an animal with a brain lesion is possibly no longer the same animal in many respects apart from the obvious lack of this brain region. The brain represents a complex feedback system. When a component is absent, the whole system, strictly speaking, is modified. The correlation between an eliminated brain area and a change of behavior must therefore be evaluated carefully, just as we have already discussed for electrostimulation experiments. The connections of the eliminated areas with the rest of the brain can be studied by means of neuroanatomical degeneration and tracing techniques (cf. p. 286).

We have discussed critically the two main methods for investigation of behaviorally relevant brain structures. In fact the predominant part of our present knowledge of relationships between substrate and function in the brain has been obtained from experiments which utilized these two techniques. Considered by themselves these techniques do not represent panaceas. These approaches must be continuously re-examined in relation to the problems addressed and must be validated in combination with other experimental methods and paradigms.

9.2.3 Neuroanatomical Mapping of Brain Activity

In order to find out which brain structures are involved in a particular behavior pattern, the deoxyglucose method may be utilized. It was developed for the neurochemical mapping of functional activity among neuronal populations [155]. The principle appears to be simple. During a stimulation experiment in the awake animal radioactive [^{14}C]-2-deoxy-D-glucose, abbreviated [^{14}C] DG, is injected intravenously over 10–15 s at a dose of 150–200 µCi/kg bodyweight. [^{14}C] DG crosses the blood brain barrier by the same carrier by which glucose is transported; the difference is not recognized by the cell membrane. Thus, [^{14}C] DG is taken up mainly by the activated neurons and will be phosphorylated to [^{14}C] DG-6-P, but not metabolized further. Hence, the level of [^{14}C] DG-6-P increases in the active neurons.

At the end of the experiment (lasting about 45 min) the brain is removed from the barbiturized animal and is frozen (to about $-100°C$). The total time from sacrificing the animal to freezing the brain in blocks will be about 15 min. Brain sections (of 20–40 µm) are exposed to X-ray film for periods of at least 2–3 weeks. The [^{14}C] DG deposit is a quantitative indicator of the energy metabolism. The density of the label can be scanned and processed by a computer and be transformed into a color code or a digitized output.

This method provides a picture of functional activity in the brain, for example, in response to behaviorally relevant stimuli. Thus, [^{14}C] DG is not only a method to study neuroanatomy, but more importantly a tool to investigate differential regional activity in the brain. Of course, control experiments are necessary for proper evaluation of the effects. Provided that the resolution is sufficiently high and the background activity is sufficiently weak, the [^{14}C] DG method may be developed as some kind of "chemical microprobe": not detecting a *neuronal element,* but indicating the activity of *neuronal assemblies* responsible for a particular task or function – and that, too, at different central levels. The [^{14}C] DG method *labels functional systems* and provides the neuroethologist with "landmarks" for quantitative, electrophysiological single cell recording studies.

9.3 Methods for Recording and Measuring Neuronal Activity

With the aid of the [^{14}C] DG method, the stimulation, and lesion techniques indications of certain "behaviorally active" areas in the CNS may be obtained. It is an aim of the neuroethologist to reveal quantitative correlations between behaviorally relevant stimuli, behavior patterns, and neuronal responses in the corresponding neural structures.

9.3.1 Recording of Summated Potentials

Receptor Unit Responses. The simplest way of obtaining information on responses of receptors is *multi-unit* recording (Fig. 156). By application of a relatively large surface electrode to an entire receptor area the receptor potentials are recorded as a mass response against a grounded reference electrode. The summated potential is amplified and visualized on an oscilloscope. From the amplitude and slope of the potential, definite data about the type and the state of sense organs can be obtained (Fig. 156 A and B). Multi-unit recordings are especially suitable for measurements of stimulus-response thresholds.

Responses of Brain Areas. Preliminary clues to possible correlates between external signals and the activity of the brain may be obtained by recording an electroencephalogram, EEG. As for the summated receptor potentials, here too, relatively large surface recording electrodes are used. They are introduced *into the brain* area to be examined or applied *to the surface* of the particular brain area – or even attached directly *to the scalp* with electrolytically conductive paste (Fig. 156C). EEG's elicited by peripheral stimulation are called *evoked potentials.* They are not action potentials, but graded slow potential fluctuations whose origin has not been completely clarified. There are indications that slow brain waves correspond to summated glial potentials (p. 36) and/or postsynaptic potentials of neuronal populations from the vicinity of the recording electrode. Since the evoked potentials are always recorded from those brain areas where the particular stimulus is processed, indications on the spatial processing path can be gained by this method. The amplitude and slope of the potential yield data on functional neural sensitivities.

In addition to these slow potentials, appearing in *response* to external stimuli, there are also slow, *spontaneous* brain rhythms. They have been classified according to their characteristic frequency ranges and designated by Greek letters:

Wave/s: $< 4 - 8 - 13 <$
Wave type: $\delta \quad \Theta \quad \alpha \quad \beta$

In man the wave band recorded in the EEG depends on the state of development of the brain. Whereas babies and infants show predominantly slow δ- and Θ-waves, in adults α- and β-rhythms predominate. The frequency of brain rhythms is also correlated to the arousal state of the brain. While awake, the "tranquil" state of the brain is expressed by α-rhythm. When stimulated, e.g., by illumination of the eyes or intensive

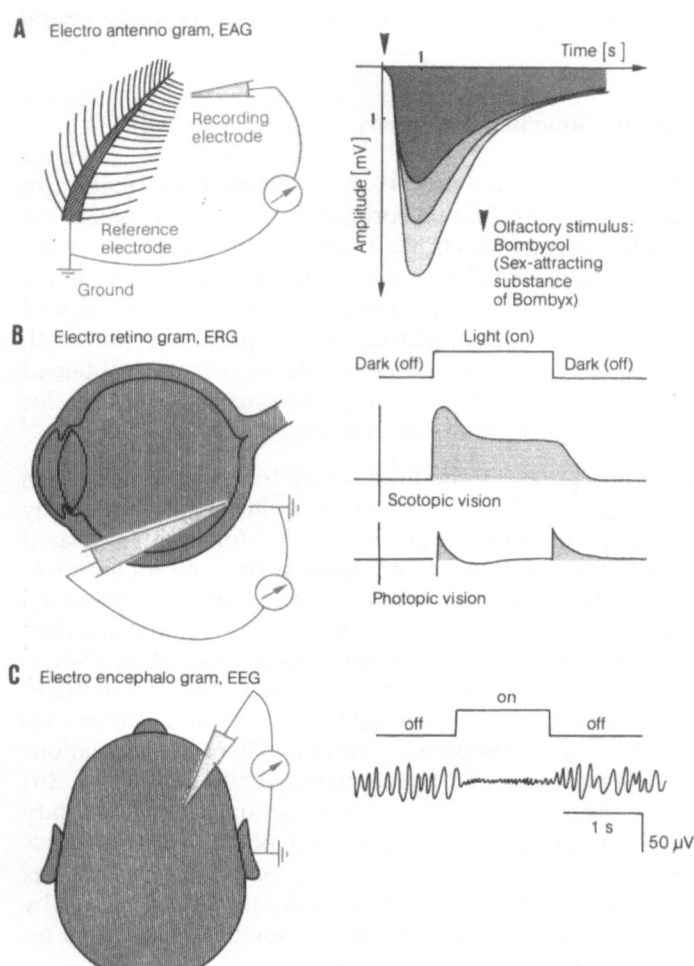

Fig. 156 A–C. Recording of summated potentials. **A** *EAG* Electro-antenno-gram, i.e., *EOG* Electro-olfacto-gram. [Modified from Boeckh et al., 1965]. **B** *ERG* Electro-retino-gram. The recording electrode is usually applied to the cornea. **C** *EEG* Electro-encephalo-gram. [Modified from Keidel W D (1973) Kurzgefaßtes Lehrbuch der Physiologie. Thieme, Stuttgart]

cogitation, α-waves are replaced by β-waves; they are characterized by higher frequencies and smaller amplitudes (Fig. 156C; cf. also Fig. 143). An interesting variation of the EEG is shown in sleep. When falling asleep, slow δ- and Θ-waves predominate. During transition to the dreaming phase – the so-called paradoxical sleep – the frequency increases, resembling the excited awake state.

The EEG may be used for various studies in learning psychology, pharmacological tests, and medical diagnostic purposes. Frequency spectrum analysis (Fourier analysis) is used for quantitative evaluation.

9.3.2 Recording of Action Potentials of Single Neurons

The study of the EEG definitely provides an important orientation to larger functional complexes in the brain. In such an instance the EEG may serve a certain indicator function, but is unable to provide a detailed picture of the underlying brain functions. Information about *neuronal* processes taking part in the control of behavior can only be obtained with recourse to microelectrode techniques that permit structural and functional analyses. For this purpose recordings of action potentials from individual receptor cells, postsynaptic neurons, interneurons, and motor neurons are required.

Microelectrodes. In principle recordings from a single neuron may be performed intra- or extracellularly. In the first case the recording electrode is plunged into the cell; in the latter case it is only pushed into the close vicinity of a neuron. Intracellular recordings provide the most exact results, namely data on the course of *postsynaptic potentials* and *action potentials.* In practice such recordings often meet with great difficulties; today they are still the domain of highly skilled neurophysiologists. Since the "language" of the neuron appears to be coded in the frequency of action potentials, the extracellular recording of these potentials is sufficient for the analysis of many problems and does provide a preliminary picture.

There are innumerable variations in electrode design that are suitable for extracellular recordings. Electrodes may be purchased at high cost, but they may also be prepared at low cost when the material is available and the technique is acquired. Why is there no universal electrode?

Selection of a Suitable Electrode. Recordings from certain types of neurons or from particular structures of a neuron – soma, axon, dendrite – often require electrodes with special characteristics such as diameter, tip form, and resistance characteristics. For instance, fine glass micropipettes are suitable for recordings from the soma. They are pulled to pointed capillaries with the aid of an electrode-pulling apparatus and then filled with a conducting solution, for example, KCl or K-citrate. The tip diameter may be ~1 μm and the AC resistance 10–15 MΩ at 1000 Hz. For recordings from fibers, pipettes with tip diameters of 2–3 μm can be used. These may be filled with a low melting metal alloy (Indium-lead) which solidifies quickly on cooling. Their resistance can be decreased to less than 200 kΩ by galvanic platinization of the tip with the formation of a fine button. Tungsten electrodes are also very suitable for extracellular recordings or cell of fiber potentials. They are electrolytically sharpened in a solution of KOH and NaNO$_2$ by connecting them to a source of alternating current, and finally insulated from the outside by dipping in a special electrode lacquer. They should have a free tip of ~7 μm length and ~1 μm diameter. Their resistance is similar to that of the electrolyte-filled glass micropipettes.

Criteria for Quality. With many microelectrodes the reasons for their success or failure are difficult to determine. Apart from resistance measurements and a microscopic evaluation we have hardly any other selection criteria. In these microranges, physical boundary phenomena play a role that may not immediately be overcome during laboratory preparations. Unfortunately in some cases, the unsuitability of an electrode is discovered during the experiment. Once the electrode preparation technique has been worked out successfully, it is important that as many parameters as possible of this process be held invariant.

The quality of a microelectrode is determined not only by the tip diameter, but also by the shape of the tip. While concave tips are frequently unsuitable, flat tips often give very good results (Fig. 157). Tip angles of 20°–30° seem to be particularly favorable. When preparing metal electrodes from stainless steel insect needles, this angle may be reproduced by using a "trick". Usually the insect needle is dipped with the tip down into a dilute HCl solution and sharpened by electrolytic etching. However, since the etching begins mainly at the tip itself – i.e., where the current density is highest – the tendency for irregularities is rather great. The situation is different if the needle is dipped in the opposite direction, i.e., with the head down (Fig. 157). Now the current

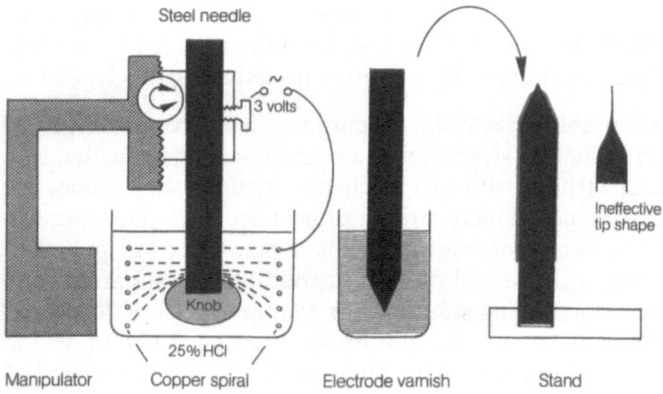

Procedures: Electrode sharpening – coating – drying

Fig. 157. Procedure for preparing a simple steel recording microelectrode (highly schematized). Details in text

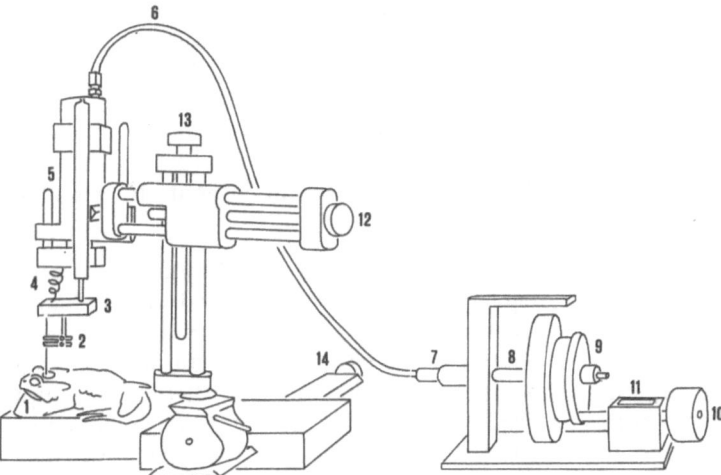

Fig. 158. Micromanipulator with hydraulic drive. *1* Experimental animal, *2* electrode with holder, *3* operational amplifier (impedance converter), *4* cable for signal conduction and power supply, *5* hydraulics for microelectrode drive, *6* connecting tube containing hydraulic fluid, *7* and *8* barrel, *9* coarse drive, *10* fine drive, *11* indicator for electrode advance in μm, *12–14* three-dimensional adjustments

density is highest in the boundary between the plastic head and the metal shaft. Since the density distribution changes in a relatively regular way, electrolysis will act differentially on the metal. This sharpening of the electrode is complete when the head drops off. All such electrodes usually have a uniform optimal tip shape. They are subsequently dipped in the special lacquer and dried vertically at room temperature, with the tip up. When flowing down, the lacquer is spread relatively equally and may leave a free 1–2 μm diameter tip of ~6 μm length; if not, the tip must be "opened" mechanically. The resistance of the electrode is in the range of 10–15 MΩ.

9.3.3 Recordings in the Immobilized Animal

The fundamental investigations on single receptor or nerve cell responses to defined stimuli are usually carried out in acute experiments in the immobilized animal. This has certain advantages: (1) Defined stimuli may be reproduced; (2) the electrode may be kept in the same position close to a neuron for several hours. The experimental animal is either anesthetized or immobilized (awake) by the injection of a neuromuscular blocking agent. In most of the cases artificial respiration is necessary.

Preparative Procedures. Before starting recording experiments a thorough neuroanatomical study of the substrate to be investigated is essential. Once the subject has been mastered, the surgical exposure of a sensory organ or brain surface may be undertaken. The recording electrode is positioned by means of a three-dimensionally movable micromanipulator, then inserted into the neuropile and finally moved close to a neuron by advancing it a few microns at a time with the aid of a fine hydraulic drive (Fig. 158).

Basic Electrophysiological Equipment. Monopolar recordings are frequently made against a grounded platinized metal plate in contact with the body of the animal and serving as a reference electrode. The recorded action potentials are conducted via an impedance converter, preamplifier, and frequency filter to the vertical (y) input of an oscilloscope and displayed on its screen (Fig. 159). Spike sequences may be filmed by a cine camera. Simultaneously, the potentials may be rendered audible as short clicks in a loudspeaker or head phone, stored on tape and counted by an electronic counter.
It is often important to record simultaneously special stimulus parameters – in the simplest case its start *(on)* and its termination *(off)* – after conversion to appropriate electrical signals in other channels of the oscilloscope. Synchronization of stimulus registration and recording of neuronal responses is achieved by electronic trigger circuits.

Screening Out External Noise. The experimental animal and the electrode have a high resistance. In high impedance recordings there is, in principle, a susceptibility to pick up noise from alternating electric fields. In order to avoid these effects, it has previously been necessary to build a Faraday cage round the animal and to carefully screen the cables coming from the electrode. When large instruments were required to

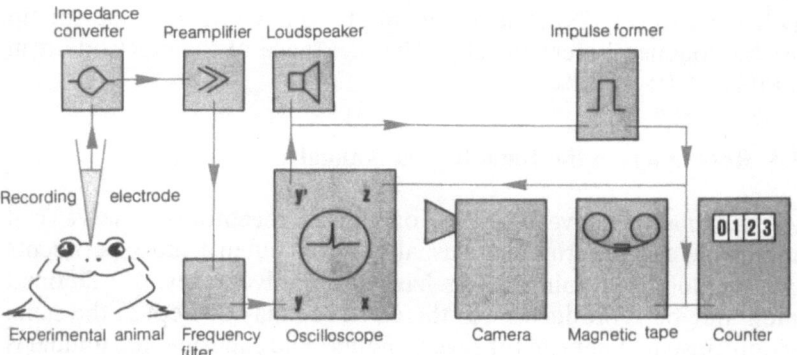

Fig. 159. Standard electrophysiological set-up for the recording of action potentials from single neurons. *Red,* Signal flow. Further details in text

generate stimulation programs, the cage finally assumed the dimensions of a laboratory. Thanks to recent advances in electronics we can largely dispense with such screenings. Over a short distance of 1–2 cm the electrode is directly connected to a miniature operational amplifier (Fig. 159). (An IC-model with a field effect transistor (FET) input stage, weighs less than 1 g). The amplifier is used as a unity-gain impedance converter: with an input resistance of more than 10^{12} Ω, and an output resistance of only a few Ω. Noise interference is thus considerably reduced. Thin, unshielded leads may be used as connections between the operational amplifier (impedance converter) and the more remote preamplifier.

What other electrical interfering effects may still be observed? Each amplifier and each resistance have a "proper" noise of their own which may be superimposed on the signal to be measured. By an appropriate choice of frequency filters (band pass filters) such broad-banded noise may be reduced (Fig. 159).

How Can the Responses of Several Neurons be Isolated from Each Other? When the electrode is inserted into a brain area and advanced into a fiber or cell layer from which responses can be recorded, it is by no means certain that an *individual* neuron has been reached. In many cases recording will include action potentials of more or less closely adjacent neurons of the same type. The amplitude of the action potentials will decrease with increasing distance between electrode and neuron (Fig. 160 B and C). By carefully shifting the position of the electrode in small steps of a couple of microns, a differentiation of the amplitudes of the various spikes is frequently possible (Fig. 160 C, a). Since, as far as possible, only the neuron with the largest potential should be included in the analysis, it is reasonable to separate its APs from the rest of the spike train. This may be achieved by adding another component to the circuit (Fig. 159). The action potentials recorded are transformed by an impulse-former into rectangular impulses of uniform height and width and fed into the intensity-(z)-axis of the oscilloscope. By adjusting the threshold of the impulse-former in such a way that it responds only to the potentials with the highest amplitude, these will appear brighter on the oscilloscope screen, outshining all others (Fig. 160 A, b). Only these APs are to be included in the analysis by film, tape, and counter (Fig. 159).

What Are the Possibilities of Orienting the Recording Electrode in the Brain? It may at first appear incomprehensible to the layman that it is possible to know from which kind of neuron in the brain a recording is derived. After all, in most cases the neurons are invisible; the recording electrode is brought to a neuron by a blind approach. Orientation in the brain is empirical, based on many factors: (1) When the neuronal discharges are audible in the loudspeaker as clicks (Fig. 159), the mere loudness indicates whether the electrode is approaching an "active" neuronal layer or moving away from it. The acoustical impression may then be verified by the recording on the oscilloscope screen. (2) Potentials recorded from certain neuron types differ to some degree. These differences are also noticeable acoustically. (3) Action potentials can be recorded from the soma, the dendrites, or from the axon of a

A

Channel a)

Channel b)

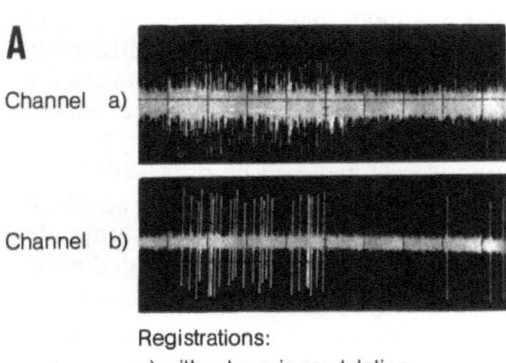

Registrations:
a) without z-axis modulation
b) with z-axis modulation

B

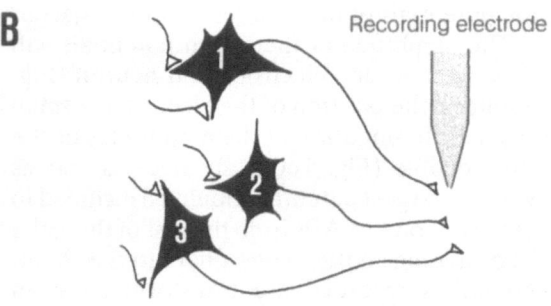

Schema of a record from three fibers

C

Channel a)

3 1 2 1 113 1 2 1 3 2

Channel b)

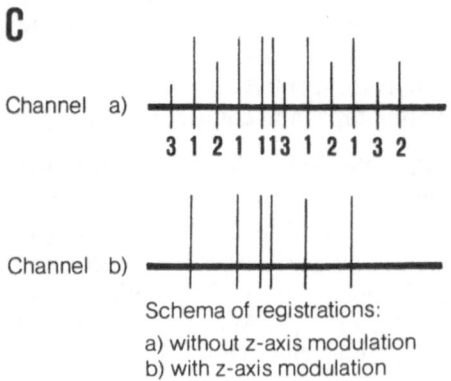

Schema of registrations:
a) without z-axis modulation
b) with z-axis modulation

Fig. 160A–C. Principle of z-axis modulation of the oscilloscope for the registration of action potentials from a single neuron. Further details in text. **A** *a* Original record of multiple unit activity without brightness control; *b* same record with brightness control emphasizing only one neuron. **B** Schematic diagram of recording from three fibers *(1), (2), (3)*. **C** *a* Diagram of multiple unit activity recorded from *(1), (2)* and *(3)*; *b* same record with brightness control emphasizing only neuron *(1)*

neuron. These potentials have characteristic time courses. (4) When the electrode is being pushed through a cell layer, it may easily injure the soma of a cell; the dying neuron now starts abrupt spontaneous discharges whose amplitudes decay completely within a few seconds. This phenomenon may also provide orientation clues. (5) When the electrode is advanced, a change from active to silent zones may further aid in establishing "landmarks". (6) For various mammals a stereotaxic atlas and a corresponding stereotaxic apparatus are available, and these enable one to predict the location of a given brain area.

Localization and Classification of Neurons. The initial experiments may be to determine the responses of individual neurons to various stimulus parameters. In the simplest case, the neuron's response to switching *on* or *off* of a stimulus or to sustained stimulation might be examined. If neurons with distinct response characteristics are discovered, one has to investigate whether they are representatives of different classes of neurons. With this information we may arrive at certain tentative conclusions regarding the basic neuronal information-processing steps. As soon as definite neuron types are recognized, they need to be localized histologically (Fig. 10A and B). For this purpose, the site of recording is marked, at the end of the experiment, by passing an anodal direct current (5 µA for 5 s) through the recording electrode. The *electrolytic lesion* is made visible subsequently in a histological section. If steel electrodes were used, the *Prussian blue* method already referred to may be applied (p. 264). If micropipettes are used, special dyes (e.g., *fast green*) can be used for recording as well as marking.

Let us keep in mind a few essential points:

1. The identification of neuronal types and their localization in certain brain areas may be deduced from a list of criteria.

2. According to the different response characteristics of neuron types preliminary clues on certain information-processing steps may be derived. They can be further explored in quantitative experiments.

3. By systematic modifications of parameters affecting the behavior mode of an animal, the corresponding activity patterns of individual neurons are established.

4. On the basis of the quantitative criteria gained thereby a further classification of neuron types may be formulated and clues to their function in triggering or controlling of a behavior mode may emerge.

5. The probability that a unit can be frequently recorded may depend on various factors, such as: (1) quality and particular features of the recording electrode; (2) physiological state of the animal after

preparation; (3) experimental conditions, for example: restrained, unrestrained, immobilized, anesthetized animal; (4) motivational state of the animal, cf. 3; (5) season of the year (in some animals). During recordings in the paralyzed animal no positive correlation between *abundance ratio* and *behavioral relevance* of a neuronal class can be expected.

9.3.4 Recordings in the Freely Moving Animal

Immobilization and Motivation. If a toad, for example, is motivated to catch prey, it snaps at a moving worm. Would it *want* to react similarly when immobilized? The animal is conscious but unable to move because of the blockage of neuromuscular transmission. But can we be certain that a toad thus constrained, on seeing the worm is in fact motivated to catch it? Only then would its brain send out the motor impulse pattern to the muscles involved in the prey-catching reaction. This can hardly be assumed, judging from what we know about the lability and sensitivity of the "internal condition" of an animal. At present, this question cannot be answered with confidence.

This dilemma reflects a common difficulty in neuroethological research. As Erich v. Holst correctly stated: "A precondition for measurements in each case is a calm, comfortable mood of the experimental animal in the resting state". The experienced ethologist may often recognize it in an animal. A restful state might be characterized in a chicken, for instance, by relaxed plumage preening, feeding, roosting, etc. An immobilized animal, on the other hand, does not indicate its mood.

Neuroethological Consequences. For a behavior program to be activated, response thresholds must be exceeded in the brain and their level may depend on the motivation of the experimental animal. In an immobilized animal, neurons of decisive significance for the release of a behavior mode may have a level of response activity below the threshold for triggering a behavioral response. Of course, this would not restrict the usefulness of the data obtained, e.g., on stimulus–response relationships. But certain neurons might be also inhibited by the physical immobilization and may therefore be overlooked by the investigator in the recording experiment. Information on such questions may be gathered in studies where the responses of individual neurons are recorded in an awake animal which is unrestrained (freely moving) – say, during a phase of behavior (Fig. 164).

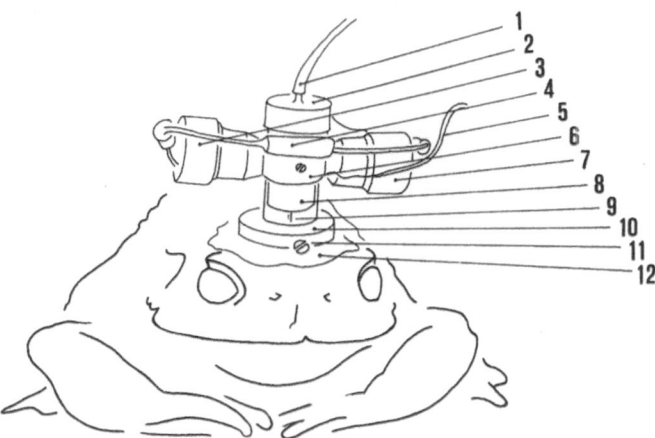

Fig. 161. Head holder for single-cell recordings from the brain of a freely moving toad *Bufo bufo. 1* Inlet tubing of hydraulics for electrode micro drive, *2* cylinder, *3* IC operational amplifier (impedance converter) for recording electrode, *4* resistance, *5* thin cables for signal conduction and power supply, *6* locking ring on cylinder, *7* IC operational amplifier for reference electrode, *8* barrel with microelectrode, *9* microelectrode, *10* holding ring cemented to cranium, *11* adjusting screw, *12* dental cement. Further details in text. [Modified from Ewert JP, Borchers HW (1974) J Comp Physiol 92:117–130]

Solution of Methodological Problems in Single Cell Recordings in the Freely Moving Animal

The following difficulties must be overcome:

1. When the animal moves, the tip of the recording electrode must not move relative to the site of recording. To avoid this, head holders, which are firmly screwed to the skull above an uncovered part of the brain,

have been developed for various animals. The head holder may consist of a miniature hydraulic drive by which the recording electrode is advanced vertically in the brain and its tip positioned in the vicinity of a neuron. This technique increases in difficulty with decreasing size of the experimental animal [156] (Figs. 161, 162). The head holder can be screwed to the skull (relatively large animals) or fixed onto it with special kinds of dental cement (small animals).

2. When the experimental animal moves, the recording electrode also picks up muscle potentials; they may be superimposed on the responses recorded from the neuron. This kind of interference may be avoided by differential recording. For this purpose two microelectrodes, insulated from each other, are positioned in the brain. One of these electrodes is connected to the noninverting input of the amplifier and the second is connected to the inverting input. The noninverting electrode (n) is used for recording and the inverting electrode (i) is placed in "silent brain

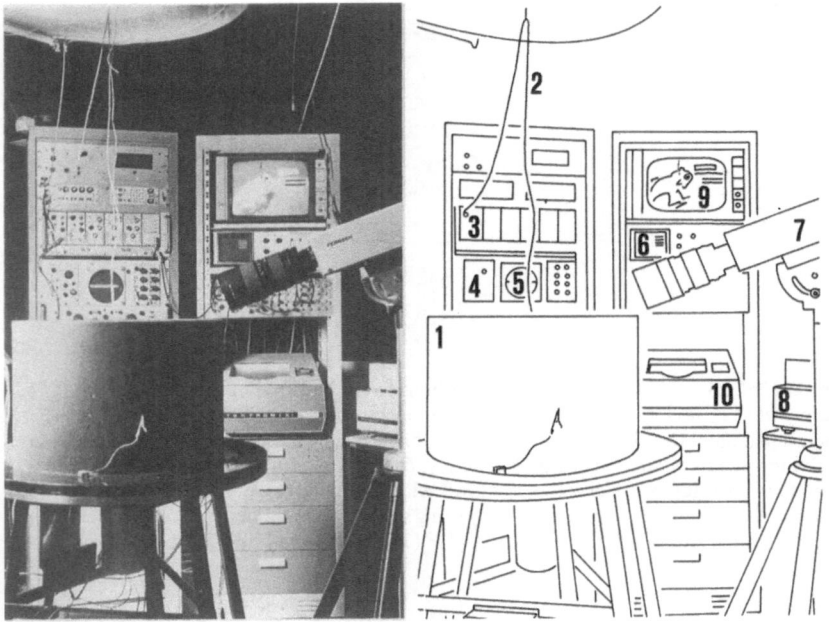

Fig. 162. Experimental set-up for recording: (1) a visual stimulus pattern (prey dummy), (2) the behavior response of the experimental animal (prey-catching of the toad), and (3) the discharges of a single neuron from the brain (of a freely moving toad). *1* Experimental arena with moving prey dummy, *2* signal conduction of the action potentials recorded from a neuron in the toad's visual system, *3* preamplifier, *4* amplifier which feeds into oscilloscope, *5* oscilloscope screen, *6* "mixer" and "scan converter", *7* TV-camera, *8* video tape, *9* monitor, *10* special dry-copying apparatus [By Ewert and Borchers, from Ewert JP (1976) Neuro-Ethologie. Springer, Berlin Heidelberg New York]

tissue". Each of the recorded potentials U_n and U_i is passed through a miniature operational amplifier (impedance converter) – also fixed on the head holder (Fig. 161, 3 and 7) – and then conducted by thin, unshielded wires to a more distant differential amplifier; from here, connections lead to the oscilloscope. The principle of this circuit is that muscle potentials U_m recorded by both electrodes (n and i) are eliminated by subtraction: $(U_n + U_m) - (U_i + U_m)$. Thus, essentially, the potential difference $U_n - U_i$ is registrated as the neuronal response. The head holder is relatively light (less than 5% of body weight) and does not appear to bother the animal. Usually the electrode keeps its position, even if the animal is jumping [157].

3. In some experiments the connections leading to the animal may be cumbersome, e.g., the electrical wirings and a tubing for the electrode advancement, if a hydraulic microdrive is used. In animals having sufficiently large skull dimensions to carry a larger head piece, the efferent connections may be dispensed with. Special electronic equipment fixed to the head of the animal allows positioning of the electrode in the brain without tubing, by remote control, and transmitting the responses of a single neuron to a distant receiver [158].

9.3.5 Recording and Evaluation of Experimental Results

a) Collecting Data

What are the different methods of recording neuronal events?

Standard Methods. The principal recording device is the oscilloscope. It is suitable for the recording of slow as well as fast processes. If it is equipped with storage capacity, a whole series of action potentials may be stored on the screen and may, if desired, be photographed with a Polaroid camera. It is also possible to film the potentials shown on the oscilloscope screen with a cine camera. This procedure is appropriate for experiments of long duration. The data may simultaneously be stored on magnetic tape during the experiment, thus permitting a subsequent "off-line" computer analysis. For the recording of slow potentials – e.g., brain potentials – electromagnetic paper recorders are adequate. They have the advantage that tracings of long-lasting processes are immediately visible. Their disadvantage is that they are too sluggish to record fast processes, such as action potentials.

Special Methods of Recording. Data collection is not quite so simple when we have to deal with recordings from the freely moving animal. As an example we again choose the visually controlled prey-catching behavior of the toad. During such an experiment, the following processes have to be recorded simultaneously: (a) the movement of the stimulus pattern (prey dummy), (b) responses of a neuron from the visual system in the brain, and (c) the movement of the toad during prey-catching.

This may be achieved in the following manner (Fig. 162): Toad and prey dummy (1) are recorded with a video camera (7). The video-signal of the camera (7) together with the neuronal responses (2–5) are fed into a mixer (6) and the subsequent signal (5 and 7) is stored on video tape (8). For evaluation purposes the course of an experiment may now be reproduced and reconstructed, frame by frame, on a video monitor (9) (cf. Figs. 163 and 164). Each frame then yields data about the space/time positions of the stimulus (prey dummy) and the animal (toad); moreover, the stored sequence of the neuronal discharges occurring at that moment may be added to the record. Frame sequences (9) which appear to be important for further analysis in space/time diagram may be delivered singly as half-tone reproductions with the aid of a special dry-copy device (10) (Figs. 162 and 163).

Other methodological problems arise when, for example, the stridulatory leg movements and the song of the grasshopper have to be recorded simultaneously. For plotting the leg movements the following technique may be used: a small, double Hall-generator is fixed on the back of the awake animal whose legs have been partially amputated (Fig. 165). A small permanent magnet is fastened on the femur of the hindleg. During

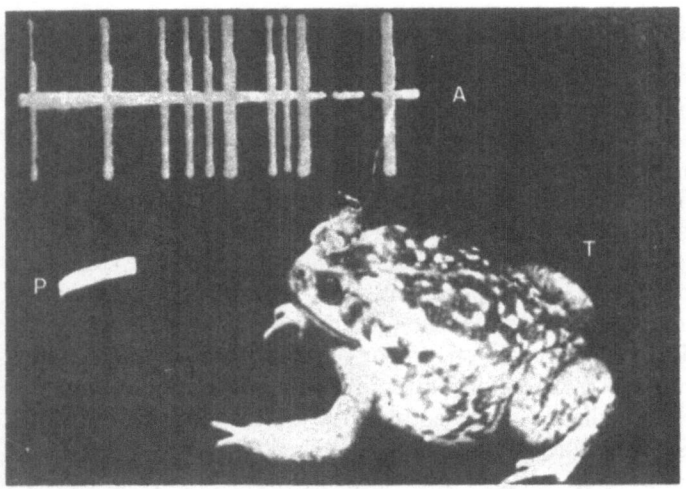

Fig. 163. An original dry-copy from an experimental program performed as in Figure 162 *(10)*. It shows the x-y position of a prey dummy *P*, the movement phase of the toad *Bufo marinus T*, and the action potentials *A* of a neuron from the toad's visual system elicited from the start of stimulation up to this moment [By Ewert and Borchers, from Ewert JP (1976) Neuro-Ethologie. Springer, Berlin Heidelberg New York]

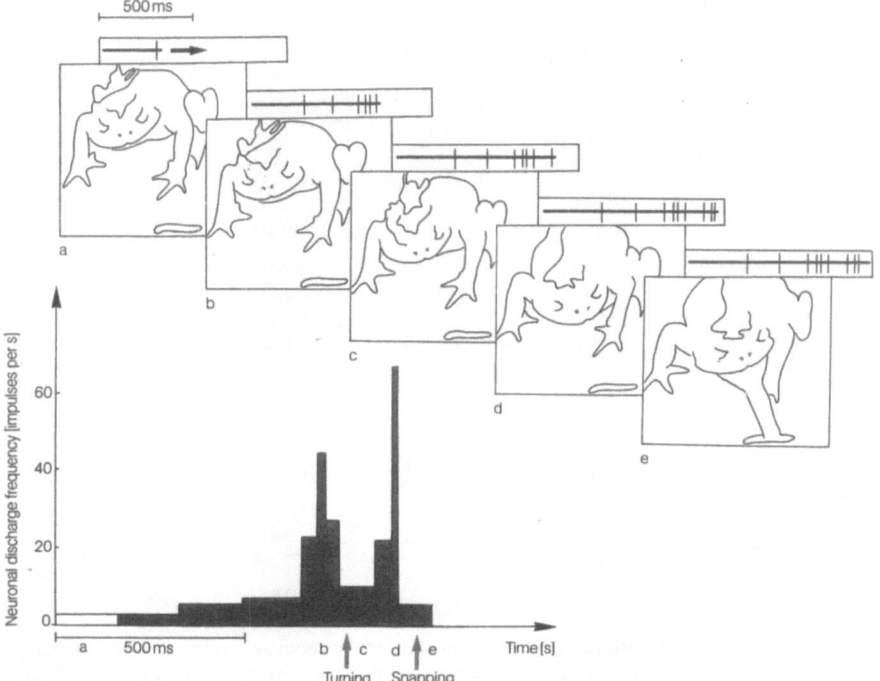

Fig. 164. Example of simultaneous recording of different neurobiological events: (1) prey stimulus, (2) prey-catching behavior pattern of the toad *Bufo bufo*, and (3) spike activity of a single neuron in the optic tectum. The diagram represents a segmentation of the events in space and time domains according to techniques described in Figures 161–163. *a* The prey object, a mealworm, crossing the visual field of the toad, has just stopped moving; the neuron was not activated during that time; the toad is sitting still. *b* Prey has not moved; neuron just starts firing; toad is sitting still. *c* Prey has not moved, discharge rate declines, toad starts turning its head toward prey. *d* Prey is still; discharge rate of the neuron increases again; toad approaches prey slightly. *e* Prey is still; discharge rate declines sharply; toad snaps at prey. Stimulus was stationary for at least 700 ms (frames *a–e*). Time interval between maximal neuronal burst (frame *d*) and snapping (frame *e*) is approximately 20 ms. Note that this neuron is not simply "visually driven". Results indicate that the neuronal activity precedes and "predicts" the toads subsequent motor response [Measurements by Borchers and Ewert, 1977]

the stridulation, it moves along the length of the Hall-generator. The output of the Hall-generator is correlated with the leg movement.

b) Methods of Spike Train Analysis

Average Neuronal Discharge Frequency. The language of a neuron appears to be coded, in the number of action potentials per unit of time.

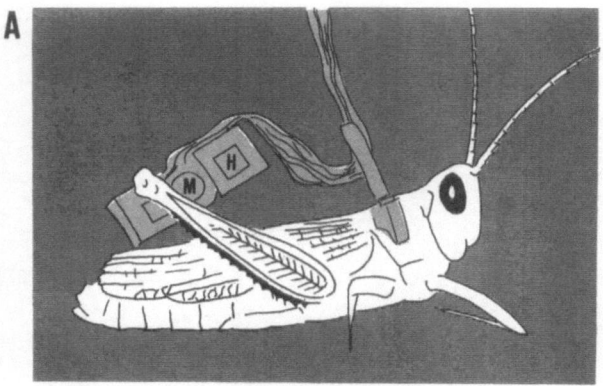

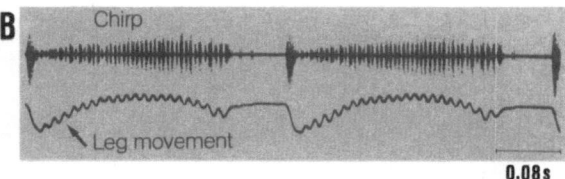

Fig. 165. A Experimental arrangement for the simultaneous recording of leg movements and acoustical signals of a stridulating grasshopper *Chorthippus mollis* using the Hall-effect. *H* Double Hall-generator fixed on the back of the freely moving animal, *M* permanent magnet fastened on the femur of the hindleg in such a way that during stridulation, it moves along the length of the Hall-generators. **B** The magnetic flux through the Hall-generators and therefore the Hall-voltage depends on the position of the leg. Simultaneous records of sound and leg movement events. Further details in text. [Modified from Elsner N (1970) Z Vergl Physiol 68:417–428]

Information on the mean discharge rate is particularly useful when the time intervals between successive spikes do not excessively deviate from each other. The frequencies are most simply measured on the recording stripe. Given the total number of impulses, n, the speed of the recording film, v [mm/s], and the distance on the stripe from the first to the last spike recorded, l [mm], the mean discharge rate is $R = n \cdot v \cdot l^{-1}$ [discharges/s]. Electronic counters are also available for recording impulse numbers or for directly displaying its frequency.

The Dwell-Histogram, DH. Information on the *mean* discharge frequency is less convenient and reliable for events showing very irregular time intervals between spikes. It may then be necessary to carry out a continuous frequency recording. For its analysis the reciprocal of each time interval $t_{n+1} - t_n$ between two successive spikes (n and $n + 1$) is plotted against the time t_{n+1} (Fig. 166 B). The dwell time is a measure

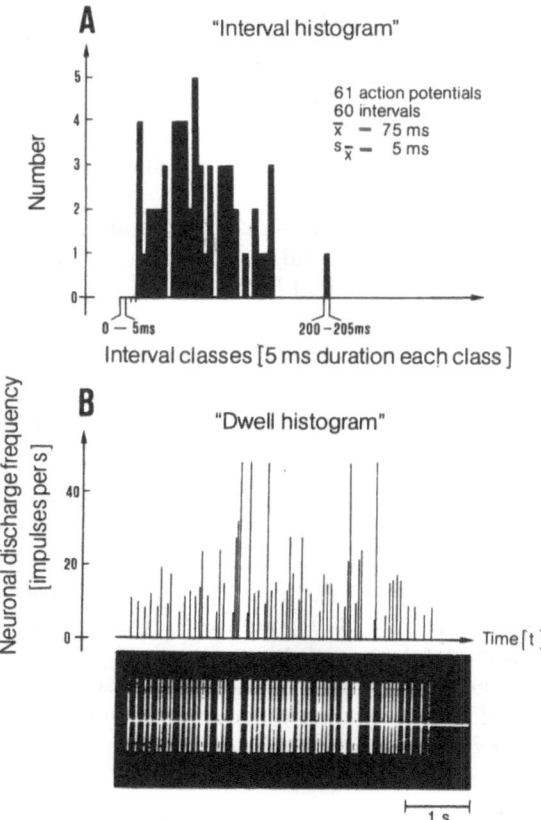

Fig. 166. A Interval histogram and **B** Dwell-histogram of a neuronal impulse pattern *(below).* Two computer analyses. Explanations in text. Discharge pattern: Original record from a single ganglion cell (class R2) of the toad retina. A black square with an edge of 4° visual angle served as stimulus moving through the excitatory receptive field against a white background at a visual angular velocity of 1°/s. [Measurements by Borchers 1976]

of the time resolution capacity when measuring intervals between successive impulses.

The Post-Stimulus-Time Histogram, PST. The PST histogram indicates the frequency course of spikes evoked at the beginning or at the end of a stimulus. The time axis is here divided continuously into equal intervals. Impulses arriving per unit of time ("bin") are counted; the width of the intervals ("bin-width") may be chosen arbitrarily.

Since the measurement has a starting point defined by the stimulus, the procedure is particularly suitable for averaging during repeated events.

The Interval Histogram, IH. As a supplement to the dwell-histogram, data on the interspike interval distribution may be important for some problems. The interval histogram shows the frequency distribution of different classes of interval (Fig. 166 A). The class dimension may be chosen arbitrarily. A variant is the *frequency histogram* (FH). It represents an impulse frequency distribution, analogous to the interval histogram.

These often quite time-consuming analyses may be performed simply, in most cases, with the aid of a suitable computer, in an "on line" operation during the experiment [160].

9.4 Investigation of Anatomical Connections in the Brain

It has been mentioned before that neurons with identical or similar functions are often located in clusters or in laminated areas. Information processing involves the interaction of neurons within such areas and between areas of distinct regions, connected via axons and dendrites. Typical forms of *physiological* interaction are established by the fundamental computations described, on the basis of convergence and divergence principles including feedback loops. What are the possibilities of testing whether there are *anatomical* connections between certain areas in distinct brain regions?

9.4.1 Neuroanatomical Methods

Various anatomical micromethods are known for accurately tracing the course of a nerve fiber in the brain (cf. Fig. 6). There are also methods for determining whether two neurons from different parts of the brain are in synaptical contact with each other. For these investigations, practically all methods make use of either the functional properties of different structures of a neuron or axoplasmic transport [159].

a) Degeneration Techniques

The neuron represents a trophic unit. When the fiber is cut (axotomy), various morphological alterations (degenerative changes) occur in the entire neuron. These changes can be used for the investigation of neuroanatomical problems, e.g., the question of interconnecting pathways between distinct cell groups in the brain. Basically we proceed in the following way: (1) destroy a brain area in the anesthetized animal; (2) let the animal survive for 8–14 days (depending on species, age, and temperature); (3) in anesthetized animal perfuse brain with formalin; (4) prepare brain for histological analysis of degeneration products.

Chromatolysis in the Soma. Franz Nissl (1860–1919) suggested that the pattern of the Nissl bodies, named after him, represents an "equivalent picture" of the nerve cell. Indeed, this pattern resembles to some degree the physiological state of the neuron. *Nissl bodies* change their appearance in relation to various physiological conditions, such as rest, fatigue, or exhaustion. Nissl substance corresponds to the endoplasmic reticulum with attached ribosomes – the sites of protein synthesis. These basophilic structures, rich in RNA, can be demonstrated histologically by staining with toluidine, cresylviolet, or other basic aniline dyes. The size, form, and distribution of Nissl bodies may vary in different neuron types: fine and dustlike, small and scarce, coarse or striated. Normally the substance is spread over the entire soma and dendrites.

However, when the axon of a neuron is sectioned (Fig. 167, top), changes occur in the appearance of the Nissl substance after a period of several days or weeks (Fig. 167, bottom): Nissl bodies are concentrated at the periphery of the soma and they disappear in the center. The cell body itself rounds off and the nucleus is displaced to the periphery. The dissolution of Nissl substance is called *chromatolysis* and the histological cell picture is named *"fish eye state"*.

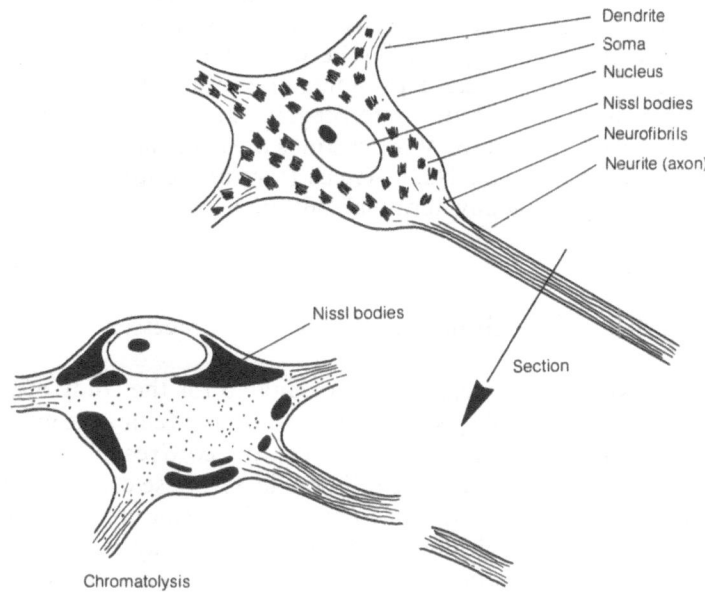

Fig. 167. Schematic representation showing retrograde degeneration after axotomy (severing the axon) of a neuron. Explanation in text

Changes in the Myelin Sheath. Axotomy causes cell changes in both directions. (1) Retrograde (proximally): Chromatolysis in the soma as described above. (2) Anterograde (distally): degeneration and death of the fiber beyond the site of the injury, while the rest of the fiber at the soma remains intact. The anterograde fiber degeneration first leads to changes in the myelin sheath, then progresses to the axon cylinder and finally to the nerve terminals (Fig. 168). Before the disintegration of the axon, the internodal myelin sheath segments assume a globular form. The myelin lipids are finally broken down to neutral fats. In histological sections, they may then be visualized by lipid staining methods, e.g., Sudan III (scarlet red).

Degeneration of the Terminal Buttons. At a particular stage of degeneration the axonal terminal buttons swell up. They can then be visualized in the histological picture by silver impregnation methods that provide a picture of the approximate sites of synapses. The technique is suitable for myelinated as well as unmyelinated nerve fibers.

Application of the Degeneration Techniques for Labeling of Connections Between Brain Areas. Let us carry out the following experiment (Fig. 168): The question arises as to whether the neurons in a nuclear area A are directly connected to those of another area B. Assuming that the fibers in the intermediate zone represent the axons from somata localized in A, the following degenerations should occur after axotomy:
1. Chromatolysis in area A (Nissl method: staining with cresylviolet).
2. Myelin sheath degeneration between the site of section and area B (Marchi method: Sudan III staining or osmium impregnation).
3. Fiber and (or) terminal button degenerations, recognizable as dustlike particles within area B. Various kinds of silver staining methods can be used: (a) Nauta and Nauta-Glees (terminal buttons and fibers), (b) Nauta-Gygax (mainly terminal buttons), (c) Fink-Heimer (terminal buttons).

b) Labeling of Individual Cells and Their Processes

The degeneration techniques are suitable for the study of connecting tracts between whole populations of neurons. Meanwhile micromethods have been developed enabling us: (1) to *record* responses from a single neuron and to classify it according to its response characteristics, (2) to *mark* the living neuron, and (3) to *reconstruct* the course of its processes in the brain histologically (Fig. 6). The marker may be delivered through a recording micropipette either by (a) intracellular injection, (b) axonal diffusion or (c) extracellular diffusion. The marker is administered either iontophoretically, by hydrostatic pressure, or by diffusion. All of these methods make use of *axoplasmic transport*.

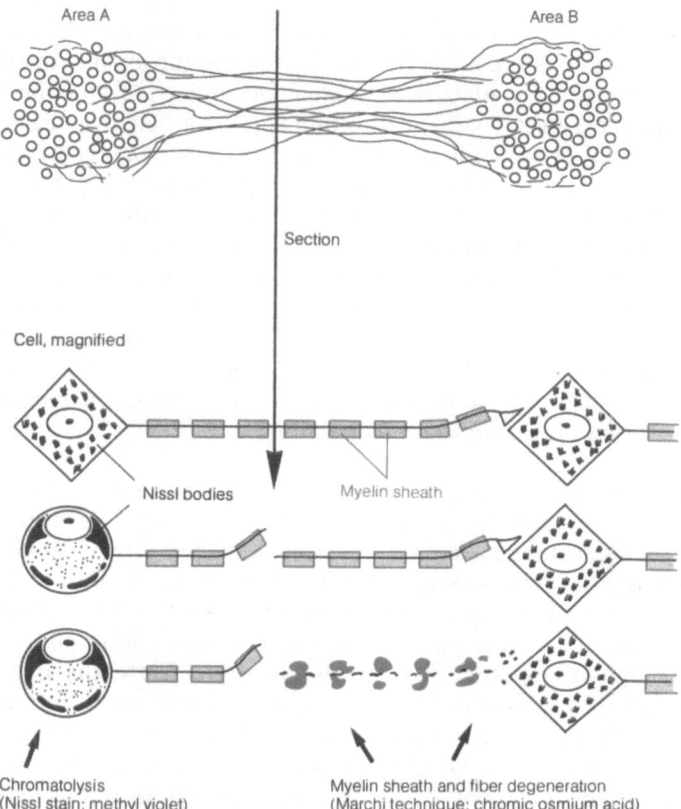

Fig. 168. Schematic representation of the appearance of degeneration in soma and fiber after axotomy as a method for neuroanatomical investigations of fiber projections from nucleus *A* to nucleus *B*. Axotomy may also lead to transneuronal degeneration (e.g., chromatolysis in *B*), which is not being considered in this example. [Modified and schematized from Glees P (1957) Morphologie und Physiologie des Nervensystems. Thieme, Stuttgart]

Transport of matter takes place in the axon as well as in the dendrites. There are, however, differences, e.g., with regard to the organelles carried in the neuroplasmic flow. There is a slow anterograde transport of 1–4 mm/day, representing the major component and corresponding in time to the fiber regeneration after axon section. The transported substance is mainly soluble protein. There is also a fast anterograde transport (> 100 mm/day), being an undercurrent within the slow-moving plasma column and dependent on the presence of intact neurotubuli. The precursors of organelles are the principal components transported. Apart from *anterograde* axonal transport systems, there is another fast transport mechanism in the opposite direction, from the axonal terminal to the cell body *(retrograde)*. *Transsynaptic* transport of material has also been described.

Whereas degeneration techniques use *pathological changes* in the neuron, cell labeling techniques utilize *physiological processes* [159].

Radioactive Labeling. The long-lived neuron is maintained by continuous protein synthesis performed in the ribosomes. Originating in the soma, a constant plasma transport occurs in the axon in the direction of the synapse. By placing a radioactive amino acid in the immediate vicinity or, better still, in the interior, of a cell by means of a thin, drawn-out glass capillary, the amino acid may be made to participate in protein synthesis and be transported to the synapse by plasma flow. Here is an example: when a goldfish is injected with ³H-histidine in the vitreous of an eye, the amino acid is taken up by the ganglion cells of the retina and transported within their axons to the projection fields in the brain (e.g., optic tectum). Its progress may be visualized autoradiographically in histological sections. ³H-leucine can be used as marker for axonal *terminals*, ³H-proline serves for *transsynaptic* marking. The survival time of the animal is 1–2 days. All of these tracer methods label the axon in the *anterograde* direction.

Injection of Dyes. Again the question should be asked where a neuron – whose response characteristics are known to us from previous neurophysiological experiments – sends its axon to. This may now be investigated by injecting a dye into the cell or close to it with a finely drawn-out micropipette whose tip diameter is <1 μm. Since the dyes have polar groups, the injection may be done iontophoretically with rectangular impulses (10 nA, 20–30 min). Thus, the neuron can be categorized both neurophysiologically and neuroanatomically.

Recently pipettes have been developed containing two microcompartments: they are called Θ-electrodes, their tip diameter is smaller than 1 μm. One chamber contains potassium citrate: it is used for the recording of neuronal responses. The other chamber contains the dye to be injected into the cell iontophoretically. Suitable dyes are, for example, *cobalt chloride* or *Procion yellow*. They are transported in part up to the finest dendritic processes (0.2 μm diameter in Purkinje cells). Their progress may later be visualized microscopically in sections by means of certain histochemical reactions (cobalt as CoS) or by fluorescence (Procion yellow; Fig. 6). Both Procion yellow and cobalt chloride may, as electrolytes, also be used in a single-chamber micropipette for the recording of neuronal responses.

Horseradish Peroxidase. As a marker, HRP has the advantage that it is transported mainly in retrograde direction. HRP can be injected into the cell body. There is also extracellular uptake by pinocytosis at all parts of the neuron. HRP is also easily picked up by injured regions of a cell. Thus, "injury staining" of adjacent neurons or passing fiber tracts may lead to misinterpretations. HRP labels the axon generally in both directions. The two possibilities can be distinguished by means of different chemical treatment, where *retrograde* labeling occurs as the "brown reaction" and *anterograde* labeling as "blue reaction". For resolving particular neuroanatomical questions HRP is at present the most frequently used marker.

9.4.2 Combinations with Electrophysiological Micromethods

If exact information on *functional connections* between various neurons is desired, a combined application of electrophysiological stimulation and recording techniques is necessary. A close collaboration between neurophysiologists and neuroanatomists is a prerequisite. The elucidation of the complicated circuits of the cerebellum by Sir John C. Eccles and John Szentágothai is a classical example. Before discussing the integrative functional properties of a special neuronal network in detail, some preliminary remarks on the structure of the cerebellar cortex are in order.

Five different neuron types contribute to the structure of the cerebellar cortex: *Purkinje cells, Golgi cells, granular cells, basket cells,* and *stellate cells* (cf. Fig. 8). They all form, in terms of their interconnections, a three-dimensional network (Fig. 9), organized according to the principles of convergence and divergence (Fig. 169). Functionally, they constitute a complex inhibitory system receiving its information from the rest of the brain and spinal cord via fiber tracts; these are the *mossy* and *climbing fibers.* After the information has been processed, it leaves the network by fibers of a specific type of neuron: the *Purkinje cells;* they form inhibitory synapses with neurons of the deep cerebellar nuclei.

The Purkinje cells are the place where – for a final accounting, as it were – *excitatory* and *inhibitory* influences converge (Fig. 169; also cf. Fig. 9). Since the spatial distribution of excitation in the Purkinje cells appears to have a significance for cerebellar functions, it could be important for the brain to know the degree of activity or inhibition prevailing in each individual Purkinje cell at any given moment. This is possibly attained by a *"reading mechanism".* Here a brief burst of impulses is sent, as a *test signal,* to the deep cerebellar neurons and to the Purkinje cells (Fig. 169). The cerebellar neuron may now compare the excitatory post-synaptic potential (EPSP), generated by the test signal, with the corresponding inhibitory postsynaptic potential (IPSP) induced by the Purkinje cell. The test signal itself originates in another brain area, the inferior olivary nucleus; the axons conducting it are called climbing fibers.

What is the experimental evidence in support of this hypothesis? Let us first consider the morphological data (Fig. 169). Each climbing fiber is in contact with a Purkinje cell by innumerable synapses. Thus, we might expect that transmission of information is insured by a high safety factor. This is borne out by the neurophysiological experiment (Fig. 169): by

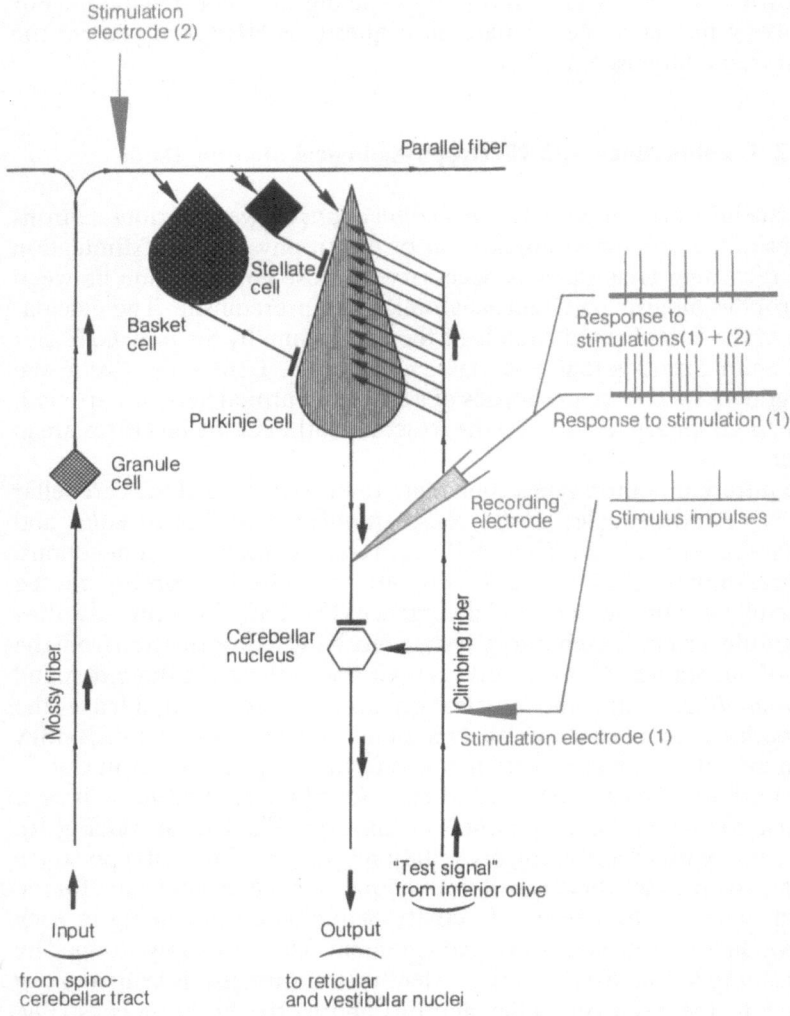

Fig. 169. Combined electrical stimulation and recording microtechniques *(red)* for the analysis of functional connections in a possible cerebellar "reading mechanism". *Arrows* designate excitatory, *lines with cross bars* inhibitory synapses. Explanations in text. [Modified from Eccles J C (1967) Naturwiss Rundsch 20:139–151; cf. also Ito M (1979) TINS 2: 122–126]

electrically stimulating the olivary nucleus, and therefore the climbing fibers (*stimulation electrode 1:* test signal), 3–4 spikes per stimulus impulse may be recorded from a Purkinje axon; they have the remarkably high discharge frequency of 300–400 Hz (cf. *recording electrode:* response to stimulation *1*). If the excitatory state of the

Purkinje cells is now changed, the response to the test stimulus should also be changed. It may be diminished experimentally by increasing the inhibitory effects of basket and stellate cells which can be excited via the parallel fibers *(stimulation electrode 2)*. The response to the simultaneously arriving test signal recorded from the Purkinje axon is now reduced (cf. *recording electrode:* response to stimulations *1 + 2*).

9.5 Statistical and Systems Theoretical Analyses

9.5.1 Correlation Analyses

Autocorrelation. There are numerous biological processes showing different time courses of frequency and amplitude. Such phenomena are as diverse as the spectra of vocal calls and the slow brain waves mentioned above. When analyzing such time-dependent processes the question of periodicity is often of interest. For statistical analysis the autocorrelation procedure can be used. We illustrate the principle for any process changing with time (Fig. 170 A, above). Its time course can be described by the function $x(t)$. With respect to the *internal connection* between the momentary values, we displace the corresponding curve by the time interval τ; then the time course is given by $x(t \pm \tau)$. At those points where periodicities occur, the curves $x(t)$ and $x(t \pm \tau)$ will differ only slightly. Information on the correlation of periodicities may be obtained by multiplying both curves for all time intervals τ (Fig. 170 A, below). In principle then the autocorrelation function is expressed as follows:

$$\Phi_{xx}(\tau) = \lim_{T \to \infty} \frac{1}{2T} \int_{-T}^{+T} x(t) \cdot x(t + \tau) \, dt.$$

By integration we obtain an averaging of the momentary values. Therefore, if periodicity is present, it is clearly brought into focus by the graphical representation of the autocorrelation function. The averaging eliminates all nonperiodical processes.

Cross Correlation. Suppose we apply a stimulus and record the brain potentials from two different points X and Y of the scalp. The question is whether there is a correlation between the two recordings of the responses or whether they are independent of each other. To answer this question we proceed in principle as we did in the case of autocorrelation. The curve of the responses $x(t)$ is temporally displaced in relation to that of the responses $y(t)$ by the value τ. By multiplication and integration (Fig. 170 B) the cross correlation function is obtained:

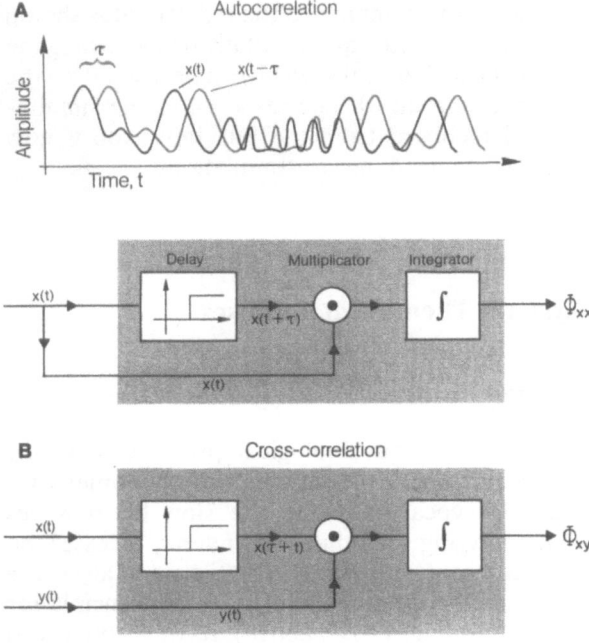

Fig. 170 A and B. Schematic representation for explaining autocorrelation **(A)** and cross-correlation analysis **(B)**. Explanations in text. [Modified from Oppelt W O (1964) Kleines Handbuch technischer Regelungsvorgänge. Verlag Chemie, Weinheim]

$$\Phi_{xy}(\tau) = \lim_{T \to \infty} \frac{1}{2T} \int_{-T}^{+T} x(t) \cdot y(t + \tau) \, dt.$$

Its course indicates whether there is a correlation between $x(t)$ and $y(t)$.

9.5.2 Possibilities for Systems Analysis

The functional properties of a neuron (or a population of neurons) in response to stimuli are determined by information processing in neurons that are presynaptic and postsynaptic to it. Together they form a system which can be considered as a *black box* (Fig. 171 A). The system properties for stimulus response correlations – the object of our inquiry – are given by a function, the so-called *weighting function* g(t). The response of the neuron under observation is determined by this function. If the function is known, the responses of this neuron toward all possible adequate stimuli may – from the systems-theory point of view – be predicted.

A   Test signals

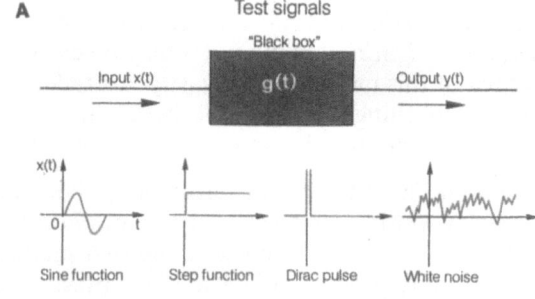

B   Systems analysis

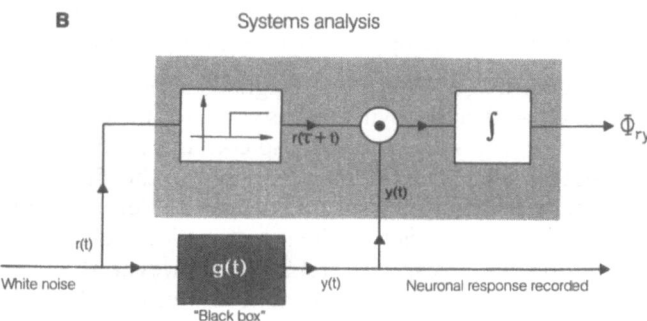

Fig. 171 A and B. Systems theoretical methods for "black box" analyses. **A** Calculations of the weighting function g(t) using different test signals. **B** Set-up for calculation of the weighting function g(t) of a framework component by evaluating the cross-correlation between the response course y(t) and the process r(t) of white noise. [Modified from Oppelt (1964), cit. in Fig. 170]

There are several ways of finding out this function. One way is to determine the *optimal stimulus* (e.g., key-stimulus) and carry out a *retrocalculation* of the system with the aid of *model networks.* As model systems we may use homogeneous two-dimensional nerve nets which are coupled by the mechanisms of lateral inhibition and lateral facilitation. By comparing the measured biological responses with the model system responses to test stimuli, approximate values for various *parameters* may be obtained [161].

Another method consists of using, in wide frequency band, test signals of the time course x(t) as input (to the sense organ) of the system to be analyzed and finding the corresponding output y(t) as system response (e.g., records of a neuron from the CNS). Signals with various time functions may, in principle, be used (Fig. 171 A):

Sine Wave. Sinusoidal stimuli of various frequencies (ω) are used as input of the system and the frequency response of the corresponding output is analyzed (Fig. 171 A). The result is a *transfer function*, $f(\omega)$, from which the *weighting function* $g(t)$ may be calculated. A considerable drawback of this method is the vastly protracted measuring period required for establishing this function; these amounts of time are usually not available when neuronal responses are recorded.

Dirac Pulse $\delta(t)$. The δ-function consists of a pulse of infinite height and infinitely narrow width, with an area of 1 (Fig. 171 A). Such a test signal has several advantages compared with the first method: (1) the Dirac pulse contains all frequencies. (2) The Dirac pulse used as input of the system immediately yields the *weighting function* $g(t)$ as system response. (3) The time required is extremely short. Drawback of the method: Owing to the enormously high intensity of the pulse, measurements in the upper ranges are distorted by nonlinearities.

Step Function $\sigma(t)$. This is a test signal leaping from zero to the value of 1, at time $t = 0$, and staying at this value for $t > 0$ (Fig. 171 A). As in the Dirac pulse, the step function also contains all frequency components. Advantages of this test signal compared to the signals described above are: brief periods of measurements and no deformation by nonlinearities. A disadvantage is that the step function yields as system response only the *transient response* $f(t)$. Since, however, the *weighting function* is the time derivative of the transient response, $g(t) = f'(t)$, the problem may be solved by calculation.

White Noise. A systems-analytical method used very often in recent times consists of applying white noise as a statistically known input signal (Fig. 171 A and B). All frequencies at equal intensity are present in equal proportions in its spectrum. In applying this method of analysis, a cross correlation $\Phi_{ry}(\tau)$ is established between the process of white noise $r(t)$ and the measured time course of the response $y(t)$ (e.g., from a neuron; Fig. 171 B). The *weighting function* may now be calculated by insertion of the known *autocorrelation function of white noise* $\Phi_{rr}(\tau)$:

$$\Phi_{ry}(\tau) = \int_0^\infty g(t) \cdot \Phi_{rr}(\tau - t)\, dt.$$

This procedure has many advantages: (1) The measurement does not consume much time. (2) Statistical contaminations are eliminated by averaging. (3) The system properties are analyzed over a wide range of frequencies [162].

9.5.3 Models

At certain levels of investigation it may be useful to elaborate a working hypothesis about the system studied and to work out a model. Models should help to plan new experiments, to predict results and to test predictions. Thus, a concept can be either supported, extended, or rejected. There are various ways of modelling ranging from global functional models to quantitative neuronal network models using either homogeneous or heterogeneuos networks with linear or nonlinear transformation properties [161–166]. Again, the investigation of the cerebellar cortex provides a good example of how a preliminary speculative approach [167] yielded meaningful predictions that could be tested by physiological experiments [168], finally leading to the concept of neuron networks as specific "neuronal machines" [169].

Special References

1. Bullock TH (1977) Introduction to nervous systems. Freeman and Co, San Francisco – Hubel DH, Stevens CF, Kandel ER, Nauta WJH, Feirtag M, Cowan WM, Iversen LL, Hubel DH, Wiesel TN, Evarts EV, Geschwind N, Kety SS, Crick FHC (1979) The brain. Sci Am 241 (3):44–232, Freeman Co, San Francisco – On the logic structure of nervous systems cf. Zettler F (1974) J Comp Physiol 95:123–167
2. Holst E v (1939) Ergeb Physiol 13:228–306 – Tinbergen N (1951) The study of instinct. Clarendon Press, Oxford – Segaar J (1961) Vakbl Biol Okt 185–195; Segaar J (1962) Acta Morphol Neerl Scand 5:49–64 – Brown JL, Hunsperger RW (1963) Anim Behav 11:439–448 – Hoyle G (1970) Adv Insect Physiol 7:349–444
3. Kuczka H (1956) Z Tierpsychol 13:185–207
4. Franzisket L (1953) Z Vergl Physiol 34:525–538
5. Heuser JE, Reese TS (1973) J Cell Biol 57:315–344
6. Axelrod J (1976) In: Thompson RF (ed) Progress in psychobiology. Readings from Scientific American. Freeman Co, San Francisco – Hallucinogenic drugs, such as mescaline and lysergic acid diethylamide *LSD*, show a strong structural resemblance to the monoamine transmitters dopamine, norepinephrine, and serotonin. They may exert their effects on consciousness by mimicking these natural transmitters at synaptic receptors in the brain. For details on neurotransmitter receptors see Snyder SH (1978) TINS 1:123–125
7. Robison GA, Nahes GG, Trinker L (eds) (1971) Cyclic AMP and cell function. Ann N Y Acad Sci 185
8. Singer SJ, Nicolson G (1972) Science 175:720–731 – Capaldi RA (1974) Sci Am 230:26–33
9. Sutherland EW, Robison GA (1966) Pharmacol Rev 18:145
10. Burrows M (1978) Verh Dtsch Zool Ges 71:68–79
11. Florey E (1970) Verh Dtsch Zool Ges 64:186–201 – The influence of ionic environment on modulation of transmitter release is discussed by Erulkar SD, Weight FF (1979) TINS 2:298–300
12. Barker JL, Nicoll RA (1973) J Physiol (London) 228:259 – Ryall RW (1978) TINS 1:164–166
13. Neuronal correlates for response decrement of the flexion reflex have been investigated by Groves PM, Thompson RF (1970) Psychol Rev 77:419–450; cf. also Spencer WA, Thompson RF, Neilson DR Jr (1966) J Neurophysiol 29:221–239
14. Schleidt W (1962) Z Tierpsychol 19:697–722
15. Hailman JP (1967) Behaviour Suppl 15:1–196 – (1969) Sci Am 221:98–106
16. Curio E (1976) The ethology of predation. Zoophysiology and ecology, vol VII. Springer, Berlin Heidelberg New York
17. Ewert JP, Borchers HW (1974) J Comp Physiol 92:117–130
18. Autrum H (1959) Naturwissenschaften 46:435
19. Schipperheyn JJ (1965) Acta Physiol Pharmacol Neerl 13:231–277
20. Hinsche G (1935) Zool Anz 111:113–122 – For preliminary data on prey-catching and avoidance behaviors in anuran amphibians see Eibl-Eibesfeldt I (1951) Behaviour 4:1–35 – Schneider D (1954) Biol Zbl 73:225–281

21. Autrum H (1968) Naturwissenschaften 55:10–18; Autrum H, Thomas I (1973). In: Autrum H et al (eds) Handbook of sensory physiology, vol VII/3. Springer, Berlin Heidelberg New York, pp 661–692
22. For corresponding results in frogs *Rana pipiens* see Ingle D (1968) Brain Behav Evol 1:500–518. For evidence on "size-constancy" in breeding behavior see Kondrashev SL (1976) Acad Nauk Zool J 55:1576–1579
23. Collett T (1977) Nature (London) 267:349–351 – For depth vision see also Ingle D (1976). In: Fite KV (ed) The amphibian visual system: A multidisciplinary approach. Academic Press, NY – Lock A, Collett T (1980) Proc R Soc Lond B 206:481–487
24. Birukow G, Meng M (1955) Naturwissenschaften 42:652–653
25. Eikmanns KH (1955) Z Tierpsychol 12:229–253
26. Various experimental concepts and strategies, originally developed in this context for the toad by J.-P Ewert and coworkers, have been adopted recently by W. Himstedt and G. Roth to investigate corresponding questions in salamanders. Surprisingly, in salamanders prey recognition (worm-preference) is supposed to fail if the dummy stimuli are moved at relatively high velocity [cf. Himstedt W, Roth G (1978) Naturwissenschaften 65:657]. For evidence of *velocity invariance* in configurational (worm/antiworm) pattern recognition in common toads cf. Borchers HW, Burghagen H, Ewert JP (1978) J Comp Physiol 128:189–192
27. Burghagen H, Ewert JP (1980) in preparation
28. Brzoska J, Schneider H (1978) Behav Process 3:125–136
29. Kondrashev SL (1976) Acad Nauk Zool J 55:1576–1579
30. Birukow G (1938) Z Vergl Physiol 25:92–142
31. Lock A, Collett T (1979) J Comp Physiol 131:179–189; cf. also Ingle D (1971) Vision Res Suppl 3:447–456
32. "Disinhibition" of prey-catching (i.e., failure of configurational prey-selection) following lesions of caudal dorsal thalamus (T) and pretectal (P) region has been first described in *Bufo bufo* [Ewert JP (1967) Z vergl Physiol 57:263–298; (1968) Z vergl Physiol 61:41–70]. The results were reconfirmed in *B. marinus* (Ingle, pers. comm. Belmont 1971), *Rana pipiens* [Ingle D (1973) Science 180:422–424], *R. temporaria*, *R. esculenta* (Schürg-Pfeiffer unpublished data 1979), and in *Salamandra salamandra* (Finkenstädt 1980). Whereas escape behavior fails to occur after complete TP lesions in *B. bufo*, corresponding lesions in *R. pipiens* suggest some residual avoidance capacity to large approaching disks (Ingle, pers comm Waltham, 1979). Preliminary results indicate that disinhibition in prey-catching can be produced either (1) by separation of posterior thalamic "nucleus" (neuropile) from the adjacent postero-lateral *pl* nucleus, or (2) by separation of rostral optic tectum from *pl* by a tiny knife cut, or (3) by suction of both *pl* and postero-central *pc* nuclei. No disinhibition occurs after suction of only central portions of *pc*. Most recently it was found that disinhibition of prey-catching also occurs after focal lesions in the area of the "corpus" (neuropile) geniculatum thalamicum and/or its lateral nucleus (Schürg-Pfeiffer, Finkenstädt, Ewert in preparation 1980). It is most interesting to note that disinhibition produced by unilateral lesions in the geniculate region concerns mainly visual objects moving in the frontal visual field, whereas complete unilateral lesions of the thalamic-pretectal region disinhibits prey-catching toward objects moving in any part of the visual field of the contralateral eye. Furthermore, the effects produced in the former case do not last as long as those in the latter case. For reviews on neuroanatomy of frog's central visual system see Scalia F (1976) and Székély G, Lázár G (1976), both in: Frog neurobiology (Llinás R, Precht W eds) Springer, Berlin Heidelberg New York – Fite KV, Scalia F (1976). In: The amphibian visual system, a multidisciplinary approach (Fite KV ed). Academic Press, New York San Francisco London
33. Comer Ch, Grobstein P (1978) Brain Res 153:217–221
34. Ewert JP, Borchers HW (1971) Z Vergl Physiol 71:165–189
35. Ingle D (1977) pers commun Belmont Mass – Following TP lesions frogs show

dramatic deficits in avoiding stationary barriers, cf. Ingle D (1979) Soc Neurosci Abstr 5:790 – From the posterior dorsal thalamus of *Bufo americanus* so-called class TH10 neurons have been recorded that respond selectively to stationary (not to moving) visual stimuli. Some neurons have binocular inputs; thus, stimulation of both eyes is necessary to elicit activity. Other neurons of this class receive also cutaneous inputs; in this case tactile stimulation facilitates the response to a subsequent stationary visual stimulus; cf. Ewert (1971) Z Vergl Physiol 74:81–102

36. For results in frogs see Stevens RJ (1973) Brain Res 49:309–321; studies in toads cf. Ewert JP, Hock FJ, Wietersheim A v (1974) J Comp Physiol 92:343–356 – A regulatory function of ACh receptors in maintenance of retino-tectal synapses is discussed by Freeman JA (1977) Nature (London) 269:268–269

37. Evidence in frogs see Shen SC, Greenfield P, Boell EJ (1955) J Comp Neurol 102:717–743

38. Székely G, Setalo G, Lázar G (1973) J Hirnforsch 14:189–225

39. Lettvin JY, Maturana HR, McCulloch WS, Pitts WH (1959) Proc IRE 47:1940–1951

40. For results in frogs see Gaze RM (1958) Q J Exp Physiol Cogn Med Sci 43:209–314

41. Grüsser OJ, Grüsser-Cornehls U, Bullock Th H (1964) Pflügers Arch Ges Physiol 279:88–93

42. Henn V, Grüsser OJ (1968) Vision Res 9:57–69

43. Keating MJ, Gaze RM (1970) Q J Exp Physiol 55:129–142

44. Corresponding results in frogs with disk shaped stimuli cf. Butenandt E, Grüsser OJ (1968) Pflügers Arch Ges Physiol 298:283–293

45. For corresponding results in frogs see Grüsser OJ, Grüsser-Cornehls U, Licker MD (1968) Vision Res 8:1173–1185 – Grüsser OJ, Grüsser-Cornehls U (1970) Verh Dtsch Zool Ges 64:201–218 – Varjú D, Pickering SG (1969) Kybernetik 6:112–119

46. Ewert JP, Siefert D (1974) J Comp Physiol 94:177–186

47. Fite KV (1969) Exp. Neurol 24:475–486 – For results on cellular and synaptic architecture of the frog's optic tectum see Székely G, Lázár G (1976). In: Llinas R, Precht W (eds) Frog neurobiology. Springer, Berlin Heidelberg New York

48. For classes of neurons in the posterior thalamus *(TH)* see Ewert JP (1971) Z Vergl Physiol 74:81–102 – Nomenclature of neurons in the optic tectum *(T)* cf. Grüsser OJ, Grüsser-Cornehls U (1970) Verh Dtsch Zool Ges 64:201–218

49. Using neuroanatomical techniques the same projection diagrams have been described by Scalia F, Fite KV (1974) J Comp Neurol 158:455–478

50. How could an AND-gate system function? "Stimulus moving anywhere in the visual field" (T4) *and* "object recognized as prey, located n degrees outside the fixation area" (T5·2) yields the command ORIENT. "Stimulus moving in the frontal visual field" (T2·2) *and* "prey located near the fixation area" (T5·2)* *and* "stimulus far afield" (T1·1) yields the command APPROACH. "Prey near the fixation area" (T5·2)* and "stimulus close to the toad" (T1·2) yields the command FIXATE. "Prey inside fixation area" (T5·2)** *and* "both retinae simultaneously stimulated" (T1·3) *and* "stimulus within snapping distance" (T3) yield the command FIXATE and SNAP. – It is to some extent possible to test the proposed "AND" conditions. Thus, the command AVOID in response to a visual predator cannot be executed, if either T5·1 neurons (by tectal lesions) or TH3 neurons (by lesions of the caudal dorsolateral thalamus) are eliminated. *Each motor command* of the prey-catching sequence requires prey recognition as indicated by the output of T5·2 neurons. Reducing T5·2 acitivity by replacing the prey stimulus by an object with nonprey configuration decreases the probability of executing *any* command in the entire *multiple action system.* Elimination of visual input from one eye to binocular class T1 neurons prolongs the execution and impairs the performance of the commands. However, the commands are not abolished, and this is obviously due to various kinds of functional neuroplasticity. For example, an "AND" condition between two command elements subserving similar functions might be altered to an "OR"

condition based on learning or due to the influence of other modulatory elements. Furthermore, after total elimination of the tectal visual neurons, cells sensitive to cutaneous input may mediate snapping. No command in the multiple action system of prey-catching can be issued, when the motivation of the animal is not appropriate. Those influences might enter a command system at the level of a command element, e.g., the T5·2 neurons.

The hypothesis on the neural basis of prey recognition in common toads (Fig. 68 on p. 118) developed by Ewert JP (1974) Sci Am 230 (3):34–42 (quantitative data cf. Ewert JP, Wietersheim v A (1974) J Comp. Physiol 92:131–148) was criticized – on the basis of theoretical arguments – by Grüsser OJ and Grüsser-Cornehls U (1976) In: Frog neurobiology (Llinás R, Precht W eds) p. 373. Springer, Berlin Heidelberg New York. Those arguments were refuted by empirical data, cf. Ewert JP, Wietersheim A v, Borchers HW (1978) J Comp Physiol 126:43–47; Wietersheim A v, Ewert JP (1978) J Comp Physiol 126:35–42; Borchers HW, Ewert JP (1979) Behav Processes 4:99–106; Ewert JP, Borchers HW, Wietersheim A v (1979) 132:191–201. In agreement with our earlier results, our current experimental investigations allow the following conclusions: (1) configurational prey recognition is performed through *interaction of neuronal networks* (Ewert and v. Seelen, 1974; Ewert and v. Wietersheim, 1974; Ewert *et al.,* 1974); (2) the corresponding *decision-making process (prey/nonprey) precedes the goal oriented behavioral response* (Ewert, 1968, 1969; Ewert and Kehl, 1978); (3) each motor command of the prey-catching sequence requires *activation of command elements subserving recognition (classification) and localization (strategy of catching)* (Ewert, 1973). These contradict the hypothesis of Grüsser and Grüsser-Cornehls (1970; 1976) which claims that prey recognition does *not* precede a motor response (e.g., orienting), and therefore *interprets prey recognition as an entire sequence of motor responses* (starting with orienting and terminating with snapping) – each behavioral component, so to speak, representing part of the recognition process. Such an assumption is without any experimental support. Cf. also review by Ewert JP (1980). In: Neurology of the optic tectum (Vanegas H, ed) Plenum Press, New York

51. Ewert JP, Borchers HW, Wietersheim A v (1979) J Comp Physiol 132:191–201 – In the frog *Rana temporaria* three types of T5 neurons have been distinguished recently. Class T5 (1) neurons are optimally activated by moving square(S) stimuli of 8° edge length, and show almost no discrimination between moving configurational worm (W) and antiworm (A) stimuli; *R. temporaria:* neuronal response to S≫W≈A; *B. bufo:* neuronal response to S>W>A; for stimuli of 4–20° edge length. In T5 (2) neurons the worm-antiworm discrimination ("worm preference") is not as sharp as that in common toads; *R. temporaria:* S≈W>A in the range of 4–8°; W>S>A in the range of 10–20°; *B. bufo:* W>S>A in the range of 4–20°. The response characteristics of T5 (3) neurons – which have been recorded recently also from the toad's optic tectum – resemble those of TH3 neurons; *R. temporaria:* S>A>W; *B. bufo:* S>A>W; in the range of 4–20°. In both *Bufo* and *Rana* the activity patterns of T5 (2) neurons in response to moving configurational stimuli resemble the probability that the same stimuli fit the prey category; cf. Schürg-Pfeiffer E, Ewert JP (1980) J Comp Physiol (in preparation)
52. Himstedt W (1977) pers commun Konstanz
53. Pathways between hypothalamus and optic tectum have been identified recently by Wilczynski W, Northcutt RG (1977) J Comp Neurol 173:219–229
54. For evidence in frogs see Trachtenberg MC, Ingle D (1974) Brain Res 79:419–430 – Wilczynski W, Northcutt RG (1977) J Comp Neurol 173:219–229
55. Tecto-thalamic projections have been described in the frog by Brown WT, Ingle D (1973) Brain Res 59:405–409
56. Results in frogs are reported by Ingle D (1973) Science 180:422–424
57. Northcutt RG, Wilczynski W (1978) pers commun St Louis; cf. also Wilczynski W, Northcutt RG (1977) J Comp Neurol 173:219–229; Wilczynski W (1978)

Connections of the midbrain auditory center in the bullfrog *Rana catesbaiana*. Ph D Dissertation Univ Michigan Ann Arbor – For organization of the nonolfactory telencephalon see Kicliter E, Ebbesson SOE (1976). In: Llinás R, Precht W (eds), Frog neurobiology. Springer, Berlin Heidelberg New York

58. Gruberg ER, Ambros VR (1974) Exp Neurol 44:187–197 – Halpern M (1972) Brain Behav Evol 6:42–68 – Kicliter E, Northcutt RG (1975) J Comp Neurol 161:239–254 – Scalia F, Colman DR (1974) Brain Res 79:496–504

59. Ewert JP, Seelen W v (1974) Kybernetik 14:167–183. A comprehensive model on prey/predator interaction in the visuomotor system of frogs and toads was recently elaborated by Lara R, Arbib MA, Cromarty A (1979) Soc Neurosci Abstr 5:469

60. The mammalian visual cortex is organized in functional units of cell assemblies which are called "sensory columns" [Mountcastle VB (1957) J Neurophysiol 20:408–434 – Hubel DH, Wiesel TN (1962) J Physiol 160:106–154]. A comparable anatomical organization has been assumed for the anuran optic tectum [Székely G, Lázár G (1976). In: Llinas R, Precht W (eds) Frog neurobiology. Springer, Berlin Heidelberg New York]. Neurons of the frog's and toad's optic tectum which appear to mediate the main efferent outflow are the large pyramidal cells and the large ganglionic cells; presumably they correspond to class T5 neurons. Recent neurophysiological recording studies from toad T5 neurons suggest similarities of some functional properties with cat cortical cells, such as: movement specificity, directional selectivity, Gestalt sensitivity and selectivity, invariance classes in Gestalt perception with regard to stimulus velocity and movement direction – to name two examples [Ewert JP, Borchers HW, Wietersheim A v (1979) J Comp Physiol 132:191–201

61. It appears that the properties of the innate releasing mechanism for feeding in common toads are temporally preprogrammed for the different special environmental situations in water and on land. Obviously the corresponding neuronal systems undergo some kind of development to be available when they are needed at a certain age. Whereas the differentiation of *ventral thalamus* takes place during larval life and is finished long before the end of metamorphosis, differentiation of *dorsal thalamus* starts before metamorphosis and is completed six months to one year after metamorphosis. In anurans the formation of dorsal thalamus proceeds in two steps: (1) edification of area dorsomedialis; (2) by cell migration shortly before middle of metamorphosis, this area gives rise to an area dorsolateralis. The differentiation of the dorsomedial area is obviously related to the establishment of tecto-thalamic optic connections. It develops only in the presence of the rostral optic tectum. The late differentiation of the area dorsolateralis – outlasting the end of metamorphosis – might be correlated with the maturation of (a) configurational prey-selection, and (b) estimation of absolute prey size, as described on p. 73 and 79. For development of the visual system see Straznicki K, Gaze RM (1972) J Embryol exp Morph 28:87–115; Clairambault P (1976). In: Llinas R, Precht W (eds) Frog neurobiology. Springer, Berlin Heidelberg New York

62. Schäfer KP (1970) Brain Behav Evol 3:222–240

63. Haseltine E, Kass L, Hartline PH (1977) Soc Neurosci Abstr 3:90

64. Schneider GE (1977). In: Sweet WH et al (eds) Neurosurgical treatment in psychiatry, pain and epilepsy. Univ Park Press, Baltimore London Tokyo – For plasticity in the nervous system see Székely G (1979) TINS 2:245–248 – Sara VR, Hall K (1979) TINS 2:263–265 – 'How a brain gets wired for adaptive functions' cf. Sperry RW (1971). In: Tobach E, Aronson LR, Shaw E (eds) The biopsychology of development. Academic Press, New York

65. Cited in Schneider GE, Jhaveri SR (1974). In: Stein DG et al (eds) Plasticity and recovery of function in the central nervous system. Academic Press, London New York; cf. also Teuber HL (1976) Neuronal Plasticity: extent and limits. 9th Winter Conf Brain Res, Keystone Colorado – For recovery of function after brain injury in man see Teuber HL (1975). In: Tower DB (ed) The nervous system, vol 2. Raven Press, New York – Teuber HL (1975). In: Ciba Foundation Symp 34 (new series):

Outcome of severe damage to the nervous system. Elsevier-Excerpta Medica, Amsterdam

66. Collewijn H (1975) J Neurobiol 6:3–22 – Simpson JI, Soodak RE (1978) Soc Neurosci Abstr 4:645 – For neuroanatomical details of the anuran accessory optic system see Montgomery N, Fite KV, Bengston L (1979) Soc Neurosci Abstr 5:798
67. Hoffmann KP (1976) Brain Res 115:150–153
68. Michael Ch R (1969) Sci Am 220:104–114 – West RW (1976) J Comp Neurol 168:355–378
69. For naming of neurons see Hughes A (1979) Brain Behav Evol 16:52–64; cf. also Rowe MH, Stone J (1979) Brain Behav 16:65–80
70. Dowling JE (1970) Invest Ophthalmol 9:655–680 – Recently a sixth type of retinal neurons has been identified, the *interplexiform cell,* which uses dopamine as neurotransmitter and may play a role in regulating the strength of lateral inhibition and center surround antagonism throughout the retina as a function of adaptive state; cf. Dowling JE, Ehinger B, Hedden WL (1978) Proc R Soc (London) 201:7–55
71. Wickelgren BG, Sterling P (1969) J Neurophysiol 32:16–23
72. Sprague JM, Berlucchi G, DiBerardino A (1970) Brain Behav Evol 3:285–294 – Snyder M, Diamond T (1968) Brain Behav Evol 1:244–288 – Schneider GE (1970) Brain Behav Evol 3:295–323 – For new results on participation of cat striate-peristriate visual cortex in form and pattern processing see Hughes HC, Sprague JM (1979) Soc Neurosci Abstr 5:789 – For subcortical vision in man see Perenin MT, Jeannerod M (1979) TINS 2:204–207
73. Hubel DH, Wiesel TN (1962) J Physiol 160:106–154 – (1968) J Physiol 195:215–243 – (1965) J Neurophysiol 28:229–289; cf. also Lee BB, Cleland BG, Creutzfeldt OD (1977) Exp Brain Res 30:527–538 – Camarda RM (1979) Exp Brain Res 36:191–194 – For nomenclature and classification of cells in the striate cortex see Henry GH (1977) Brain Res 133:1–28 – Kato H, Bishop PO, Orban GA (1978) J Neurophysiol 41:1071–1095 – Results on central representation of "depth" in monkey visual cortex are reviewed by Poggio GF (1979) TINS 2:199–201 – For visual cortical areas and their connections see Mishkin M (1972). In: Karczmar AG, Eccles JC (eds) Brain and human behavior. Springer, Berlin Heidelberg New York – New results on thalamo-cortical connections are reported by Peters A (1979) TINS 2:183–185 – Transformations of complex stimulus patterns by different types of geniculate and cortical neurons have been investigated by Creutzfeldt OD, Nothdurft HC (1978) Naturwissenschaften 65:307–318
74. There are neurons in the monkey's infero-temporal cortex responding most vigorously to the sight of a monkey hand, cf. Gross CP, Rocha-Miranda CE, Bender DB (1972) J Neurophysiol 35:96–111 – Most recently in the temporal lobe of alert rhesus monkey neurons have been recorded responding selectively to faces (human or rhesus monkey, 3-D, or projected). These neurons were unresponsive to gratings, simple geometrical or complex visual patterns, or stimuli of other sensory modalities. Some of the cells required eyes in a face model, some hair, some the mouth. Some cells responded to each of these features, and others responded best when such component features of faces were combined. The response property of some neurons was invariant with respect to changes of (1) orientation, (2) color, (3) size, and (4) distance of the face; cf. Perret DI, Rolls ET, Caan W (1979) Neurosci Lett Supl 3:358
75. For results on pattern discrimination in cats following lesions in striate and peristriate visual cortex see Sprague JM (1966) Science 153:1544–1547 – Spear PD, Braun JJ (1969) Exp Neurol 25:331–348 – For visual mechanisms beyond the striate cortex see Mishkin M (1966). In: Russel RW (ed) Frontiers in physiological psychology. Academic Press, New York – Functional recovery after lesions of the nervous system is reviewed by Eidelberg E, Stein DG (1974) Neurosci Res Prog 12:191–303
76. Schneider GE (1969) Science 163:895–902
77. Goodale MA, Murison RCC (1975) Brain Res 88:243–261 – Ingle D (1978) pers commun NATO Adv Study Inst, Waltham Mass

78. Hoffmann KP, Heitländer H, Sireteanu R (1978) Arch Ital Biol 116:452–462 – On functional significance of X and Y cells in normal and visually deprived cats cf. Sherman SM (1979) TINS 2:192–195 – For results on functional plasticity in the immature striate cortex of the monkey shown by the [^{14}C] Deoxyglucose method see DesRosiers MH, Sakurada O, Jehle J, Shinohara M, Kennedy C, Sokoloff L (1978) Science 200:447–449 – Effects of visual deprivation on the visual cortex of cat and monkey are summarized by Hubel DH (1978) The Harvey Lectures 72:1–51. Academic Press, New York; cf. also Hubel DH, Wiesel TN, LeVay S (1977) Phil Trans R Soc (London) B 278:377–409

79. Pettigrew JD, Kasamatsu T (1978) Nature (London) 271:761–763

80. Hirsch HVB, Spinelli DN (1971) Exp Brain Res 12:509–527

81. Held R (1965) Sci Am 213:84–94

82. Grobstein P, Chow KL (1975) Science 190:352–358 – Hubel DH, Wiesel TN, LeVay S (1977) Phil Trans R Soc (London) B 278:377–409 – Hubel DH (1978) The Harvey Lectures 72:1–51. Academic Press, New York

83. Rosenzweig MR, Bennet EL, Diamond MC (1972) Sci Am 226:22–29 – Recent investigations in monkeys and mice provide evidence that physical and social interactions with the environment during development significantly influence the morphology (spiny branchlets, soma size) of cerebellar Purkinje cells; cf. Floeter MK, Greenough WT (1979) Science 206:277–229 – Physh JJ, Weiss GM (1979) Science 206:230–231

84. For neurophysiological results on the visual Wulst of birds – which is an analog of the mammalian visual cortex – see Pettigrew JD, Konishi M (1976) Science 193:675–678 – Neural correlates of imprinting in the domestic chick are reported by Bateson P (1978) Nature (London) 273:659–660 – Martin JT (1978) Science 200:565–566 – Salzen EA, Parker DM, Williamson AJ (1978) Exp Brain Res 31:107–116 – Horn G (1979) TINS 2:219–222

85. For neuronal correlates of sound production in grasshoppers see Elsner N (1974) J Comp Physiol 88:67–102 – (1975) J Comp Physiol 97:291–322

86. Huber F (1960) Z Vergl Physiol 44:60–132 – Otto D (1971) Z Vergl Physiol 74:227–271

87. Hirth C, Elsner N (1978) Verh Dtsch Zool Ges 71:261

88. For transmission of acoustic information in the auditory system of locusts see Adam LJ (1969) Z Vergl Physiol 63:227–289 – Rheinlaender J (1975) J Comp Physiol 97:1–53 – Kalmring K (1975) J Comp Physiol 104:103–159

89. Gross morphology of the AIAA neuron is similar to that of the "interneuron 1" described by Casaday GB, Hoy RR (1977) J Comp Physiol 121:1–13 – Evidence of a *temporal pattern* as cue for recognition of species-specific calling song in crickets is given by Pollack GS, Hoy RR (1979) Science 204:429–432 – *Time intensity trading* in locust auditory interneurons as a principle of sound localization is discussed by Rheinlaender J, Mörchen A (1979) Nature (London) 281:672–674 – A hypothesis of the mechanisms underlying conspecific song recognition at the ventral cord level in bush crickets was recently presented by Kühne R, Lewis B, Kalmring K (1980 in press) Behav Process – Ethological concepts and neurobiological results in crickets are discussed by Huber F (1978) Anim Behav 26:969–981

90. For results in tree frogs see Gerhardt HC (1974) J Exp Biol 61:229–241 – (1976) Nature (London) 261:692–694 – (1978) J Exp Biol 74:59–73

91. Feng AS, Narins PM, Capranica RR (1975) J Comp Physiol 100:221–229

92. Gerhardt HC, Mudry KM (1978) pers commun Columbia Mo.

93. Gerhardt HC (1978) Science 199:992–994; cf. also Hubl L, Mohneke R, Schneider H (1977) Behav Process 2:305–314. Temperature coupling has been also described in electric fish and fireflies: Feng AA (1976) Comp Biochem Physiol A 55:99; Carlson A, Copeland J, Raderman R, Bulloch A (1976) Anim Behav 24:786

94. Gerhardt HC (1976) Nature (London) 261:692–694

95. Capranica RR (1968) Behaviour 31:302–325. – The morphology and physiology of

the anuran auditory system has been recently reviewed by Capranica RR (1976). In: Llinas R, Precht W (eds) Frog neurobiology. Springer, Berlin Heidelberg New York – Capranica RR (1977). In: Taylor DH, Guttman SI (eds) The reproductive biology of amphibians. Plenum Press, New York – For neural correlates of frog calling see Schmidt RS (1971) Behaviour 39:288–317 – (1976) J Comp Physiol 108:99–113 – (1978) J Comp Physiol 126:49–56

96. For neural bases of auditory communication in lower and higher vertebrates see Creutzfeldt OD, Scheich H, Schreiner C (eds) (1979) Hearing mechanisms and speech. Springer, Berlin Heidelberg New York

97. Simmons JA, Fenton M, O'Farrel MJ (1979) Science 203:16–21 – Simmons JA (1973) J Acoust Soc Am 54:157–173

98. Jen Ph HS (1978) Nature (London) 275:143–144

99. Schnitzler HU (1973) J Comp Physiol 82:79–92

100. Suga N, O'Neill WE (1978) TINS 1:35–38

101. For tonotopic organization in the bat auditory cortex see Suga N, Jen Ph HS (1976) Science 194:542–544 – For peripheral auditory tuning see Schnitzler HU, Suga N, Simmons JA (1976) J Comp Physiol 106:99–110 – Pollack G, Henson Jr OW, Johnson R (1979) J Comp Physiol 131:255–266 – For auditory processing in the inferior colliculus see Schuller G (1979) J Comp Physiol 132:39–46 – Schuller G, Pollack G (1979) J Comp Physiol 132:47–54

102. Suga N, O'Neill WE, Manabe T (1979) Science 203:270–274

103. Bell CC, Bradbury J, Russel CJ (1976) J Comp Physiol 110:65–88

104. Heiligenberg W (1974) J Comp Physiol 91:223–240

105. Bullock TH (1969) Brain Behav Evol 2:85–118

106. Matsubara J, Heiligenberg W (1978) J Comp Physiol 125:285–290

107. Scheich H (1974) Science 185:365–367

108. Heiligenberg W, Baker C, Matsubara J (1978) J Comp Physiol 127:267–286 – The behavior of *Eigenmannia* in response to multiple jamming has been investigated by Partridge B, Heiligenberg W (1979) Soc Neurosci Abstr 5:470

109. Efferent fibers in the optic nerve of frogs have been described by Maturana HR (1958) J Anat (London) 92:21–27

110. Suga N, Schlegel P (1972) Science 177:82–84

111. Müller-Preuss P (1978) Neurosci Lett Suppl 1:7

112. Kandel ER (1976) Cellular basis of behavior. WH Freeman Co, San Francisco

113. Hoyle G (ed) (1977) Identified neurons and behavior of arthropods. Plenum Press, New York London

114. For new results on the neuronal network involved in escape swimming in *Tritonia* see Getting PA (1977) Soc Neurosci Abstr 3:382 – Getting PA, Lennard PR, Hume RI (1979) Soc Neurosci Abstr 5:495 – For neuronal generation of swimming movements in the leech see Stent GS, Kristan WB, Friesen WO, Ort CA, Poon M, Calabrese RL (1978) Science 200:1348–1356 – A reexamination on "bursting neural networks" underlying cyclic behaviors is presented by Russel DF, Hartline DK (1978) Science 200:453–455

115. The term "command" for single neurons capable of eliciting coordinated movements in crayfish was introduced by Wiersma CAG, Ikeda K (1964) Comp Biochem Physiol 12:509 – The "behavior of neurons" is reviewed by Wiersma CAG (1974). In: Schmitt FO (ed) The neurosciences, 3rd study program. MIT Press, Cambridge Mass – Recent investigations of command neurons responsible for the rhythmic feeding behavior in the marine mollusc *Pleurobranchea californica* indicate that these neurons receive synaptic feedback from the motor network they excite. Furthermore, the activity of feeding command neurons correlates well with two aspects of behavioral plasticity: (1) They are activated by food stimuli, but are inhibited by those food stimuli which the animal was trained to avoid; (2) feeding command neurons are inhibited in response to food if the naive animals are satiated; cf. Gillette

R, Kovac MC, Davis WJ (1978) Science 199:798–801 – Davis WJ, Gillette R (1978) Science 199:801–803

116. Kimmel Ch B, Powell SL, Eaton RC (1978) Soc Neurosci Abstr 4:362
117. Mountcastle VB (1976) Neurosci Res Program (NRP) 14:1–47 – The activity of cortical *motor neurons* in primates during the performance of hand and finger movements has been investigated recently by Lemon RN, Kuypers HGJM (1979) Neurosci Lett Suppl 3:114 – For concepts of motor organization see Miles FA, Evarts EV (1979) Ann Rev Psychol 30:327–362 – Wurtz RH, Albano JE (1980) Ann Rev Neurosci 3, in press
118. The command neuron concept was recently redefined (see italics) and discussed by Kupferman I, Weiss KR (1978) Behav Brain Sci 1:3–39
119. Doty RW (1976) The concept of neural centers. In: Fentress JC (ed) Simpler networks and behavior. Sinauer Assoc Inc, Sunderland Mass
120. The term drive instead of motivation is nowadays used only occasionally in the literature of animal behavior. According to the arguments of many ethologists, however, the two terms are not totally equivalent in meaning. For detailed discussion of this complex concept see Hinde RA (1970) Animal behaviour: synthesis of ethology and comparative physiology. McGraw-Hill, New York – Concepts of *drive* were discussed recently by Deutsch JA (1979) TINS 2:240–244
121. Holst E v, Saint Paul U v (1962) Sci Am 206:50–59
122. Ruiter L de, Wiepkema PR, Reddingius J (1969) Ann NY Acad Sci 157:1204–1216 – Hypothalamic regulation of feeding was recently reviewed by Grossman SP (1979) Annu Rev Psychol 30:209–242; cf. also Wayner M, Oomura Y (1975) Pharmacol Biochem Behav Suppl 1
123. The role of the noradrenergic fiber system in behavior has been discussed recently by Mason ST (1979) TINS 2:82-83 – For axon sparing cell lesions in *LH* by microinjection of kainic acid (p. 265) cf. Grossman SP, Dacey D, Halaris AE, Collier T, Routtenberg A (1978) Science 202:537–539 – Peterson GM, Moore RY (1979) Soc Neurosci Abstr 5:222
124. Anand BK, Chhina GS, Sharma KN, Dua S, Singh B (1964) Am J Physiol 207:1146–1154 – Oomura Y, Ono T, Ooyama H, Wayner MJ (1969) Nature (London) 222:282–284 – Hunger in humans can be induced by 2-deoxy-D-glucose which inhibits intracellular glucose utilization (p. 266), cf. Thompson DA, Campbell RC (1977) Science 198:1065
125. Gold thiomalate, gold thiogalacton or gold thioglycerol produce no comparable damage in the VMH. Mayer J, Thomas DW (1967) Science 156:328–337
126. Kennedy GC (1966) Br Med Bull 22:216–220
127. Oomura Y, Ooyama H, Yamanoto T, Naka F (1968) Physiol Behav 2:97–115 – Grossman SP (1968) Fed Proc 27:1349–1360
128. Grossman SP (1967). In: Kare MR, Maller O (eds) The chemical senses and nutrition. Johns Hopkins Press, Baltimore – The effects of central cholinergic and adrenergic stimulation on eating and drinking in rats were investigated recently by Finkelstein JA, Chance WT (1979) Soc Neurosci Abstr 5:217 – For effects of testosterone implants in *VMH* on food-intake see Nunez AA, Siegel LI, Gray JM, Wade GN (1979) Soc Neurosci Abstr 5:222
129. Baxter BL (1967) Exp Neurol 19:412–432; (1968) Exp Neurol 21:1–10 – Grant LD, Jarrad LE (1968) Brain Res 10:392–401
130. The effect of angiotensin II on thirst is well known in mammals and birds and has also been reported for the few reptiles and fish that have been investigated so far; Fitzsimons JT (1975) Prog Brain Res 42:215–233 – The influence of angiotensin II depends on different central sites of action; cf. Simpson JB, Reed M, Keil LC, Thrasher TN, Ramsay DJ (1979) Soc Neurosci Abstr 5:459
131. For effects of diencephalic and cortical lesions on mating behavior in rats see Heimer L, Larsson K (1967) Brain Res 3:248–263 – Larsson K (1964) J Exp Zool 155:203–214 – Lisk RD (1969) J Reprod Fertil 19:353–356 – In male rats the

following pathway is known to facilitate male sexual behavior: Olfactory bulb → corticomedial amygdala → medial preoptic area (via stria terminalis) → anterior medial hypothalamus. *Lesions* of the medial preoptic area (and anterior medial hypothalamus) disrupt male sexual behavior; *electrical point stimulation* elicits, and *testosterone implants* in castrated males reactivate male sexual behavior in rats (as well as in fish, frogs and birds). Following medial preoptic lesions functional recovery occurs if surgery was performed prepuberally and rats had socially interacted with peers; cf. Twiggs DG, Popolow HB, Gerall AA (1978) Science 200:1414–1415 – Lesion of area septalis facilitates lordosis both in male and female rats. Electrical stimulation of medial preoptic or septal area inhibits lordosis in females. Estrogen sensitive neurons were identified in female rats in the medial preoptic area, the medial and ventromedial hypothalamus and the area septalis. Effects of estradiol implants in medial preoptic area on the lordotic and ejaculatory behavior in male rats have been reported recently by Dohanich GP, Ward IL (1979) Soc Neurosci Abstr 5:443 – For reviews on central pharmacological, and hormonal control of reproductive behavior see Meyerson BJ, Eliasson M (1977). In: Iversen LL, Iversen SD, Snyder SH (eds) Handbook of psychopharmacology, vol 8. Plenum Press, New York – Malsbury CW, Pfaff DW (1974). In: DiCara L (ed) Limbic and autonomic nervous systems research. Plenum Press, New York – Pfaff DW, Diakow C, Zigmond RE, Kow LM (1974). In: Schmitt FO, Worden FG (eds) The neurosciences, 3rd study program. MIT Press, Cambridge Mass – Bermant G, Davidson JM (1974) Biological basis of sexual behavior. Harper and Row, New York

132. There are indications that the embryonic brain of the rat is undifferentiated. Female sexual differentiation may require "estrogen imprinting" with *low* estrogen concentrations of placental origin. *High* levels of estrogen, after conversion from testicular androgens, stimulate the male type of brain circuitry. During masculinization of the brain, androgens may act synergistically with estrogens; cf. Döhler KD (1978) TINS 1:138–140 – MacKinnon PCB (1978) TINS 1:136–138 – For steroid hormon receptors in the brain see Greenstein BD (1978) TINS 1:4–6 – As to the mechanism by which hormones regulate sexual difference, it seems that there are important variations between primates (Rhesus monkeys, *Homo sapiens)* and rodents. Herbert I (1977) Exp. Brain Res (EBBS Abstr) R 59

133. Anatomical comparison of normal adult male and female song systems in the brains of zebra finches reveals that all types of cells and identified connections are *generally* present in both sexes. Physiological experiments suggest that early exposure to estradiol may exert a specific inductive effect on the telencephalic song neurons which renders them physiologically competent to respond to testosterone in the adult. Thus, *testosterone* treatment activates song in adult female zebra finches which were implanted with 17 β-*estradiol* when chicks, but fails to activate song in those females which were implanted with dihydrotestosterone. More specifically, estradiol treatment of genetically female chicks organizes malelike cytoarchitectonic differentiation of the *telencephalic* song nuclei RA, HVC, MAN and X (Fig. 139), whereas testosterone induces masculinization of the *brain stem* song nuclei N. XII and DM; cf. Gurney ME, Konishi M (1979) Soc Neurosci Abstr 5:446 – For hearing, single-unit analysis, and vocalization in songbirds see Konishi M (1969) Science 166:1178–1181 – For unit responses to species-specific sound in birds see Leppelsack HJ (1978) Fed Proc 37:2336–2341 – Scheich H (1977). In: Bullock TH (ed) Dahlem workshop on recognition of complex acoustic signals. Dahlem Konferenzen, Berlin – Scheich H, Bonke D, Langner G (1979). In: Creutzfeldt O, Scheich H, Schreiner C (eds) Hearing mechanisms and speech. Springer, Berlin Heidelberg New York – For space and frequency orientation in the owl auditory system cf. Knudson EI, Konishi M (1978) J Neurophysiol 41:870–884

134. Flynn J, Vanegas H, Foote W, Edwards S (1970). In: Whalen RF et al (eds) The neural control of behavior. Academic Press, London New York – For influence of

mesencephalic lesions on mouse killing in rat see Chaurand JP, Schmitt P, Karli P (1973) Physiol Behav 10:507–515

135. Fernandez de Molina A, Hunsperger RW (1962) J Physiol 160:200–213 – The neural bases of aggression and sexual behavior in rhesus monkey have been reviewed by Perachio AA, Alexander M (1975). In: Bourne GH (ed). The rhesus monkey I: anatomy and physiology. Academic Press, New York – A review on the neural bases of threat and attack in cats is presented by Flynn JP (1976). In: Grenell RG, Gabay S (eds) Biological foundations of psychology. Raven Press, New York – Relationships between anxiety and anxiolytic drugs are discussed by File SE (1978) TINS 1:9–11 – For results on brain serotonin, body weight and isolation-induced aggression in mice see Walters JK, Lavooy M, Posch R (1979) Soc Neurosci Abstr 5:666 – Lithium has been shown in mice to suppress social aggression (inter-male attacks) but not maternal aggression (attack by lactating female on intruding male); cf. Brain PF, Al-Maliki S (1979) Neurosci Lett Suppl 3:29 – For pharmacology of aggression and sex see Miczek KA, Barry H (1976). In: Glick SD, Goldfarb J (eds) Behavioral pharmacology. CV Mosby, St. Louis – For behavioristic concepts of aggression see Hutchinson RR (1972). In: Cole JK, Jensen DD (eds) Nebraska symposium on motivation. University of Nebraska Press, Lincoln Nebr – Cahoon DD (1972) Psychol Rec 22:463–476

136. Valenstein ES (1973) Brain control. Wiley, New York – Psychosurgery is also discussed by Bridges PK (1978) TINS 1:108–110; Robin A (1978) TINS 1:111–112

137. Pribram KH, Broadbent DE (1970) Biology of memory. Academic Press, London New York – On the search for engrams see Lashley KS (1950) Symp Soc Exp Biol 4:454–481 – The role of neocortex in learning has been discussed by Oakley DA (1979) TINS 2:149–152

138. Burns BD (1958) The mammalian cerebral cortex. Arnold, London

139. Mishkin M (1978) NATO Adv Study Inst, Waltham Mass – From the orbito-frontal cortex of rhesus monkey neurons have been recorded responding selectively to sight or taste of a particular food (e. g. banana). Many neurons showed visual responses to particular rewarding or aversive food stimuli. There are also neurons exhibiting specific responses to frustrating events (e. g. removal of food); cf. Thorpe SJ, Maddison S, Rolls ET (1979) Neurosci Lett Suppl 3:77 – The role of frontal and temporal association areas in the context of *combined* amygdalo-hippocampal lesions in the monkey are discussed by Mishkin M (1978) Nature (London) 273:297–298; cf. also results on chronic blindness following nonvisual cortical lesions in monkeys reported by Nakamura RK, Mishkin M (1979) Soc Neurosci Abstr 5:800

140. Bridges PK (1978) TINS 1:108–110

141. Olds J, Milner P (1954) J Comp Physiol Psychol 47:419–427 – Results on biochemistry and behavior have been reviewed by Groves PM, Rebec GV (1976) Ann Rev Psychol 27:91–127

142. The availability of DNA strands for transcribing messenger RNA depends on histone and nonhistone proteins which are attached to the DNA. Whereas histone proteins generally inhibit RNA synthesis, nonhistone proteins (which can bind to histone proteins) specifically facilitate this process in relation to their degree of phosphorylation; cf. Stein GS, Stein JS, Kleinsmith LJ (1975) Sci Am 232:46–57 – Entingh D, Dunn A, Glassman E, Wilson JE, Hogan E, Damstra (1975). In: Gazzaniga MS, Blakemore C (eds) Handbook of psychology. Academic Press, London New York

143. Zippel HP (ed) (1973) Memory and transfer of information. Plenum Press, New York London

144. Ungar G (1972) Naturwissenschaften 59:85–91 – Ungar G, Desiderio DM, Parr W (1972) Nature (London) 238:198–202 – For memory transfer through cannibalism in *Planaria* see Hartry AL, Keith-Lee P, Morton WD (1964) Science 146:274–275

145. The effect of synthetic scotophobin on the light tolerance of teleost fish was studied by

Thines G, Domagk GF, Schonne E (1973). In: Zippel HP (ed) Memory and transfer. Plenum Press, New York London

146. Horn E, Greiner B, Horn I (1979) J Comp Physiol 131:129–135

147. de Wied D (1973). In: Zippel HP (ed) Memory and transfer. Plenum Press, New York London – In mammals the action of a new group of putative neurotransmitters, called the "neuropeptides", is being discovered. They differ from previously identified transmitters in that they mediate complex phenomena, such as: differentiation and survival of neurons (*nerve-growth factor*, NGF), modulation of neuronal communication *(prostaglandins)*, female sexual behavior *(luteinizing hormone releasing hormone)*, olfaction related information *(carnosine)*, secretion of insulin and glucagon *(somatostatin)*, neurogastrointestinal actions *(bombesin)*, thirst *(angiotensin* II), pain related information (*substance* P), perception of pain and emotional experience (*endorphins, enkephalins,* both exhibiting similarity with morphine, thus binding with opiate receptors), improvement of memory *(vasopressin)* – For effects of nerve growth factor, NGF, on sympathetic neuron development see review by Black IB (1978) TINS 1:101–104 – For carnosine as principal neurotransmitter in the mammalian olfactory pathway see Margolis FL (1978) TINS 1:42–44 – Effects of bombesin on thermoregulation, hyperglycaemia, and satiety are reviewed by Brown M, Vale W (1979) TINS 2:95–97 – The relationship between thirst, angiotensin II, substance P and enkephalin is discussed by de Caro G, Micossi LG, Venturi F (1978) Nature (London) 277:51–53 – The role of β-endorphin and ACTH–opiate peptides in the regulation of behavior has been reviewed recently by Miller RJ (1978) TINS 1:29–31, Jacquet YF (1979) TINS 2:140–143 – Experiments in mice show that a deficiency in opiate receptors is paralleled by poor electroacupuncture analgesia; cf. Peets JM, Pomeranz B (1978) Nature (London) 273:675–676 – Painful memories appear to be mediated by an opiate system; cf. Kapp BS, Gallagher M (1979) TINS 2:177–180 – For opiate receptor mechanisms see Snyder SH, Matthysse S (1975) Neurosci Res Prog 13:3–166 – Endorphins and acupuncture analgesia are discussed by Nathan PW (1978) TINS 1:21–23 – For Neurobiology of peptides see Iversen LL, Nicoll RA, Vale WW (1978) Neursci Res Prog 16:211–370 – Liebeskind JC, Dismukes RK (1978) Neurosci Res Prog 16:490–635 – Aspects of hypothalamic regulation of the pituitary gland are reviewed by Schally AV (1978) Science 202:18–28

148. Martin U, Martin H, Lindauer M (1978) J Comp Physiol 124:193–201

149. Holst D v (1969) Z Vergl Physiol 63:1–58

150. Calhoun JB (1962) Sci Am 206:139–148; Calhoun JB (1971). In: Essor AH (ed) Behavior and environment. Plenum Press, New York London

151. Quote from the lecture by Dietrich von Holst: *Social stress in animal and man.* Academy of Sciences of Rhineland Westphalia in Münster, Westph (N 253)

152. Definite problems can also be studied on the basis of genetic mutants and structural anomalies within the nerve nets related to them. For corresponding investigations in Siamese cats and various mammalian albinos see Hubel DH, Wiesel TN (1971) J Physiol 218:33–62 – Guillery RW (1974) Sci Am 230:44–54

153. Cerebral utilization of glucose during and after cortical spreading depression has been analyzed by Shinohara M, Dollinger B, Brown G, Rapoport S, Sokoloff L (1979) Science 203:188–190 – For axon sparing brain lesioning techniques using kainic acid see Coyle JT, Schwarcz R (1976) Nature (London) 263:244–246 – Simson EL, Gold RM, Standish LJ, Pellett PL (1977) Science 198:515–517 – Merker B (1978) Science 200:1417

154. For chemical lesions using 6-hydroxy-dopamine see Sachs S, Jonsson G (1975) Biochem Pharmacol 24:1–8 – Killing of single neurons by injection of proteolytic enzymes is reported by Parnas I, Bowling D (1977) Nature (London) 270:626–628 – For killing of "lucifer yellow"-filled and irradiated cells see Miller JP, Selverston AI (1979) Soc Neurosci Abstr 5:496 – A method for reversible inactivation of small

regions of brain tissue using injections of the local anesthetic lidocaine is described by Malpeli JG, Schiller PH (1979) J Neurosci Methods 1:143–151

155. Kennedy C, Des Rosiers MH, Jehle JW, Reivich M, Sharpe F, Sokoloff L (1975) Science 187:850–853

156. Cooley RK, Vanderwolf CH (1975) Brain Res Bull 3:175–179 – Borchers HW (1975) Methodische Grundlagen für die Erforschung des visuellen Systems. GHK, Kassel – Skydell JL, Capranica RR (1975) Electroencephalogr Clin Neurophysiol 38:325–328 – Borchers HW (1979) Neurosci Lett Suppl 3:291 – For parallel recording of single unit *electrical* activity see Kuperstein M, Wittington D (1979) Soc Neurosci Abstr 5:495 – Pickard RS (1979) TINS 2:259–261 – Techniques for recording of electro*chemical* events are discussed by Adams RN (1978) TINS 1:160–163

157. Rossetto MA, Vandercar DH (1972) Physiol Behav 9:106–109

158. McElligott JG (1973). In: Phillips MI (ed) Brain unit activity during behavior. Ch C Thomas, Springfield

159. Recent advances in neuroanatomical methodology have been summarized by Jones EG, Hartman BK (1978) Ann Rev Neurosci 1:215–296 – Degeneration methods in neurobiology are discussed by Leonard CM (1979) TINS 2:156–159 – For transsynaptic staining with procion yellow see Kuhnt U, Kelly MJ, Schaumberg R (1979) Exp Brain Res 35:371–385 – On recent developments in intracellular staining cf. Mason CA, O'Shea M (1979) TINS 2:76–79 – For histochemical localization of HRP see Olsson Y, Malmgren LT (1978) TINS 1:105–107 – The characteristics of dendritic and axonal transport are decribed by Schubert PE (1976) In: Gispen WH (ed) Molecular and functional neurobiology. Elsevier Publ Comp, Amsterdam – Double labeling techniques to detect branching axons (i. e., a neuron projecting to more than one target) are described by Kuypers HGJM, Catsman-Berrevoets CE, Padt RE (1977) Neurosci Lett 6:127–133; cf. also Cesaro P, Nguyen-Legros J, Berger B, Alvarez C, Albe-Fessard D (1979) Neurosci Lett 15:1–7 – For reconstruction of brain circuitries by neural implants see Björklund A, Stenevi U (1979) TINS 2:301–306

160. For analysis of neuronal spike trains see Perkel DH, Gerstein GL, Moore GP (1967) Biophys J 7:391–440 – Muschaweck LG, Loevner D (1978) Intern J Neurosci 8:51–60

161. Seelen W v in Ewert JP (1976). In: Fite KV (ed) The amphibian visual system. Academic Press, London New York – Seelen W v (1968) Kybernetik 5:133–148

162. Seelen W v, Hoffmann KP (1976) Biol Cybernet 22:7–20 – Marmarelis PZ, Marmarelis VZ (1978) Analysis of physiological systems: The white-noise approach. Plenum Press, New York

163. Arbib MA (1972) Commun ACM 15:521–527

164. Mittelstaedt H (1977). In: Kybernetik 1977. Oldenbourg Verlag, München Wien – Reichardt W, Poggio T (1976) Quart Rev Biophys 9:311–438

165. Bullock TH (1965) Amer Zool 5:745–755

166. Szentágothai J, Arbib MA (1974) Neurosci Res Bull 12:307–510

167. Szentágothai J (1963) Magy Tud Akad Biol Orv Tud Osztl Kozl 6:217–227 – (1965) Prog Brain Res 14:1–32 – (1967) In: Lissák K (ed) Recent developments of neurobiology in Hungary, vol 1; Results in neuroanatomy, neurochemistry, neuropharmacology and neurophysiology. Akadémiai Kiadó, Budapest – (1968) Proc IEEE 56:960–968

168. Eccles JC (1965) Perspect Biol Med 8:289–310

169. Eccles JC (1967) Proc Nat Acad Sci 58:336–343 – Eccles JC, Ito M, Szentágothai J (1967) The cerebellum as a neuronal machine. Springer, Berlin Heidelberg New York

Suggested Reading

Ethology

Altman I (1966) Organic foundations of animal behavior. Holt, Rinehart and Winston Inc, New York

Altman I (1975) The environment and social behavior. Brooks Cole Publ Co, Monterey Calif

Aronson LR, Tobach E, Lehrman DS, Rosenblatt JS (eds) (1970) Development and evolution of behavior. WH Freeman Co, San Francisco

Baerends GP, Beer C, Manning A (eds) (1975) Function and evolution in behavior. Clarendon Press, Oxford

Bandura A (1973) Aggression. A social learning analysis. Pentrice Hall, New York

Bandura A, Walters RH (1963) Social learning and personality development. Holt, Rinehart and Winston, New York

Barton Brown L (ed) (1974) Insect behaviour. Springer, Berlin Heidelberg New York

Bateson P, Klopfer PH (1978) Perspectives in ethology. Plenum Press, New York

Beach FA (ed) (1965) Sex and behavior. John Wiley, New York

Bell PR (ed) (1959) Darwin's biological work. Cambridge Univ Press, Cambridge

Berkowitz L (1962) Aggression. A social psychological analysis. McGraw-Hill, New York London

Bermant G, Davidson JM (1974) Biological bases of sexual behavior. Harper and Row, New York

Bower TG (1977) A primer of infant development. WH Freeman Co, San Francisco

Breland K, Breland M (1966) Animal behavior. Macmillan, New York

Buchholtz Ch (1973) Das Lernen bei Tieren. G Fischer, Stuttgart

Burkhardt D (Hrsg) (1966) Signale der Tierwelt. B Moos, München

Byrne WL (ed) (1970) Molecular approaches to learning and memory. Academic Press, London

Chance MRA, Larsen RR (eds) (1976) The social structure of attention. John Wiley, London

Colgan PW (1979) Quantitative ethology. John Wiley and Sons, Chichester Sussex

Corning WC, Dyal JA, Willows AOD (1973) Invertebrate learning. Plenum Press, New York

Dethier VG, Stellar E (1964) Animal behavior. Its evolutionary and neurological basis. Prentice Hall, Englewood Cliffs New Jersey

Eibl-Eibesfeldt I (1979) Ethology, the biology of behavior. Holt, Rinehart and Winston Inc, New York

Esser AH (1971) Behavior and environment. The use of space by animals and men. Plenum Press, New York

Etkin W (ed) (1964) Social behavior and organization among vertebrates. Chicago Univ Press, Chicago

Flechtner HJ (1976) Gedächtnis und Lernen in psychologischer Sicht. S Hirzel, Stuttgart

Flugel JC (1950) Probleme und Ergebnisse der Psychologie. Hundert Jahre Psychologischer Forschung. Klett, Stuttgart

Foppa K (1966) Lernen, Gedächtnis, Verhalten. Ergebnisse und Probleme der Lernpsy-
    chologie. Kiepenheuer und Witsch, Köln Berlin
Foss BM (ed) (1965) Determinants of infant behavior. Methuen Press, London
Freemon FR (1972) Sleep research: A critical review. Charles C Thomas, Springfield
    Illinois
Fuller JL, Thompson WR (1960) Behavior genetics. John Wiley, New York
Garattini S, Sigg EB (eds) (1969) Aggressive behavior. Elsevier-Excerpta Medica,
    Amsterdam
Griffin DR (1958) Listening in the dark. Yale Univ Press, New Haven
Guthrie ER (1952) The psychology of learning. Harper and Row, New York
Hailman JP (1977) Optical signals. Indiana Univ Press, Bloomington London
Hartmann EL (1973) The functions of sleep. Yale Univ Press, New Haven
Hess EH (1973) Imprinting: Early experience and the developmental psychobiology of
    attachment. Nostrand, New York
Hess EH, Petrovich SB (eds) (1977) Imprinting. Benchmark Papers in animal behavior,
    vol 5. Academic Press, London New York
Heymer A (1977) Ethological dictionary. Parey Verlag, Berlin Hamburg
Hilgard ER (1956) Theories of learning. Appleton-Century, New York
Hilgard ER, Bower GH (1973) Theorien des Lernens I und II. Klett, Stuttgart
Hinde RA (1970) Animal behaviour: A synthesis of ethology and comparative
    psychology. McGraw-Hill, New York
Hinde RA, Stevenson-Hinde J (eds) (1973) Constraints on learning. Academic Press,
    London
Hirsch J (1967) Behavior – genetic analysis. McGraw-Hill, New York
Holloway RL (ed) (1974) Primate aggression, territoriality, and xenophobia. Academic
    Press, London
Holst E von (1969/70) Zur Verhaltensphysiologie bei Tieren und Menschen. Gesammelte
    Abhandlungen Bd I und II. Piper, München
Immelmann K (Hrsg) (1974) Verhaltensforschung. Ergänzungsband zu Grzimeks
    Tierleben. Kindler, Zürich
Immelmann K (1975) Wörterbuch der Verhaltensforschung. Kindler, Zürich
Jürgens U, Ploog D (1974) Von der Ethologie zur Psychologie. Kindler, Zürich
Kerr FWL, Casey KL (eds) (1978) Pain. NRP*, vol 16, no 1. MIT Press, Cambridge Mass
Klopfer PH (1962) Aspects of ecology. Pentrice Hall Inc, New Jersey
Klopfer PH (1968) Ökologie und Verhalten. G Fischer, Stuttgart
Lamprecht J (1975) Verhalten. Grundlagen-Erkenntnisse, Entwicklungen der Ethologie.
    Studio Visuell. Herder, Freiburg
Laudien H (1977) Physiologie des Gedächtnisses. Uni-Taschenbücher 707. Quelle und
    Meyer, Heidelberg
Lehrman DS, Hinde RA, Shaw E (eds) (1970) Advances in the study of behavior.
    Academic Press, New York
Lorenz K (1965a) Über tierisches und menschliches Verhalten. Aus dem Werdegang der
    Verhaltenslehre. Gesammelte Abhandlungen, Bd I und II. Piper, München
Lorenz K (1965b) Evolution and modification of behavior. Chicago Univ Press, Chicago
Lorenz K (1978) Vergleichende Verhaltensforschung: Grundlagen der Ethologie.
    Springer, Berlin Heidelberg New York
Manning A (1979) An introduction to animal behaviour, 3rd edn. Springer, Berlin
    Heidelberg New York
Marler PR, Hamilton WJ (1966) Mechanisms of animal behavior. John Wiley and Sons,
    New York
Marler PR, Vandenbergh JG (eds) (1979) Social behavior and communication. Plenum
    Press, New York
McGaugh JL, Herz MJ (1972) Memory consolidation. Albion, San Francisco

* NRP = Neuroscience Research Program Bulletin.

McLaughlin B (1971) Learning and social behavior. Free Press, New York
Michael RP, Crook JH (eds) (1973) Comparative ecology and behavior of primates. Academic Press, London
Money J, Ehrhardt AA (1972) Man and woman, boy and girl: The differentiation and dimorphism of gender identity from conception to maturity. Johns Hopkins Univ Press, Baltimore
Montagu MFA (1968) Man and aggression. Oxford Univ Press, New York
Müller-Schwarze D, Mozell MM (eds) (1977) Chemical signals in vertebrates. Plenum Press, New York
Ploog D (1974) Die Sprache der Affen und ihre Bedeutung für die Verständigung des Menschen. Kindler, München
Ploog D, Gottwald P (1974) Verhaltensforschung (Instinkt, Lernen, Hirnfunktion). Urban und Schwarzenberg, München
Ploog D, Melnechuk T (eds) (1969) Primate communication. NRP, vol 7, no 5. MIT Press, Cambridge Mass
Ploog D, Melnechuk T (eds) (1971) Are apes capable of language? NRP, vol 9, no 5. MIT Press, Cambridge Mass
Premack D (1976) Intelligence in ape and man. John Wiley, New York
Pribram KH (ed) (1969) On the biology of learning. Harcourt, New York
Pribram KH, Broadbent DE (eds) (1970) Biology of memory. Academic Press, London New York
Razran G (1971) Mind and evolution. An east-west-synthesis of learned behavior and cognition. Houghton Mifflin, Boston
Rosenblum LA (ed) (1975) Primate behavior. Academic Press, London
Scott JP (1960) Aggression. Chicago Univ Press, Chicago
Shorey HH (1976) Animal communication by pheromones. Academic Press, London
Skinner BF (1953) Science and human behavior. Macmillan, New York
Sluckin W (1965) Imprinting and early learning. Aldine, Chicago
Smith FV (1971) Purpose in animal behavior. Hutchinson Univ Library, London
Stamm RA, Zeier H (eds) (1978) Lorenz und die Folgen, Bd VI. In: Die Psychologie des 20. Jahrhunderts. Kindler, Zürich
Stellar E, Sprague JM (eds) (1971) Progress in Psychology. Academic Press, London New York
Tembrock G (1971) Biokommunikation. Informationsübertragung im biologischen Bereich. Bd I und II. Akademie-Verlag, Berlin
Tembrock G (1977) Grundlagen des Tierverhaltens. Akademie-Verlag, Berlin
Thorpe WH (1963) Learning and instinct in animals. Methuen Co, London
Thorpe WH (1979) The origin and rise of ethology: the science of the natural behavior of animals. Holt, Rinehart and Winston Inc, New York
Thorpe WH, Zangwill OL (eds) (1961) Current problems in animal behavior. Cambridge Univ Press, Cambridge
Tinbergen N (1951) The study of instinct. Clarendon Press, Oxford
Tinbergen N (1953) Social behaviour in animals, with special reference to vertebrates. Methuen Press, London
Tinbergen N (1963) The herring gull's world, 3rd edn, Collins, London
Tobach E, Aronson LR, Shaw E (eds) (1971) The biopsychology of development. Academic Press, London
Uexküll J v (1921) Umwelt und Innenwelt der Tiere, 2. Aufl Berlin
Uexküll J v (1928) Theoretische Biologie. J Springer, Berlin
Wickler W (1968) Mimikry. Signalfälschung in der Natur. Kindler, München
Wickler W (1970) Stammesgeschichte und Ritualisierung. Zur Entstehung tierischer und menschlicher Verhaltensmuster. Piper, München
Wilson EO (1971) The insect societies. Belknap Press of Harvard Univ, Cambridge Mass
Wilson EO (1975) Sociobiology, the new synthesis. Belknap Press of Harvard Univ, Cambridge Mass

Young WC (ed) (1961) Sex and internal secretions, 3rd edn, Williams and Wilkins, Baltimore

Sensory Physiology and Neurophysiology

Aidley DJ (1978) The physiology of excitable cells. 2nd edn. Cambridge Univ Press, Cambridge
Ali MA (ed) (1978) Sensory ecology, review and perspectives. NATO Adv Stud Inst series A, vol 18. Plenum Press, New York
Autrum H, Jung R, Loewenstein WR, MacKay DM, Teuber HL (eds) (1971) Handbook of sensory physiology. Springer, Berlin Heidelberg New York
Beidler LM (ed) (1971) Chemical sens. Part 1: Olfaction. Handbook of sensory physiology, vol IV. Springer, Berlin Heidelberg New York
Beidler LM (ed) (1971) Chemical senses. Part 2: Taste. Handbook of sensory physiology, vol IV. Springer, Berlin Heidelberg New York
Beidler LM, Reichardt WE (eds) (1970) Sensory transduction. NRP, vol 8, no 5. MIT Press, Cambridge Mass
Békésy GV (1960) Experiments in hearing. McGraw-Hill, New York
Bennet MVL (ed) (1974) Synaptic transmission and neuronal interaction. Raven Press, New York
Bennet MVL, Goodenough DA (1978) Gap junctions, electronic coupling, and intracellular communication. NRP, vol 16, no 3. MIT Press, Cambridge Mass
Boeckh J (1975) Nervensysteme und Sinnesorgane der Tiere. Studio Visuell. Herder, Freiburg
Bourne GH (ed) (1972) The structure and function of neurons. Academic Press, London New York
Brazier MAB (1977) The electrical activity of the nervous system. 4th edn. Williams and Wilkins, Baltimore
Bullock TH (1977) Introduction to nervous systems. WH Freeman Co, San Francisco
Bullock TH (ed) (1977) Dahlem workshop on recognition of complex acoustic signals. Dahlem Konferenzen, Berlin
Burkhardt D (1971) Wörterbuch der Neurophysiologie. VEB G Fischer, Jena
Capranica RR (1965) The evoked vocal response of the bullfrog. A study of communication by sound. MIT Press, Cambridge Mass
Carthy JD, Newell GE (1968) Invertebrate receptors. Academic Press, London New York
Cowan WM, Hall ZW, Kandel ER (eds) (1979) Annual review of neuroscience, vol 1–2. Annual Review Inc, California
Dartnall HJA (ed) (1972) Photochemistry of vision. Handbook of sensory physiology, vol VII/1. Springer, Berlin Heidelberg New York
Dawson H (1972) The physiology of the eye. Academic Press, London New York
Eccles JC (1964) The physiology of synapses. Springer, Berlin Heidelberg New York
Fessard A (ed) (1974) Electroreceptors and other specialized receptors in lower vertebrates. Handbook of sensory physiology, vol III/3. Springer, Berlin Heidelberg New York
Field J (ed) (1959/60) Handbook of physiology. Section 1: Neurophysiology, vol. 1–3. American Physiological Society, Washington
Fite KV (ed) (1976) The amphibian visual system: A multidisciplinary approach. Academic Press, London New York
Fuortes MG (ed) (1972) Physiology of photoreceptor organs. Handbook of sensory physiology, vol VII/2. Springer, Berlin Heidelberg New York
Ganong WF (1977) The nervous system. Lange Medical Publication, Los Altos Ca

Gauer OH, Kramer K, Jung R (Hrsg) (1971) Physiologie des Menschen, Bd 4: Muskel, Urban und Schwarzenberg, München

Gauer OH, Kramer K, Jung R (Hrsg) (1971) Physiologie des Menschen, Bd 10: Allgemeine Neurophysiologie. Urban und Schwarzenberg, München

Gauer OH, Kramer K, Jung R (Hrsg) (1971) Physiologie des Menschen, Bd 11: Somatische Sensibilität, Geruch und Geschmack (Sinnesphysiologie I). Urban und Schwarzenberg, München

Gauer OH, Kramer K, Jung R (Hrsg) (1971) Physiologie des Menschen, Bd 12: Hören, Stimme, Gleichgewicht (Sinnesphysiologie II). Urban und Schwarzenberg, München

Granit R (1955) Receptors and sensory perception. Yale Univ Press, New Haven

Gulick WL (1971) Hearing: Physiology and psychophysics. Oxford Univ Press, New York

Hammes GG, Molinoff PB, Bloom FE (eds) (1973) Receptor biophysics and biochemistry. NRP, vol 11, no 3. MIT Press, Cambridge Mass

Held R, Richards W (eds) (1972) Perception: Mechanisms and models. Readings from Scientific American. WH Freeman, San Francisco

Hoppe W, Lohmann W, Markl H, Ziegler H (eds) (1977) Biophysik. Springer, Berlin Heidelberg New York

Horridge GA (1968) Interneurons. WH Freeman Co, San Francisco

Horridge GA (ed) (1974) The compound eye and vision of insects. Clarendon Press, Oxford

Hubbard JI, Llinas R, Quastel DMJ (1969) Electrophysiological analysis of synaptic transmission. Edward Arnold, London

Ingle DJ, Mansfield JW, Goodale MA (eds) (1980) Advances in the analysis of visual behavior. MIT Press, Cambridge Mass

Jameson D, Hurvich LM (eds) (1972) Visual psychophysics. Handbook of sensory physiology, vol VII/4. Springer, Berlin Heidelberg New York

Jung R (ed) (1973) Central processing of visual information, Part A: Integrative functions and comparative data. Handbook of sensory physiology, vol VII/3. Springer, Berlin Heidelberg New York

Jung R (ed) (1973) Central processing of visual information. Part B: Visual centers in the brain. Handbook of sensory physiology, vol VII/3. Springer, Berlin Heidelberg New York

Kandel ER (ed) (1977) Cellular biology of neurons. Handbook of physiology. The nervous system vol 1. Williams and Wilkins, Baltimore

Katz B (1966) Nerve, muscle, and synapse. McGraw-Hill. New York

Keidel WD (Hrsg) (1975) Physiologie des Gehörs. Thieme, Stuttgart

Kerr FWL, Casey KL (eds) (1978) Pain NRP, vol 16, no 1. MIT Press, Cambridge Mass

Kuffler SW, Nicholls JG (1977) From neuron to brain. Sinauer Associates Inc, Sunderland Mass

Lindauer M (Hrsg) (1973) Orientierung der Tiere im Raum. Teil 1: Sinnes- und neurophysiologische Grundlagen. Teil 2: Intraspezifische Kommunikation. G Fischer, Stuttgart

Lindauer M (1975) Verständigung im Bienenstaat. G Fischer, Stuttgart

Llinas RR, Heusser JE (eds) (1977) Depolarization-release coupling systems in neurons. NRP, vol 15, no 4. MIT Press, Cambridge Mass

Llinas R, Precht W (eds) (1976) Frog neurobiology. Springer, Berlin Heidelberg New York

Loewenstein WR (ed) (1971) Principles of receptor physiology. Handbook of sensory physiology, vol I. Springer, Berlin Heidelberg New York

MacKay DM (ed) (1969) Evoked potentials as indicators of sensory information processing. NRP, vol 7, no 3. MIT Press, Cambridge Mass

Masterton RB (ed) (1978) Sensory integration. Handbook of behavioral neurobiology, vol 1. Plenum Press, New York

McLennan J (1970) Synaptic transmission. Saunders, Philadelphia

Møller AR (ed) (1973) Basis mechanism in hearing. Academic Press, London New York

Motokawa K (1970) Physiology of colour and pattern vision. Springer, Berlin Heidelberg
New York
Noback CR (ed) (1978) Sensory systems of primates. Plenum Press, New York
Nystrom RA (1973) Membrane physiology. Prentice Hall, Englewood Cliffs New Jersey
Pappas GO, Purpura DP (1972) Structure and function of synapses. Raven Press, New
York
Perkel DH, Bullock TH (eds) (1963) Neural coding. NRP, vol 6, no 3. MIT Press,
Cambridge Mass
Pfenninger KH (1973) Synaptic morphology and cytochemistry. G Fischer, Stuttgart
Pöppel E, Held R, Dowling JE (eds) (1977) Neuronal mechanisms in visual perception.
NRP, vol 15, no 3. MIT Press, Cambridge Mass
Porter R (ed) (1978a) Studies in neurophysiology – presented to AK McIntyre.
Cambridge Univ Press, London
Porter R (ed) (1978b) International review of physiology, vol 17, Neurophysiology III.
Univ Park Press, Baltimore
Quarton G, Melnechuk T, Schmitt FO (1967) The neurosciences: A study program.
Rockefeller Univ Press, New York
Rahmann H (1976) Neurobiologie. Uni-Taschenbücher 557. Eugen Ulmer, Stuttgart
Rockstein M (ed) (1974) The physiology of insecta. Academic Press, London New York
Rosenblith WA (ed) (1961) Sensory communication. MIT Press, Cambridge Mass
Ruch TC, Patton HD, Woodburg JW, Towe AL (1965) Neurophysiology. Saunders,
Philadelphia
Sarnat HB, Netsky MG (1974) Evolution of the nervous system. Oxford Univ Press, New
York
Schmidt RF (ed) (1978) Fundamentals of sensory physiology. Springer, Berlin Heidelberg
New York
Schmidt RF (ed) (1978) Fundamentals of neurophysiology. Springer, Berlin Heidelberg
New York
Schmitt FO (ed) (1970) The neurosciences: Second study programm. Rockefeller Univ
Press, New York
Schmitt FO, Worden FG (eds) (1974) The neurosciences: Third study program. MIT
Press, Cambridge Mass
Schmitt FO, Worden FG (eds) (1978) The neurosciences: Fourth study program. MIT
Press, Cambridge Mass
Shahani M (ed) (1976) The motor system: Neurophysiology and muscle mechanisms.
Elsevier Sci Publ Co, Amsterdam Oxford New York
Shepherd GM (1979) The synaptic organization of the brain. 2nd edn. Oxford Univ Press,
New York
Smith BH, Kreutzberg GW (eds) (1976) Neuron-target cell interactions. NRP. vol 14, no
3. MIT Press, Cambridge Mass
Snyder AW, Menzel R (eds) (1975) Photoreceptor optics. Springer, Berlin Heidelberg
New York
Stanford AL (1975) Foundations of biophysics. Academic Press, London New York
Stevens Ch F (1966) Neurophysiology. John Wiley and Sons, New York
Sybesma Ch (1977) An introduction to biophysics. Academic Press, London New York
Ten Bruggencate G (1972) Experimentelle Neurophysiologie. Goldmann, München
Thompson RF (1976) Progress in psychobiology: Readings from Scientific American.
WH Freeman Co, San Francisco
Trendelenburg W (1961) Der Gesichtssinn. Springer, Berlin Göttingen Heidelberg
Uttal WR (1973) The psychobiology of sensory coding. Harper and Row, New York
Wehner RG (ed) (1972) Information processing in the visual system of arthropods.
Springer, Berlin Heidelberg New York
Wiersma CAG (ed) (1967) Invertebrate nervous systems. Their significance for
mammalian neurophysiology. Univ Chicago Press, Chicago London

Worden FG, Galambos R (eds) (1972) Auditory processing of biologically significant sounds. NRP, vol 10, no 1. MIT Press, Cambridge Mass
Zettler F, Weiler R (eds) (1976) Neuroprinciples in vision. Springer, Berlin Heidelberg New York
Zusne L (1970) Visual perception of form. Academic Press, London New York

CNS and Behavior

Adey WR, Bawin SM (eds) (1977) Brain interactions with weak electric and magnetic fields. NRP, vol 15, no 1. MIT Press, Cambridge Mass
Adey WR, Birch LC, Lindsley DB, Maddox J, May RM, Olds J, O'Neil WM (1974) Brain mechanisms and the control of behavior. Heinemann Educational Books, London
Aladjalova NA (ed) (1964) Slow electrical processes in the brain. Elsevier Sci Publ Co, Amsterdam Oxford New York
Anochin PK (1967) Das funktionelle System als Grundlage der physiologischen Architektur des Verhaltens. Abhandlungen aus dem Gebiet der Hirnforschung und Verhaltensphysiologie, Bd 1. VEB G Fischer, Jena
Ashby WR (1960) Design for a brain; the origin of adaptive behavior. Science Paperbacks. Assoc Book Publ, London
Bach-y-Rita P (1972) Brain mechanisms in sensory substitution. Academic Press, London New York
Balagura S (1973) Hunger: A biopsychological analysis. Basic Books, New York
Bargmann W, Schadé JP (eds) (1964) Progress in brain research. Vol. 5. Lectures on the diencephalon. Elsevier Sci Publ Co, Amsterdam Oxford New York
Baust W (Hrsg) (1970) Ermüdung, Schlaf und Traum. Wiss Verlagsgesellschaft, Stuttgart
Beach FA, Hebb DO, Morgan CT, Nissen HW (eds) (1960) The neurophysiology of Lashley. McGraw-Hill, New York
Begleiter H(ed) (1979) Evoked brain potentials and behavior. Plenum Press, New York
Bermant G, Davidson JM (1974) Biological basis of sexual behavior. Harper and Row, New York
Biesold D, Matthies H (1976) Neurobiologie. VEB G Fischer, Jena
Birbaumer N (1975) Physiologische Psychologie. Springer, Berlin Heidelberg New York
Boddy J (1978) Brain systems and psychological concepts. John Wiley Sons, Chichester New York
Brazier MAB (ed) (1960) The central nervous system and behavior. Naus Foundation, New York
Brazier MAB (ed) (1975) Growth and development of the brain. Raven Press, New York
Bullock TH (ed) (1966) Simple systems for the study of learning. NRP, vol 4, no 2. MIT Press, Cambridge Mass
Bullock TH, Horridge GA (1965) Structure and function of the nervous systems of invertebrates. WH Freeman Co, San Francisco
Buser PA, Rougeul-Buser A (eds) (1978) Cerebral correlates of conscious experience. INSERM Symp 6. Elsevier Biomed Press, Amsterdam
Carlson NR (1977) Physiology of behavior. Allyn and Bacon, Boston London Sydney Toronto
Chow KL, Leimann AL (eds) (1970) The structural and functional organization of the neocortex. NRP, vol 8, no 2. MIT Press, Cambridge Mass
Clara M (1959) Das Nervensystem des Menschen. JA Barth, Leipzig
Clarke E, Dewhurst K (1972) An illustrated history of brain function. Univ California Press, Berkeley Ca
Clemente CD, Lindsley DB (eds) (1967) Aggression and defense. Neural mechanisms and social pattern. Univ Calif Press, Berkeley Los Angeles Ca
Crabtree JM, Moyer KE (eds) (1977) Bibliography of aggressive behavior: A reader's guide to research literature. AR Liss Inc, New York

Creutzfeldt O, Scheich H. Schreiner Ch (eds) (1979) Hearing mechanisms and speech. EBBS-Workshop Göttingen. Springer, Berlin Heidelberg New York
DeFeudis FV, DeFeudis PAF (1977) Elements of the behavioral code. Academic Press Inc, London
Delgado J (1971) Gehirnschrittmacher. Ullstein, Frankfurt
Deutsch JA (ed) (1973) The physiological basis of memory. Academic Press, London New York
DiCara L (ed) (1974) Limbic and autonomic nervous systems research. Plenum Press, New York
Dubner R, Sessle BJ, Storey AT (1978) The neural basis of oral and facial functions. Plenum Press, New York
Drucker-Colin R, Shkurovich M, Sterman MB (eds) (1979) The function of sleep. Academic Press, New York
Eccles JC (1970) Facing reality. Springer, Berlin Heidelberg New York
Eccles JC (1977) The understanding of the brain. McGraw-Hill, New York
Eccles JC (1979) The human mystery. Springer, Berlin Heidelberg New York
Eccles JC, Gibson WC (1979)˙Sherrington – his life and thought. Springer, Berlin Heidelberg New York
Eccles, JC, Ito M, Szentágothai J (1967) The cerebellum as a neuronal machine. Springer, Berlin Heidelberg New York
Edelman GE, Mountcastle VB (eds) (1978) The mindful brain. MIT Press, Cambridge Mass
Eidelberg E, Stein DC (eds) (1974) Functional recovery after brain lesions of the nervous system. NRP, vol 12, no 2. MIT Press, Cambridge Mass
Eleftherious BE, Sprott RL (eds) (1975) Hormonal correlates of behavior. Plenum Press, New York
Epstein AN, Kissileff HR, Stellar E (1973) The neuropsychology of thirst: New findings and advances in concepts. VH Winston and Sons, Washington. Distributed by Halstead Press, Division of John Wiley and Sons
Evarts EV, Bizzi E, Burke RE, Delong M, Thach JrWT (eds) (1971) Central control of movement. NRP, vol 9, no 1. MIT Press, Cambridge Mass
Faber DS, Korn H (eds) (1978) Neurobiology of the Mauthner cell. Raven Press, New York
Fentress JC (1976) Simpler networks and behavior. Sinauer Associates Inc, Sunderland Mass
Finger S (ed) (1978) Recovery from brain damage – research and theory. Plenum Press New York
Freeman RD (ed) (1979) Developmental neurobiology of vision. NATO Adv Stud Inst series A, vol 27. Plenum Press, New York
Gazzaniga MS (ed) (1979) Neuropsychology. Plenum Press, New York
Gazzaniga MS, LeDoux EJ (1978) The integrated mind. Plenum Press, New York
Granit R (1970) The basis of motor control. Academic Press, London New York
Granit R (1977) The purposive brain. The MIT Press, London
Granit R, Pompeiano O (eds) (1979) Reflex control of posture and movement. Elsevier Biomed Press, Amsterdam
Grenell RG, Gabay S (eds) (1976) Biological functions of psychology. Raven Press, New York
Hall RD, Bloom FE, Olds J (eds) (1977) Neuronal and neurochemical substrates of reinforcement. NRP, vol 15, no 2. MIT Press, Cambridge Mass
Harnad S, Doty RW, Goldstein L, Jaynes J, Krauthammer G (eds) (1977) Lateralization in the nervous system. Academic Press, New York
Haseloff OW (Hrsg) (1971) Hirnforschung und Psyche. Colloquium Verlag, Berlin
Heiligenberg W (1977) Principles of electrolocation and jamming avoidance in electric fish. A neuroethological approach. Springer, Berlin Heidelberg New York
Held R (1976) Recent progress in perception. Readings from Scientific American. WH Freeman Co, San Francisco

Hess WR (1954) Das Zwischenhirn, Syndrome, Lokalisation, Funktionen. Schwabe und Co, Basel

Hess WR (1968) Hypothalamus und Thalamus. Experimental-Dokumente. Thieme, Stuttgart

Holst E v (1939) Die relative Koordination als Phänomen und als Methode zentralnervöser Funktionsanalyse. Ergeb Physiol 13:228–306

Horn G, Hinde RA (eds) (1970) Short-term changes in neural activity and behavior. Cambridge Univ Press, Cambridge

Hoyle G (1970) Cellular mechanisms underlying behavior – neuroethology. In: Advances in insect physiology. London

Hoyle G (ed) (1977) Identified neurons and behavior of arthropods. Plenum Press, New York

Hubel DH, Wiesel TN (1977) Functional architecture of macaque monkey visual cortex. (Ferrier lecture). Proc R Soc Lond B 198:1–59

Ingle D (ed) (1968) The central nervous system and fish behavior. Univ Chicago Press, Chicago London

Ingle D, Schneider GE (eds) (1970) Subcortical visual systems. Karger, Basel

Ingle D, Sprague JM (eds) (1975) Sensorimotor function of the midbrain tectum. NRP, vol 13, no 2. MIT Press, Cambridge Mass

Ingle D, Mansfield RJW, Goodale MA (eds) (1980) Advances in the analysis of visual behavior. MIT Press, Cambridge Mass.

Ito M (ed) (1979) Integrative control functions of the brain, vol 1. Elsevier Biomed Press, Amsterdam

John ER (1967) Mechanisms of memory. Academic Press, London New York

Kandel ER (1976) Cellular basis of behavior: An introduction to neurobiology. WH Freeman Co, San Francisco

Kandel ER (1979) Behavioral biology of *Aplysia:* A contribution to the comparative study of opisthobranch molluscs. WH Freeman Co, San Francisco

Karczmar AG, Eccles JC (eds) (1972) Brain and human behavior. Springer, Berlin Heidelberg New York

Keidel WD (Hrsg) (1973) Kurzgefaßtes Lehrbuch der Physiologie. Thieme, Stuttgart

King FA (ed) (1978) Handbook of behavioral neurobiology. Plenum Press, New York

Kinsbourne M (ed) (1978) Asymmetrical function of the brain. Cambridge Univ Press, Cambridge

Klüver H (1933) Behavior mechanisms in monkeys. Univ Chicago Press, Chicago London; New edition (1957) Phoenix Science Series, New York

Lentz TL (1968) Primitive nervous systems. Yale Univ Press, New Haven

Livingston K, Hornykiewicz O (eds) (1978) Limbic mechanisms. The continuing evolution of the limbic system concept. Plenum Press, New York

Livingston RB (ed) (1966) Brain mechanisms in conditioning and learning. NRP, vol 4, no 3. MIT Press, Cambridge Mass

Lund RD (1978) Development and plasticity of the brain. Oxford Univ Press, New York

Mark VH, Ervin FR (1970) Violence and the brain. Harper and Row, New York

Meyer KE (1971) The physiology of hostility. Markham, Chicago

Meyer KE (1976) Physiology of aggression and implications for control. Raven Press, New York

Moruzzi G, Fessard A, Jasper HH (eds) (1963) Progress in brain research, vol 1. Brain mechanisms. Elsevier Sci Publ Co, Amsterdam Oxford New York

Mountcastle VB (1979) Medical physiology vol I. Mosby Co, St Louis Missouri

Mountcastle VB (1976) The world around us: Neuronal command functions for selective attention. NRP, vol 14, Suppl. MIT Press, Cambridge Mass

Nauta WJH (ed) (1964) Some brain structure and functions related to memory. NRP, vol 2, no 5. MIT Press, Cambridge Mass

Nauta WJH, Koella WP, Quarton GC (eds) (1966) Sleep, wakefulness, dreams and memory. NRP, vol 4, no 1. MIT Press, Cambridge Mass

Novin D, Wyrwicka W, Bray GA (eds) (1970) Hunger: Basic mechanisms and clinical implications. Raven Press, New York

O'Keefe J, Nadel L (1978) The hippocampus as a cognitive map. Clarendon Press, Oxford

Olds J (1977) Drives and reinforcements. Behavioral studies of hypothalamic functions. Raven Press, New York

Penfield W, Rasmussen T (1968) The cerebral cortex of man. Hafner, New York

Philips MI (ed) (1973) Brain unit activity during behavior. Charles C Thomas, Springfield Illinois

Pincus JH, Tucker GJ (1979) Behavioral neurology, 2nd edn. Oxford Univ Press, New York

Popper KR, Eccles JC (1977) The self and its brain. Springer, Berlin Heidelberg New York

Prosser CL (1973) Comparative animal physiology. Saunders, Philadelphia

Roberts TD (1967) Neurophysiology of postural mechanisms. Butterworth, London

Roeder KD (1967) Nerve cells and insect behavior. Harvard Univ Press, Cambridge Mass

Rolls, ET (1975) The brain and reward. Pergamon Press, New York

Rüdiger W (Hrsg) (1965) Probleme der Physiologie des Gehirns. VEB Verlag Volk und Gesundheit, Berlin

Sheer DE (ed) (1961) Electrical stimulation of the brain. Univ Texas Press, Austin

Smythies JR (1970) Brain mechanisms and behaviour. An outline of the mechanisms of emotion, memory, learning and the organization of behaviour, with particular regard to the limbic system. Blackwell Scientific Publications, Oxford Edinburgh

Stellar E, Corbit JD (eds) (1973) Neural control of motivated behavior. NRP, vol 11, no 4. MIT Press, Cambridge Mass

Sternbach RA (1968) Pain: A psychophysiological analysis. Academic Press London New York

Swaab DF, Schadé JP (eds) (1974) Progress in brain research. Vol. 41. Integrative hypothalamic activity. Elsevier Sci Publ Co, Amsterdam Oxford New York

Teyler TJ (1976) A primer of psychobiology. Brain and behavior. WH Freeman Co, San Francisco

Teyler TJ (ed) (1978) Brain and learning. D Reidel Publ Comp, Dordrecht Holland

Thompson RF (1967) Foundations of physiological psychology. Harper and Row, New York

Thompson RF, Verzeano M, Weinberger NM (eds) (1970) The neural control of behavior. Academic Press, London New York

Tobach E, Aronson LR, Shaw E (eds) (1971) The biopsychology of development. Academic Press, London

Tokizane T, Schadé JP (eds) (1966) Progress in brain research. Vol. 21 A. Correlative neurosciences. Part A: Fundamental mechanisms. Part B: Clinical studies. Elsevier Sci Publ Co, Amsterdam Oxford New York

Toller van C (1979) The nervous body: An introduction to the autonomic nervous system and behavior. John Wiley Sons, Chichester New York

Valenstein ES (ed) (1973) Biology of drives. NRP, vol 6, no 1. MIT Press, Cambridge Mass

Valenstein ES (1973) Brain control. John Wiley and Sons, New York

Weiskrantz L (ed) (1978) Functions of the septo-hippocampal system. Ciba Found Symp 58 (new series). Elsevier-Excerpta Medica, Amsterdam

Wells MJ (1962) Brain and behaviour in cephalopods. Univ Press. Stanford

Whalen RF, Thompson RF, Verzeano M, Weinberger NM (eds) (1970) The neural control of behavior. New York, Academic Press

Wiersma CAG (ed) (1967) Invertebrate nervous systems. Chicago Univ Press, Chicago

Yahr MD, Purpura DP (eds) (1967) Neurophysiological basis of normal and abnormal motor activities. Raven Press, New York

Young JZ (1980) Program of the brain. Oxford University Press, New York

Neurochemistry

Agranoff BW, Aprison MH (eds) (1978) Advances in neurochemistry, vol 3. Plenum Press, New York

Barchas DJ, Berger PA, Ciaranello RD, Elliot GR (eds) (1977) Psychopharmacology: From theory to practice. Oxford Univ Press, New York

Bogoch S (1968) The biochemistry of memory. Oxford Univ Press, New York

Bunney Jr W, Murphy D (eds) (1976) The neurobiology of lithium. NRP, vol 14, no 2. MIT Press, Cambridge Mass

Burnstock G, Hökfelt T (eds) (1979) Nonadrenergic autonomic neurotransmission mechanisms. NRP, vol 17, no 3. MIT Press, Cambridge Mass

Cooper JR, Bloom FE, Roth RH (1979) The biochemical basis of neuropharmacology 3rd Edn. Oxford Univ Press, New York

Costa E, Trabucchi M (eds) (1978) The endorphins. Advances in biomedical psychopharmacology, vol 18. Raven Press, New York

Cox B, Morris ID, Weston AH (eds) (1978) Pharmacology of the hypothalamus. Macmillan Press, London and Basingstoke

Cuénod M, Kreutzberg GW, Bloom FE (eds) (1979) Development and chemical specificity of neurons. Elsevier Biomed Press, Amsterdam

Danielli JF, Moran JF, Triggle DJ (eds) (1970) Fundamental concepts in drug-receptor interactions. Academic Press, London New York

Delgado JM, DeFeudis FV (eds) (1977) Behavioral neurochemistry. Spectrum Publ Inc, New York

Donald JJ (ed) (1977) Cholinergic mechanisms and psychopharmacology. Plenum Press, New York

Donovan B (1970) Mammalian neuroendocrinology. McGraw-Hill, New York

Dörner G, Kawakami M (eds) (1978) Hormones and brain development. Elsevier Biomed Press, Amsterdam

Dunn AJ, Bondy SC (1974) Functional chemistry of the brain. Flushing SP Books, New York (1974) distributed by Halstead Press Division of John Wiley and Sons

Eigen M, DeMaeyer LCM (eds) (1964) Information storage and processing in biomolecular systems. NRP, vol 2, no 3. MIT Press, Cambridge Mass

Erlich YH, Volvka J, Davis LG, Brunngraber EG (eds) (1979) Modulators, mediators, and specifiers in brain function. Advances in experimental medicine and biology, vol 116. Plenum Press, New York

Florey E (1966) An introduction to general and comparative animal physiology. Saunders, Philadelphia

Gainer H (ed) (1977) Peptides in neurobiology. Plenum Press, New York

Ganong WF, Martini L (eds) (1969) Frontiers in neuroendocrinology. Oxford Univ Press, New York

Guillemin R (1978) The brain as an endocrine organ. NRP, vol 16, suppl. MIT Press, Cambridge Mass

Haber B, Aprison MH (eds) (1978) Neuropharmacology and behavior. Plenum Press, New York

Hall DH, Bloom FE, Olds J (eds) (1977) Neuronal and neurochemical substrates of reinforcement. NRP, vol 15, no 2. MIT Press, Cambridge Mass

Hall ZW, Hildebrand JG, Kravitz EA (1974) Chemistry of synaptic transmission. Chiron Press, Newton Mass

Hughes J (ed) (1978) Centrally acting peptides. Macmillan Press, London and Basingstoke

Iversen LL, Iversen SD, Snyder SH (eds) (1978) Biology of mood and antianxiety drugs. Handbook of psychopharmacology, vol 13. Plenum Press New York

Iversen LL, Nicoll RA, Vale WW (eds) (1978) Neurobiology of peptides. NRP, vol 10, no 2. MIT Press, Cambridge Mass

Jaenicke L (ed) (1974) Biochemistry of sensory functions. Springer, Berlin Heidelberg New York
Julien RM (1978) A primer of drug action. WH Freeman Co, San Francisco
Katz B (1969) The release of neural transmitter substances. Charles C Thomas, Springfield Illinois
Kuschinsky G, Lüllmann H (1972) Kurzes Lehrbuch der Pharmakologie. Thieme, Stuttgart
Liebeskind JC, Dismukes RK (eds) (1978) Peptides and behavior: a critical analysis of research strategies. NRP, vol 16, no 4. MIT Press, Cambridge Mass
McGeer PL, Eccles JC, McGeer EG (1978) Molecular neurobiology of the mammalian brain. Plenum Press, New York
Nalbandov AV (ed) (1963) Advances in endocrinology. Univ Illinois Press, Urbana
Palmer GO (1979) Neuropharmacology of cyclic nucleotids. Urban and Schwarzenberg, München
Porter R, O'Connor M (eds) (1970) Molecular properties of drug receptors. (A Ciba Foundation Symposium). J and A Churchill, London
Rang HP (ed) (1973) Drug receptors. Crowell Collier Macmillan, London
Ree van JM, Terenius L (eds) (1978) Characteristics and function of opioids. Elsevier Biomed Press, Amsterdam
Roberts PJ, Woodruff GN, Iversen LL (eds) (1978) Dopamine. Advances in biochemical psychopharmacology, vol 19. Raven Press, New York
Siegel GJ, Albers RW, Katzman R, Agranoff BW (eds) (1976) Basic neurochemistry. Little Brown, Boston Mass
Smith BH, Kreutzberg GW (eds) (1976) Neuron-target cell interactions. NRP, vol 14, no 3. MIT Press, Cambridge Mass
Usdin E, Bunney WE (ed) (1975) Pre- and postsynaptic receptors. Marcel Dekker, New York
Valzelli L (ed) (1978) Psychopharmacology of aggression. Modern problems of pharmacopsychiatry, vol 13. S Karger AG, Basel
Wurtman RJ (ed) (1971) Brain monoamines and endocrine function. NRP, vol 9, no 2, MIT Press, Cambridge Mass

Neuroanatomy

Bargmann W (1964) Histologie und mikroskopische Anatomie des Menschen. G Thieme Stuttgart
Bernhard CG, Schadé JP (1967) Progress in brain research, vol 26. Developmental neurology. Elsevier Sci Publ Co, Amsterdam Oxford New York
Bloom W, Fawcett DW (1978) A textbook of histology. Saunders, Philadelphia
Braitenberg V (1973) Gehirngespinste, Neuroanatomie für kybernetisch Interessierte. Springer, Berlin Heidelberg New York
Carpenter MB (1972) Core text of neuroanatomy. Williams and Wilkins, Baltimore
Carpenter MB (1976) Human neuroanatomy. Williams and Wilkins, Baltimore
Crosby EC, Humphrey T, Lauer EW (1962) Correlative anatomy of the nervous system. Macmillan, New York
Edds Jr MV, Gaze RM, Schneider GE, Louis NI (eds) (1979) Specificity and plasticity of retinotectal connections. NRP, vol 17, no 2. MIT Press, Cambridge Mass
Forssmann WG, Heym Ch (1974) Grundriß der Neuroanatomie. Heidelberger Taschenbücher 139. Springer, Berlin Heidelberg New York
Gaze RM (1970) The formation of nerve connections. Academic Press, London New York
Glees P (1957) Morphologie und Physiologie des Nervensystems. Thieme, Stuttgart
Glees P (1975) Das menschliche Gehirn. Hippokrates, Stuttgart
Gottschick J (1952) Die Leistungen des Nervensystems. VEB G Fischer, Jena

Hamilton LW (1976) Basic limbic system anatomy of the rat. Plenum Press, New York

Jacobson M (1978) Developmental neurobiology. Plenum Press, New York

Kappers CUA, Huber GC, Crosby EC (1967) The comparative anatomy of the nervous system of vertebrates, including man. Hafner, New York

Kemali M, Braitenberg V (1969) Atlas of the frog's brain. Springer, Berlin Heidelberg New York

Nieuwenhuys R, Voogd J, Huijzen C van (1978) The human central nervous system. A synopsis and atlas. Springer, Berlin Heidelberg New York

Noback CR, Demarest RJ (1972) The nervous system: Introduction and review. McGraw-Hill, New York

Peters A, Palay SL, de Webster HF (1970) The fine structure of the nervous system: The cells and their processes. Springer, Berlin Heidelberg New York

Plum, F, Gjedde A, Samson FE (eds) (1976) Neuroanatomical functional mapping by the radioactive 2-deoxy-D-glucose method. NPR, vol 14, no 4. MIT Press, Cambridge Mass

Rakic P (ed) (1975) Local circuit neurons. NRP, vol 13, no 3. MIT Press, Cambridge Mass

Robertson RT (ed) (1978) Neuroanatomical research techniques. Methods in physiological psychology, vol II. Academic Press, New York San Francisco London

Rohen JW (1971) Funktionelle Anatomie des Nervensystems. Schattauer, Stuttgart

Sandri C, von Buren JM, Akert K (1977) Progress in brain research. Vol 46. Membrane morphology of the vertebrate nervous system. A study of freeze-etch technique. Elsevier Sci Publ Co, Amsterdam Oxford New York

Sarkissow SA (1967) Grundriß der Struktur und Funktion des Gehirns. VEB Verlag Volk und Gesundheit, Berlin

Singer M, Schadé JP (eds) (1964) Mechanisms of neural regeneration. Elsevier Sci Publ Co, Amsterdam Oxford New York

Singer M, Schadé JP (eds) (1965) Degeneration patterns in the nervous system. Elsevier Sci Publ Co, Amsterdam Oxford New York

Strausfeld NJ (1976) Atlas of an insect brain. Springer, Berlin Heidelberg New York

Varon SS, Somjen GG (eds) (1979) Neuron-glia interactions. NPR, vol 17, no 1. MIT Press, Cambridge Mass

Wolstenholme GEW, O'Connor M (eds) (1968) Growth of the nervous system. Little Brown and Co, Boston

Neurogenetics

Brazier MAB (ed) (1976) Growth and development of the brain: Nutritional genetic and environmental factors. Raven Press, New York

Corner MA, Baker RE, Poll van de NE, Swaab DF, Uylings HBM (eds) (1978) Maturation of the nervous system. Elsevier Biomed Press, Amsterdam

Ehrmann L, Omen GS, Caspari E (eds) (1972) Genetics environment and behavior. Academic Press, London New York

Fieve RR, Rosenthal D, Brill A (eds) (1975) Genetic research in psychiatry. Johns Hopkins Univ Press, Baltimore Md

Gottesmann II, Shields J (1972) Schizophrenia and genetics: A twin study vantage point. Academic Press, London New York

Kaplan AR (ed) (1972) Genetic factors in "schizophrenia". Charles C Thomas, Springfield Illinois

Mednick SA, Schulsinger F, Higgins J, Bell B (eds) (1974) Genetics environment and psychopathology. American Elsevier, New York

Oliverio A (ed) (1977) Gene machinery in the brain. Elsevier Biomed Press, Amsterdam

Sidman RL, Green MC, Appel SH (1965) Catalog of the neurological mutants of the mouse. Harvard Univ Press, Cambridge Mass

Teorell T, Dedrick RL, Coudliffe PG (eds) (1974) Pharmacology and pharmacogenetics. Plenum Press, New York
Weiss P (ed) (1950) Genetic neurology. Problems of the development, growth, and regeneration of the nervous system and of its functions. Univ Chicago Press, Chicago
Worden F, Childs B, Matthysse St, Gershon ES (eds) (1976) Frontiers of psychiatric genetics. NRP vol 14, no 1. MIT Press, Cambridge Mass

Biocybernetics

Arbib, MA (1969) Theories of abstract automata. Prentice Hall, Englewood Cliffs New Jersey
Arbib MA (1972) The metaphorical brain: An introduction to cybernetics as artificial intelligence and brain theory. Wiley-Interscience, New York
Beatty J, Legewie H (eds) (1977) Biofeedback and behavior. Plenum Press, New York
Braines SN, Napalkow AW, Swetschinski WB (1971) Neurokybernetik. VEB Verlag Volk und Gesundheit, Berlin
Deutsch S (1967) Models of the nervous system. John Wiley and Sons, New York
Drischel H (1973) Einführung in die Biokybernetik. Akademie-Verlag, Berlin
Drischel H, Dettmar P (Hrsg) (1975) Biokybernetik V. VEB G Fischer, Jena
Flechtner HJ (1970) Grundbegriffe der Kybernetik. Wiss Verlagsgesellschaft, Stuttgart
Gold HJ (1977) Mathematical modeling of biological systems. John Wiley & Sons, New York London Sidney Toronto
Grodins FS (1963) Control theory and biological systems. Columbia Univ Press, New York London
Grüsser OJ, Klinke R (Hrsg) (1971) Zeichenerkennung durch biologische und technische Systeme. Springer, Berlin Heidelberg New York
Hassenstein P (1973) Information and control in the living organism. An elementary introduction. Chapman and Hall, London
Heynert H (1978) Grundlagen der Bionik. Dr A Hüthig Verlag, Heidelberg
Marko H, Färber G (1968) Kybernetik 1968. Oldenbourg, München
Marmarelis PZ, Marmarelis VZ (1978) Analysis of physiological systems: The white-noise approach. Plenum Press, New York
McFareland DJ (1972) Feedback mechanisms in animal behavior. Academic Press, London New York
McFareland DJ (1978) Motivational control systems analysis. Academic Press, London New York
Milson JH (1966) Biological control systems analysis. McGraw-Hill, New York
Mittelstaedt H (Hrsg) (1961) Regelungsvorgänge in lebenden Wesen. Oldenbourg, München
Reiss RF (1964) Neural theory of modeling. Univ Press, Stanford
Röhler R (1974) Biologische Kybernetik. Teubner, Stuttgart
Sachsse H (1971) Einführung in die Kybernetik. Vieweg, Braunschweig
Sklansky J (ed) (1973) Pattern recognition. Dowden Hutchinson and Ross, Stroudsburg Pa
Sommerhoff G (1974) Logic of the living brain. John Wiley Inc, New York
Stark L (1968) Neurological control systems: Studies in bioengineering. Plenum Press, New York
Szentágothai J, Arbib MA (1974) Conceptual models of neural organization. NRP, vol 12, no 3. MIT Press, Cambridge Mass
Varju D (1977) Systemtheorie für Biologen und Mediziner. Heidelberger Taschenbücher 182. Springer, Berlin Heidelberg New York
Wiener N (1948) Cybernetics. MIT Press, Cambridge Mass

Wiener N, Schadé JP (1965) Cybernetics of the nervous system. Elsevier Sci Publ Co, Amsterdam Oxford New York
Young JZ (1964) A model of the brain. Clarendon Press, Oxford

Experimental Techniques

Allen MB, Doetsch GS, Gindin RA, Haar FL, Yagmai F (eds) (1978) A manual of neurosurgery. Univ Park Press, Baltimore London Tokyo
Blinkov SM, Glezer II (1968) Das Zentralnervensystem in Zahlen und Tabellen. VEB G Fischer, Jena
Börnert D (1974) Leitfaden der Biotelemetrie. VEB G Fischer, Jena
Burck HC (1973) Histologische Technik. Leitfaden für die Herstellung mikroskopischer Präparate in Unterricht und Praxis. Thieme, Stuttgart
Bureš J, Petráǔ M, Zachar J (1967) Electrophysiological methods in biological research. Academic Press, London New York
Cooley RK, Vanderwolf CH (1978) Stereotaxic surgery in the rat: A photographic series. AJ Kirby Co, London Ontario
Creutzfeld OD, Probst W (1974) Neurophysiologische Technik. In: Gauer OH, Kramer K, Jung R (Hrsg) Physiologie des Menschen, Bd 10. Urban und Schwarzenberg, München
Fryer TB, Miller HA, Sandler H (eds) (1976) Biotelemetry III. Academic Press, London New York
Galigher AE, Kozloff EN (1971) Essentials of practical microtechnique. Lea and Febiger, Philadelphia
Gay WI (ed) (1974) Methods of animal experimentation I–V. Academic Press, London New York
Glaser EM, Ruchkin DS (1976) Principles of neurobiological signal analysis. Academic Press, London New York
Haensch G, Haberkamp de Antón G (1976) Dictionary of Biology. BLV Verlagsgesellschaft, München Wien
Heymer A (1977) Ethological dictionary. P Parey, Berlin Hamburg
Knapp W, Holubar K, Wick G (eds) (1978) Immunofluorescence and related staining techniques, Elsevier Biomed Press, Amsterdam
Kater SB, Nicholson Ch (eds) (1973) Intracellular staining in neurobiology. Springer, Berlin Heidelberg New York
Kerkut GA (1968–1971) Experiments in physiology and biochemistry, vol 1–4. Academic Press, London New York
Lavallée M, Schanne OF, Hébert N (eds) (1969) Glass microelectrodes. John Wiley and Sons, New York
Marks N, Rodnight R (eds) (1978) Research methods in neurochemistry, vol 4. Plenum Press, New York
McGeer EG, Olney JW, McGeer PL (eds) (1978) Kainic acid as a tool in neurobiology. Raven Press, New York
Miller H, Harrison DC (eds) (1976) Biomedical electrode technology: Theory and practice. Academic Press, London New York
Miller TA (1979) Insect neurophysiological techniques. Springer, Berlin Heidelberg New York
Moore DH (ed) (1968/69) Physical techniques in biological research. 2A and B: Physical chemical techniques. Academic Press, London New York
Muralt v A (1958) Neue Ergebnisse der Nervenphysiologie. Springer, Berlin Göttingen Heidelberg
Nastuk WL (ed) (1971) Physical techniques in biological research, vol 1: Optical techniques, vol 2: Physical chemical techniques, vol 3: Cells and tissues, vol 4: Special

methods, vol 5 and 6: Electrophysiological methods. Academic Press, London New York

Nauta WJH, Ebbesson SOE (eds) (1970) Contemporary research methods in neuroanatomy. Springer, Berlin Heidelberg New York

Neher E (1974) Elektronische Meßtechnik in der Physiologie. Springer, Berlin Heidelberg New York

Offner FF (1967) Elektronik in der Biologie. Moderne Biologie. BLV, München

Pellegrino LJ, Pellegrino AS, Cushman AJ (1979) A stereotaxic atlas of the rat brain. Plenum Press, New York

Phillips MI (ed) (1973) Brain unit activity during behavior. Ch C Thomas, Springfield

Plonsey R, Fleming DG (eds) (1969) Bioelectric phenomena. McGraw-Hill, New York

Regan D (1972) Evoked potentials in psychology, sensory physiology and clinical medicine. Chapman and Hall, London

Robertson RT (ed) (1978) Methods in physiological psychology, vol 2: Neuroanatomical research techniques. Academic Press, New York

Ryall RW, Kelly JS (eds) (1978) Iontophoresis and transmitter mechanisms in the mammalian central nervous system. Elsevier Biomed Press, Amsterdam

Schneider B, Ranft U (1978) Simulationsmethoden in der Medizin und Biologie. Springer, Berlin Heidelberg New York

Skinner JE (1971) Neuroscience: A laboratory manual. Saunders, Philadelphia

Snider RS, Niemer WT: A stereotaxic atlas of the cat brain. Univ Chicago Press, Chicago London

Spiegel A (1976) Versuchstiere. Eine Einführung in ihre Zucht und Haltung. G Fischer, Stuttgart

Stokes AW (1971) Praktikum der Verhaltensforschung. G Fischer, Stuttgart

Strausfeld NJ (1980) Neuroanatomical techniques. Springer, Berlin Heidelberg New York

Thompson RF (ed) (1973/74) Methods in physiological psychology, vol I. Bioelectric recording techniques, Thompson RF, Patterson MM (eds) Parts A–C. Academic Press, London New York

Webster WG (1975) Principles of research methodology in physiological psychology. Harper and Row, New York

Westhues M, Fritsch R (1961) Die Narkose der Tiere, Bd II. Allgemeinnarkose. P Parey, Berlin Hamburg

Subject Index

Brain stimulation (electrical)
bat   176
bull   236
cat   5, 129
chicken   9, 216
cricket   161
dog   5
fish   129
man   5, 242, 260
monkey   5, 6
moth   208
rabbit   129
rat   223
snail   204
techniques   259
toad   93
Brainstem   144
Bristling hair, see Tail-
bristling
Bufo   27, 62, 69, 131,
136, 157, 168, 174, 199,
203, 205, 246, 248, 259,
262, 277–281
Bull   236
Bullfrog   167, 173, 174
Bungarus   56

Ca^{2+} ions   31, 38, 51, 53,
54
Calling song   161, 164,
165
Calliphora   150
Calls, see Mating calls
Canary   233, 234
Captivity   257
Carbachol   223
Caricature   57, 71
Carrion receptor   150
Castration experiments
232
Cat   5, 129, 135, 138,
142, 219, 226, 237, 238,
252, 256
Categories of significance
(meaning)   57, 63, 67,
121
Caterpillar   83
Cell
architecture   19, 20
injections   14, 288
Cells, see Neuron types,
Glial cells
Center concept   214

Central
grey   95, 237
pattern generator
203–215
programs   1, 32, 93,
125, 158, 160, 203
Central nervous system,
CNS   1, 2, 7, see also
Behaviorally relevant
brain structures, Brain
Cerebellar cortex   19, 20,
135, 289, 295
Cerebellum, see Cerebellar
cortex
Cerebral cortex   4–6, 23,
138, 182, 183, 212, 218,
228, 236, 240, 242, 243
CF neurons   181
CF-FM sound   176
Change of meaning   67
Charge carrier   33
Chemical
brain stimulation, see
Brain stimulation
lesions, see Lesion tech-
niques, Techniques
Chemoreceptors   147,
206
Chicken   9, 216, 276
Chinning   253
Choice behavior   168,
171, 172, 258
Cholinergic transmission,
see Acetylcholine
Cholinesterase   38, 56, 98
Chorus, see Frog chorus
Chromatolysis   285, 287
Chronic recording   69,
199, 205, 222, 276–282
Circuitry   203
Circulating excitations
44, 240, 243
Circum striate belt   138,
139
Clasping   90
Classes of meaning, see
Categories of signifi-
cance
Classification (neurons)
46, 50, 111, 118, 122,
125, 126, 131, 135–140,
149, 164, 168, 181, 183,
193, 205, 214, 275, see
also Neuron
Climbing fibers   135, 290

Cluster   26
Coagulation   95, 264
Cobalt
chloride   288
nitrate   165
Cochlea   180
Coder neurons   121, 188,
193
Coding (synapse)   30, 37,
40
Colliculus
inferior   176, 179
superior   129, 130,
136, 137, 140
Colored perch   59, 61
Column (sensory)   26
Command
element   125, 213, 214
function   211
neuron   203–215
neuron concept   213
system   125, 214
Commissure   22
Common toad, see *Bufo*
Communication, general
1, 2, 158
auditory   158
bats   183
chemical   147, 150
crickets   159
electric fish   186, 201
frogs   1, 166
moths   150
squirrel monkeys   202,
229, 237
Comparative aspects
aggression   235
command systems   213,
214
configurational stimuli
59–66
dendrite, neurite   40
drive circuits   225, 226
echo-/electrolocation
186
frequency filter/time do-
main analysis   198
motor programs   157,
158, 208
muscle fibers   56
neuroethology   1–12
neurons   13, 17, 19
neurosurgery   239
orientation detectors/
Gestalt decoders   121

330   Subject Index

Zoophysiology and Ecology

Coordination Editor: D. S. Farner
Editors: W. S. Hoar, B. Hoelldobler,
H. Langer, M. Lindauer

A Selection

Volume 7
E. Curio
The Ethology of Predation

1976. 70 figures, 16 tables. X, 250 pages
ISBN 3-540-07720-0

"Eberhard Curio has written a good book
about the ethology of predation... It is good
because it is stimulating, exhaustive and logical. No important aspect of the subject is
missed. The author illustrates all his main
points with a multiplicity of examples drawn
from recent research. The reference list of
nearly 700 items is evidence of the thorough-
ness of the treatment and the marshalling of
examples used in explanation. Curio is an
enthusiast and conveys the excitement to be
found in much of the research on this subject; he also draws pointed attention to the
gaps in our knowledge. For all these reasons
the book is a *must* for ethologists, ecologists,
experimental psychologists and university
libraries. As a first treatment of seminal
quality the book could well become a reference classic and inspire numerous research
projects... The illustrations are clear and
relevant..."

The Quart. Review Biology

Springer-Verlag
Berlin
Heidelberg
New York

Volume 8
W. Leuthold
African Ungulates

A Comparative Review of Their Ethology and
Behavioral Ecology

1977. 55 figures, 7 tables. XIII, 307 pages
ISBN 3-540-07951-3

"...Dr. Leuthold displays a masterly command of his subject and few important papers
seem to have escaped his notice,... The work
is basically a review of published knowledge
with an original approach, enlivened by the
author's interpretations and based on his
intimate first-hand knowledge of the subject.
The first chapter, on the application of ethological knowledge to wildlife management,
covers an important area... a wealth of references is given so that the chapter provides a
useful guide to the literature. The illustrations
are good and well chosen to demonstrate
points made verbally and not first to embellish
the text. The book will provide excellent background reading for undergraduates and
research students as well as for anyone seriously interested in African wildlife. On the
whole, it can be thoroughly recomended."

J. Applied Ecology

Volume 10
H.-U. Thiele
Carabid Beetles in Their Environments

A Study on Habitat Selection by Adaptations
in Physiology and Behaviour
Translated from the German by J. Wieser

1977. 152 figures, 58 tables. XVII,
XVII, 369 pages
ISBN 3-540-08306-5

"The book is a fine synthesis of current knowledge of homeostatic aspects of ecological relationships of carabids, and it is a fitting tribute
to the man to whom it is dedicated: Carl
H. Lindroth... The material is well organized
and the text is easily readable, thanks to the
clarity of thought and expression of the author
and to the skill of an able translator."

Science

A Springer journal

Behavioral Ecology and Sociobiology

Managing Editor: H. Markl
in cooperation with a distinguished advisory
board

Behavioral Ecology and Sociobiology was
founded by Springer-Verlag in 1976 as an
international journal. Drawing on a philo-
sophy developed and nurtured for more than
half a century in the original *Zeitschrift für
vergleichende Physiologie* (now *Journal of
Comparative Physiology*), it presents original
articles and short communications dealing
with the experimental analysis of animal
behavior on an individual level and in population. Special emphasis is given to the functions, mechanisms, and evolution of ecological adaptations of behavior. Specific areas
covered include:

- orientation in space and time
- communication and all other forms of
  social and interspecific behavior
- origins and mechanisms of behavior preferences and aversions, e.g., with respect to
  food, locality, and social partners
- behavioral mechanisms of competition and
  resource partitioning
- population physiology
- evolutionary theory of social behavior.

Behavioral Ecology and Sociobiology is designed to serve as a link between researchers and
students in a variety of disciplines.

Subscription Information upon request.

Journal of Comparative Physiology · A+B

Founded in 1924 as
Zeitschrift für vergleichende Physiologie
by K. von Frisch and A. Kühn

A. Sensory, Neural, and Behavioral
   Physiology

Editorial Board: H. Autrum, R. R. Capranica,
C. Delcomyn, K. von Frisch, G. A. Horridge,
M. Lindauer, C. L. Prosser

Advisory Board: H. Atwood, S. Daan,
W. H. Fahrenbach, B. Hölldobler, Y. Katsuki,
M. Konishi, M. F. Land, M. S. Laverack, H. C.
Lüttgau, H. Markl, A. Michelsen, D. Ottoson,
F. Papi, C. S. Pittendrigh, W. Precht, J. D. Pye,
A. Roth, H. F. Rowell, D. G. Stavenga,
R. Wehner, J. J. Wine

B. Biochemical, Systemic, and Environmental Physiology

Editorial Board: K. Johansen, B. Linzen,
W. T. W. Potts, C. L. Prosser

Advisory Board: G. A. Bartholomew, H. Bern,
P. J. Butler, Th. Eisner, D. H. Evans, S.
Nilsson, O. Randall, R. B. Reeves, G. H.
Satchell, T. J. Shuttleworth, G. Somero,
K. Urich, S. Utida, G. R. Wyatt, E. Zebe

The *Journal of Comparative Physiology* pub-
lishes original articles in the field of animal
physiology. In view of the increasing number
of papers and the high degree of scientific
specialization the journal is published in two
sections.

A. Sensory, Neural, and Behavioral
   Physiology
Physiological Basis of Behavior; Sensory
Physiology; Neural Physiology; Orientation,
Communication; Locomotion; Hormonal
Control of Behavior

B. Biochemical, Systemic, and Environmen-
   tal Physiology
Comparative Aspects of Metabolism and
Enzymology; Metabolic Regulation; Respira-
tion and Gas Transport; Physiology of Body
Fluids; Circulation; Temperature Relations;
Muscular Physiology

Subscription Information upon request.

Springer-Verlag
Berlin
Heidelberg
New York